A FIRST GRADUATE COURSE
IN ABSTRACT ALGEBRA

PURE AND APPLIED MATHEMATICS

A Program of Monographs, Textbooks, and Lecture Notes

MONOGRAPHS AND TEXTBOOKS IN
PURE AND APPLIED MATHEMATICS

1. *K. Yano,* Integral Formulas in Riemannian Geometry (1970)
2. *S. Kobayashi,* Hyperbolic Manifolds and Holomorphic Mappings (1970)
3. *V. S. Vladimirov,* Equations of Mathematical Physics (A. Jeffrey, ed.; A. Littlewood, trans.) (1970)
4. *B. N. Pshenichnyi,* Necessary Conditions for an Extremum (L. Neustadt, translation ed.; K. Makowski, trans.) (1971)
5. *L. Narici et al.,* Functional Analysis and Valuation Theory (1971)
6. *S. S. Passman,* Infinite Group Rings (1971)
7. *L. Dornhoff,* Group Representation Theory. Part A: Ordinary Representation Theory. Part B: Modular Representation Theory (1971, 1972)
8. *W. Boothby and G. L. Weiss, eds.,* Symmetric Spaces (1972)
9. *Y. Matsushima,* Differentiable Manifolds (E. T. Kobayashi, trans.) (1972)
10. *L. E. Ward, Jr.,* Topology (1972)
11. *A. Babakhanian,* Cohomological Methods in Group Theory (1972)
12. *R. Gilmer,* Multiplicative Ideal Theory (1972)
13. *J. Yeh,* Stochastic Processes and the Wiener Integral (1973)
14. *J. Barros-Neto,* Introduction to the Theory of Distributions (1973)
15. *R. Larsen,* Functional Analysis (1973)
16. *K. Yano and S. Ishihara,* Tangent and Cotangent Bundles (1973)
17. *C. Procesi,* Rings with Polynomial Identities (1973)
18. *R. Hermann,* Geometry, Physics, and Systems (1973)
19. *N. R. Wallach,* Harmonic Analysis on Homogeneous Spaces (1973)
20. *J. Dieudonné,* Introduction to the Theory of Formal Groups (1973)
21. *I. Vaisman,* Cohomology and Differential Forms (1973)
22. *B.-Y. Chen,* Geometry of Submanifolds (1973)
23. *M. Marcus,* Finite Dimensional Multilinear Algebra (in two parts) (1973, 1975)
24. *R. Larsen,* Banach Algebras (1973)
25. *R. O. Kujala and A. L. Vitter, eds.,* Value Distribution Theory: Part A; Part B: Deficit and Bezout Estimates by Wilhelm Stoll (1973)
26. *K. B. Stolarsky,* Algebraic Numbers and Diophantine Approximation (1974)
27. *A. R. Magid,* The Separable Galois Theory of Commutative Rings (1974)
28. *B. R. McDonald,* Finite Rings with Identity (1974)
29. *J. Satake,* Linear Algebra (S. Koh et al., trans.) (1975)
30. *J. S. Golan,* Localization of Noncommutative Rings (1975)
31. *G. Klambauer,* Mathematical Analysis (1975)
32. *M. K. Agoston,* Algebraic Topology (1976)
33. *K. R. Goodearl,* Ring Theory (1976)
34. *L. E. Mansfield,* Linear Algebra with Geometric Applications (1976)
35. *N. J. Pullman,* Matrix Theory and Its Applications (1976)
36. *B. R. McDonald,* Geometric Algebra Over Local Rings (1976)
37. *C. W. Groetsch,* Generalized Inverses of Linear Operators (1977)
38. *J. E. Kuczkowski and J. L. Gersting,* Abstract Algebra (1977)
39. *C. O. Christenson and W. L. Voxman,* Aspects of Topology (1977)
40. *M. Nagata,* Field Theory (1977)
41. *R. L. Long,* Algebraic Number Theory (1977)
42. *W. F. Pfeffer,* Integrals and Measures (1977)
43. *R. L. Wheeden and A. Zygmund,* Measure and Integral (1977)
44. *J. H. Curtiss,* Introduction to Functions of a Complex Variable (1978)
45. *K. Hrbacek and T. Jech,* Introduction to Set Theory (1978)
46. *W. S. Massey,* Homology and Cohomology Theory (1978)
47. *M. Marcus,* Introduction to Modern Algebra (1978)
48. *E. C. Young,* Vector and Tensor Analysis (1978)
49. *S. B. Nadler, Jr.,* Hyperspaces of Sets (1978)
50. *S. K. Segal,* Topics in Group Kings (1978)
51. *A. C. M. van Rooij,* Non-Archimedean Functional Analysis (1978)
52. *L. Corwin and R. Szczarba,* Calculus in Vector Spaces (1979)
53. *C. Sadosky,* Interpolation of Operators and Singular Integrals (1979)
54. *J. Cronin,* Differential Equations (1980)
55. *C. W. Groetsch,* Elements of Applicable Functional Analysis (1980)

56. *I. Vaisman*, Foundations of Three-Dimensional Euclidean Geometry (1980)
57. *H. I. Freedan*, Deterministic Mathematical Models in Population Ecology (1980)
58. *S. B. Chae*, Lebesgue Integration (1980)
59. *C. S. Rees et al.*, Theory and Applications of Fourier Analysis (1981)
60. *L. Nachbin*, Introduction to Functional Analysis (R. M. Aron, trans.) (1981)
61. *G. Orzech and M. Orzech*, Plane Algebraic Curves (1981)
62. *R. Johnsonbaugh and W. E. Pfaffenberger*, Foundations of Mathematical Analysis (1981)
63. *W. L. Voxman and R. H. Goetschel*, Advanced Calculus (1981)
64. *L. J. Corwin and R. H. Szczarba*, Multivariable Calculus (1982)
65. *V. I. Istrătescu*, Introduction to Linear Operator Theory (1981)
66. *R. D. Järvinen*, Finite and Infinite Dimensional Linear Spaces (1981)
67. *J. K. Beem and P. E. Ehrlich*, Global Lorentzian Geometry (1981)
68. *D. L. Armacost*, The Structure of Locally Compact Abelian Groups (1981)
69. *J. W. Brewer and M. K. Smith, eds.*, Emmy Noether: A Tribute (1981)
70. *K. H. Kim*, Boolean Matrix Theory and Applications (1982)
71. *T. W. Wieting*, The Mathematical Theory of Chromatic Plane Ornaments (1982)
72. *D. B. Gauld*, Differential Topology (1982)
73. *R. L. Faber*, Foundations of Euclidean and Non-Euclidean Geometry (1983)
74. *M. Carmeli*, Statistical Theory and Random Matrices (1983)
75. *J. H. Carruth et al.*, The Theory of Topological Semigroups (1983)
76. *R. L. Faber*, Differential Geometry and Relativity Theory (1983)
77. *S. Barnett*, Polynomials and Linear Control Systems (1983)
78. *G. Karpilovsky*, Commutative Group Algebras (1983)
79. *F. Van Oystaeyen and A. Verschoren*, Relative Invariants of Rings (1983)
80. *I. Vaisman*, A First Course in Differential Geometry (1984)
81. *G. W. Swan*, Applications of Optimal Control Theory in Biomedicine (1984)
82. *T. Petrie and J. D. Randall*, Transformation Groups on Manifolds (1984)
83. *K. Goebel and S. Reich*, Uniform Convexity, Hyperbolic Geometry, and Nonexpansive Mappings (1984)
84. *T. Albu and C. Năstăsescu*, Relative Finiteness in Module Theory (1984)
85. *K. Hrbacek and T. Jech*, Introduction to Set Theory: Second Edition (1984)
86. *F. Van Oystaeyen and A. Verschoren*, Relative Invariants of Rings (1984)
87. *B. R. McDonald*, Linear Algebra Over Commutative Rings (1984)
88. *M. Namba*, Geometry of Projective Algebraic Curves (1984)
89. *G. F. Webb*, Theory of Nonlinear Age-Dependent Population Dynamics (1985)
90. *M. R. Bremner et al.*, Tables of Dominant Weight Multiplicities for Representations of Simple Lie Algebras (1985)
91. *A. E. Fekete*, Real Linear Algebra (1985)
92. *S. B. Chae*, Holomorphy and Calculus in Normed Spaces (1985)
93. *A. J. Jerri*, Introduction to Integral Equations with Applications (1985)
94. *G. Karpilovsky*, Projective Representations of Finite Groups (1985)
95. *L. Narici and E. Beckenstein*, Topological Vector Spaces (1985)
96. *J. Weeks*, The Shape of Space (1985)
97. *P. R. Gribik and K. O. Kortanek*, Extremal Methods of Operations Research (1985)
98. *J.-A. Chao and W. A. Woyczynski, eds.*, Probability Theory and Harmonic Analysis (1986)
99. *G. D. Crown et al.*, Abstract Algebra (1986)
100. *J. H. Carruth et al.*, The Theory of Topological Semigroups, Volume 2 (1986)
101. *R. S. Doran and V. A. Belfi*, Characterizations of C*-Algebras (1986)
102. *M. W. Jeter*, Mathematical Programming (1986)
103. *M. Altman*, A Unified Theory of Nonlinear Operator and Evolution Equations with Applications (1986)
104. *A. Verschoren*, Relative Invariants of Sheaves (1987)
105. *R. A. Usmani*, Applied Linear Algebra (1987)
106. *P. Blass and J. Lang*, Zariski Surfaces and Differential Equations in Characteristic $p > 0$ (1987)
107. *J. A. Reneke et al.*, Structured Hereditary Systems (1987)
108. *H. Busemann and B. B. Phadke*, Spaces with Distinguished Geodesics (1987)
109. *R. Harte*, Invertibility and Singularity for Bounded Linear Operators (1988)
110. *G. S. Ladde et al.*, Oscillation Theory of Differential Equations with Deviating Arguments (1987)
111. *L. Dudkin et al.*, Iterative Aggregation Theory (1987)
112. *T. Okubo*, Differential Geometry (1987)

113. *D. L. Stancl and M. L. Stancl,* Real Analysis with Point-Set Topology (1987)
114. *T. C. Gard,* Introduction to Stochastic Differential Equations (1988)
115. *S. S. Abhyankar,* Enumerative Combinatorics of Young Tableaux (1988)
116. *H. Strade and R. Farnsteiner,* Modular Lie Algebras and Their Representations (1988)
117. *J. A. Huckaba,* Commutative Rings with Zero Divisors (1988)
118. *W. D. Wallis,* Combinatorial Designs (1988)
119. *W. Więsław,* Topological Fields (1988)
120. *G. Karpilovsky,* Field Theory (1988)
121. *S. Caenepeel and F. Van Oystaeyen,* Brauer Groups and the Cohomology of Graded Rings (1989)
122. *W. Kozlowski,* Modular Function Spaces (1988)
123. *E. Lowen-Colebunders,* Function Classes of Cauchy Continuous Maps (1989)
124. *M. Pavel,* Fundamentals of Pattern Recognition (1989)
125. *V. Lakshmikantham et al.,* Stability Analysis of Nonlinear Systems (1989)
126. *R. Sivaramakrishnan,* The Classical Theory of Arithmetic Functions (1989)
127. *N. A. Watson,* Parabolic Equations on an Infinite Strip (1989)
128. *K. J. Hastings,* Introduction to the Mathematics of Operations Research (1989)
129. *B. Fine,* Algebraic Theory of the Bianchi Groups (1989)
130. *D. N. Dikranjan et al.,* Topological Groups (1989)
131. *J. C. Morgan II,* Point Set Theory (1990)
132. *P. Biler and A. Witkowski,* Problems in Mathematical Analysis (1990)
133. *H. J. Sussmann,* Nonlinear Controllability and Optimal Control (1990)
134. *J.-P. Florens et al.,* Elements of Bayesian Statistics (1990)
135. *N. Shell,* Topological Fields and Near Valuations (1990)
136. *B. F. Doolin and C. F. Martin,* Introduction to Differential Geometry for Engineers (1990)
137. *S. S. Holland, Jr.,* Applied Analysis by the Hilbert Space Method (1990)
138. *J. Okniński,* Semigroup Algebras (1990)
139. *K. Zhu,* Operator Theory in Function Spaces (1990)
140. *G. B. Price,* An Introduction to Multicomplex Spaces and Functions (1991)
141. *R. B. Darst,* Introduction to Linear Programming (1991)
142. *P. L. Sachdev,* Nonlinear Ordinary Differential Equations and Their Applications (1991)
143. *T. Husain,* Orthogonal Schauder Bases (1991)
144. *J. Foran,* Fundamentals of Real Analysis (1991)
145. *W. C. Brown,* Matrices and Vector Spaces (1991)
146. *M. M. Rao and Z. D. Ren,* Theory of Orlicz Spaces (1991)
147. *J. S. Golan and T. Head,* Modules and the Structures of Rings (1991)
148. *C. Small,* Arithmetic of Finite Fields (1991)
149. *K. Yang,* Complex Algebraic Geometry (1991)
150. *D. G. Hoffman et al.,* Coding Theory (1991)
151. *M. O. González,* Classical Complex Analysis (1992)
152. *M. O. González,* Complex Analysis (1992)
153. *L. W. Baggett,* Functional Analysis (1992)
154. *M. Sniedovich,* Dynamic Programming (1992)
155. *R. P. Agarwal,* Difference Equations and Inequalities (1992)
156. *C. Brezinski,* Biorthogonality and Its Applications to Numerical Analysis (1992)
157. *C. Swartz,* An Introduction to Functional Analysis (1992)
158. *S. B. Nadler, Jr.,* Continuum Theory (1992)
159. *M. A. Al-Gwaiz,* Theory of Distributions (1992)
160. *E. Perry,* Geometry: Axiomatic Developments with Problem Solving (1992)
161. *E. Castillo and M. R. Ruiz-Cobo,* Functional Equations and Modelling in Science and Engineering (1992)
162. *A. J. Jerri,* Integral and Discrete Transforms with Applications and Error Analysis (1992)
163. *A. Charlier et al.,* Tensors and the Clifford Algebra (1992)
164. *P. Biler and T. Nadzieja,* Problems and Examples in Differential Equations (1992)
165. *E. Hansen,* Global Optimization Using Interval Analysis (1992)
166. *S. Guerre-Delabrière,* Classical Sequences in Banach Spaces (1992)
167. *Y. C. Wong,* Introductory Theory of Topological Vector Spaces (1992)
168. *S. H. Kulkarni and B. V. Limaye,* Real Function Algebras (1992)
169. *W. C. Brown,* Matrices Over Commutative Rings (1993)
170. *J. Loustau and M. Dillon,* Linear Geometry with Computer Graphics (1993)
171. *W. V. Petryshyn,* Approximation-Solvability of Nonlinear Functional and Differential Equations (1993)

172. E. C. Young, Vector and Tensor Analysis: Second Edition (1993)
173. T. A. Bick, Elementary Boundary Value Problems (1993)
174. M. Pavel, Fundamentals of Pattern Recognition: Second Edition (1993)
175. S. A. Albeverio et al., Noncommutative Distributions (1993)
176. W. Fulks, Complex Variables (1993)
177. M. M. Rao, Conditional Measures and Applications (1993)
178. A. Janicki and A. Weron, Simulation and Chaotic Behavior of α-Stable Stochastic Processes (1994)
179. P. Neittaanmäki and D. Tiba, Optimal Control of Nonlinear Parabolic Systems (1994)
180. J. Cronin, Differential Equations: Introduction and Qualitative Theory, Second Edition (1994)
181. S. Heikkilä and V. Lakshmikantham, Monotone Iterative Techniques for Discontinuous Nonlinear Differential Equations (1994)
182. X. Mao, Exponential Stability of Stochastic Differential Equations (1994)
183. B. S. Thomson, Symmetric Properties of Real Functions (1994)
184. J. E. Rubio, Optimization and Nonstandard Analysis (1994)
185. J. L. Bueso et al., Compatibility, Stability, and Sheaves (1995)
186. A. N. Michel and K. Wang, Qualitative Theory of Dynamical Systems (1995)
187. M. R. Darnel, Theory of Lattice-Ordered Groups (1995)
188. Z. Naniewicz and P. D. Panagiotopoulos, Mathematical Theory of Hemivariational Inequalities and Applications (1995)
189. L. J. Corwin and R. H. Szczarba, Calculus in Vector Spaces: Second Edition (1995)
190. L. H. Erbe et al., Oscillation Theory for Functional Differential Equations (1995)
191. S. Agaian et al., Binary Polynomial Transforms and Nonlinear Digital Filters (1995)
192. M. I. Gil', Norm Estimations for Operation-Valued Functions and Applications (1995)
193. P. A. Grillet, Semigroups: An Introduction to the Structure Theory (1995)
194. S. Kichenassamy, Nonlinear Wave Equations (1996)
195. V. F. Krotov, Global Methods in Optimal Control Theory (1996)
196. K. I. Beidar et al., Rings with Generalized Identities (1996)
197. V. I. Arnautov et al., Introduction to the Theory of Topological Rings and Modules (1996)
198. G. Sierksma, Linear and Integer Programming (1996)
199. R. Lasser, Introduction to Fourier Series (1996)
200. V. Sima, Algorithms for Linear-Quadratic Optimization (1996)
201. D. Redmond, Number Theory (1996)
202. J. K. Beem et al., Global Lorentzian Geometry: Second Edition (1996)
203. M. Fontana et al., Prüfer Domains (1997)
204. H. Tanabe, Functional Analytic Methods for Partial Differential Equations (1997)
205. C. Q. Zhang, Integer Flows and Cycle Covers of Graphs (1997)
206. E. Spiegel and C. J. O'Donnell, Incidence Algebras (1997)
207. B. Jakubczyk and W. Respondek, Geometry of Feedback and Optimal Control (1998)
208. T. W. Haynes et al., Fundamentals of Domination in Graphs (1998)
209. T. W. Haynes et al., eds., Domination in Graphs: Advanced Topics (1998)
210. L. A. D'Alotto et al., A Unified Signal Algebra Approach to Two-Dimensional Parallel Digital Signal Processing (1998)
211. F. Halter-Koch, Ideal Systems (1998)
212. N. K. Govil et al., eds., Approximation Theory (1998)
213. R. Cross, Multivalued Linear Operators (1998)
214. A. A. Martynyuk, Stability by Liapunov's Matrix Function Method with Applications (1998)
215. A. Favini and A. Yagi, Degenerate Differential Equations in Banach Spaces (1999)
216. A. Illanes and S. Nadler, Jr., Hyperspaces: Fundamentals and Recent Advances (1999)
217. G. Kato and D. Struppa, Fundamentals of Algebraic Microlocal Analysis (1999)
218. G. X.-Z. Yuan, KKM Theory and Applications in Nonlinear Analysis (1999)
219. D. Motreanu and N. H. Pavel, Tangency, Flow Invariance for Differential Equations, and Optimization Problems (1999)
220. K. Hrbacek and T. Jech, Introduction to Set Theory, Third Edition (1999)
221. G. E. Kolosov, Optimal Design of Control Systems (1999)
222. N. L. Johnson, Subplane Covered Nets (2000)
223. B. Fine and G. Rosenberger, Algebraic Generalizations of Discrete Groups (1999)
224. M. Väth, Volterra and Integral Equations of Vector Functions (2000)
225. S. S. Miller and P. T. Mocanu, Differential Subordinations (2000)

226. R. Li et al., Generalized Difference Methods for Differential Equations: Numerical Analysis of Finite Volume Methods (2000)
227. H. Li and F. Van Oystaeyen, A Primer of Algebraic Geometry (2000)
228. R. P. Agarwal, Difference Equations and Inequalities: Theory, Methods, and Applications, Second Edition (2000)
229. A. B. Kharazishvili, Strange Functions in Real Analysis (2000)
230. J. M. Appell et al., Partial Integral Operators and Integro-Differential Equations (2000)
231. A. I. Prilepko et al., Methods for Solving Inverse Problems in Mathematical Physics (2000)
232. F. Van Oystaeyen, Algebraic Geometry for Associative Algebras (2000)
233. D. L. Jagerman, Difference Equations with Applications to Queues (2000)
234. D. R. Hankerson et al., Coding Theory and Cryptography: The Essentials, Second Edition, Revised and Expanded (2000)
235. S. Dăscălescu et al., Hopf Algebras: An Introduction (2001)
236. R. Hagen et al., C*-Algebras and Numerical Analysis (2001)
237. Y. Talpaert, Differential Geometry: With Applications to Mechanics and Physics (2001)
238. R. H. Villarreal, Monomial Algebras (2001)
239. A. N. Michel et al., Qualitative Theory of Dynamical Systems: Second Edition (2001)
240. A. A. Samarskii, The Theory of Difference Schemes (2001)
241. J. Knopfmacher and W.-B. Zhang, Number Theory Arising from Finite Fields (2001)
242. S. Leader, The Kurzweil-Henstock Integral and Its Differentials (2001)
243. M. Biliotti et al., Foundations of Translation Planes (2001)
244. A. N. Kochubei, Pseudo-Differential Equations and Stochastics over Non-Archimedean Fields (2001)
245. G. Sierksma, Linear and Integer Programming: Second Edition (2002)
246. A. A. Martynyuk, Qualitative Methods in Nonlinear Dynamics: Novel Approaches to Liapunov's Matrix Functions (2002)
247. B. G. Pachpatte, Inequalities for Finite Difference Equations (2002)
248. A. N. Michel and D. Liu, Qualitative Analysis and Synthesis of Recurrent Neural Networks (2002)
249. J. R. Weeks, The Shape of Space: Second Edition (2002)
250. M. M. Rao and Z. D. Ren, Applications of Orlicz Spaces (2002)
251. V. Lakshmikantham and D. Trigiante, Theory of Difference Equations: Numerical Methods and Applications, Second Edition (2002)
252. T. Albu, Cogalois Theory (2003)
253. A. Bezdek, Discrete Geometry (2003)
254. M. J. Corless and A. E. Frazho, Linear Systems and Control: An Operator Perspective (2003)
255. I. Graham and G. Kohr, Geometric Function Theory in One and Higher Dimensions (2003)
256. G. V. Demidenko and S. V. Uspenskii, Partial Differential Equations and Systems Not Solvable with Respect to the Highest-Order Derivative (2003)
257. A. Kelarev, Graph Algebras and Automata (2003)
258. A. H. Siddiqi, Applied Functional Analysis: Numerical Methods, Wavelet Methods, and Image Processing (2004)
259. F. W. Steutel and K. van Harn, Infinite Divisibility of Probability Distributions on the Real Line (2004)
260. G. S. Ladde and M. Sambandham, Stochastic Versus Deterministic Systems of Differential Equations (2004)
261. B. J. Gardner and R. Wiegandt, Radical Theory of Rings (2004)
262. J. Haluška, The Mathematical Theory of Tone Systems (2004)
263. C. Menini and F. Van Oystaeyen, Abstract Algebra: A Comprehensive Treatment (2004)
264. E. Hansen and G. W. Walster, Global Optimization Using Interval Analysis: Second Edition, Revised and Expanded (2004)
265. M. M. Rao, Measure Theory and Integration, Second Edition, Revised and Expanded (2004)
266. W. J. Wickless, A First Graduate Course in Abstract Algebra (2004)

Additional Volumes in Preparation

A FIRST GRADUATE COURSE IN ABSTRACT ALGEBRA

W. J. Wickless

University of Connecticut
Storrs, Connecticut, U.S.A.

CRC Press

Taylor & Francis Group
Boca Raton London New York

CRC Press is an imprint of the
Taylor & Francis Group, an **informa** business

CRC Press
Taylor & Francis Group
6000 Broken Sound Parkway NW, Suite 300
Boca Raton, FL 33487-2742

First issued in paperback 2019

© 2004 by Taylor & Francis Group, LLC
CRC Press is an imprint of Taylor & Francis Group, an Informa business

No claim to original U.S. Government works

ISBN-13: 978-0-8247-5627-7 (hbk)
ISBN-13: 978-0-367-39441-7 (pbk)

Visit the Taylor & Francis Web site at
http://www.taylorandfrancis.com

and the CRC Press Web site at
http://www.crcpress.com

Preface

This is intended to be a reader-friendly, accessible textbook for a first year graduate course in abstract algebra, along the lines of Elbert Walker's *Introduction to Abstract Algebra*, now out of print. The instructor should have no trouble covering the basic topics, Chapters 1-5, together with a few optional ones from Chapters 6-8, in a one year sequence. Chapters 6-8 are logically independent, so can be covered, partially or totally, in any order. Our text incorporates three pedagogical innovations: (1) the tensor product and the notion of projectivity for modules, introduced in Chapter 3, are used in several instances to simplify standard proofs, for example, see the structure theorem for finitely generated modules over a pid and the Cayley-Hamilton Theorem; (2) review material is placed in a series of remarks, directly before it is needed, rather than in a (to me unexciting) chapter at the beginning; (3) a number of simple practice exercises are inserted in the text. None of these exercises is intended to be particularly challenging; they are there as a self-test on the material just discussed. I urge the student to spend the few minutes required trying them before reading on. More difficult, including Ph.D. qualifying level, problems are in the problem sets at the end of each few sections. I have tried to include enough interesting problems for students to practice on a particular topic without overkill. As a student I always found a set of thirty to forty problems (mostly repetitive and some of them marginal to understanding the key concepts) discouraging. My motto is, if you can do one of a certain kind of problem, you can do ten. The converse is also clear.

A recurring theme in the book is the search for a set of algebraic invariants. These are usually a set of cardinal numbers that are invariants of an algebraic object and that determine it up to isomorphism. We visit this particular topic in eight sections, all titled "A structure theorem for _ _ _ _ _]."

By now, the material covered and the proofs of most of the theorems are fairly standard. I am obviously indebted to the many authors before me who have written excellent abstract algebra texts. I first learned the material from Birkhoff and MacLane's text [BML] and then from Jacobson's books. Their present versions, Jacobson's Basic Algebra I,II, provide an extremely comprehensive algebra reference. I still often look back at them for a treatment of more advanced topics, e.g. arithmetic in Dedekind domains. At a lower (advanced undergraduate/beginning graduate) level one has Fraleigh [Fr], clearly written with a large number of illustrative examples, and the always lively exposition of Rotman [R2]. I still also like the old linear algebra text of Hoffman and Kunze [HK]; it has a lot of nice examples and problems.

Occasionally, when a theorem or section has been presented so well in a previous text that it would be both presumptuous and foolish to try to improve the original presentation, I simply give a slight revision, with appropriate acknowledgment, of the original version. A few suggestions for further reading are provided at the end of each chapter.

Part 1, containing the standard topics on groups, rings, modules and vector spaces, should be enough for a first semester course. Part 2 contains the standard results on field and Galois theory, together with chapters on noncommutative rings, group extensions, and abelian groups. It should be enough to finish out the year.

This book is dedicated to my wife Cynthia, my son Miles and my daughter in law Lori, who have been my companions through good times and bad. I would like to thank my colleagues at the University of Connecticut and my algebra colleagues everywhere for their support. I'd especially like to thank Chuck Vinsonhaler, Sarah Glaz, Walt and Penny Lowrie, Vince and Cecelia Giambalvo, Bob Schor, the late Conrad Schwarz, and Carolina Herfkins for friendship beyond the call of duty. Special thanks go to Jim Hurley for generously offering to read through early versions of this manuscript. His editing improved the style-and in a few cases the content-considerably. Finally, I would like to honor the memory of Ross Beaumont and Richard Pierce, who introduced me to the pleasure of abstract algebra and inspired me by their warmth and generosity

A special word to the student: First, I envy you the pleasure of seeing most of this material for the first time. As texts go, this one is fairly short. I don't like books where you have to skim through pages which repeat the same thing over and over and try to figure out what is important. The book has the length it does because there is no extraneous filler and because I tend to say things only once. You will need to read **slowly** and **please** do the exercises and most of the problems. The problem sets are reasonably small, but almost every problem should teach you something you need to know. My goal was to present things in a straightforward, non-ponderous and (perhaps) occasionally humorous way. When you finish, you should have a solid background in a fascinating subject. I hope you like the book. Comments and suggestions for improvement are welcome. Send them to me at wickless@math.uconn.edu

W. J. Wickless

Contents

Chapter 1

Groups (mostly finite)

1.1 Definitions, examples, elementary properties

Remark 1.1.1 *We assume the reader is familiar with the standard notions and notations of set theory. A particular thing we wish to point out is that in listing the elements of a set, say $S = \{x_1, ..., x_n\}$, unless said otherwise, we make no assumption that the listed elements are distinct. When we write $A \subseteq B$, we allow the possibility that the sets A and B could be equal. To indicate that A is properly contained in B, we use the notation $A \subset B$. Recall that the* **Cartesian product** *$A \times B = \{(a, b) : a \in A, b \in B\}$. A* **function** *$f : A \to B$ is a subset $f \subseteq A \times B$ such that (a, b) and (a, b') in f implies $b = b'$. As usual, if $(a, b) \in f$ we write $b = f(a)$. The set A is called the* **domain** *and B the* **codomain** *of the function f. The function $f : A \to B$ is called* **monic** *if, whenever a, a' are distinct elements of A, it follows that $f(a) \neq f(a')$. The function f is* **epic** *if $f(A) = B$, that is each $b \in B$ is $f(a)$ for some $a \in A$. We call f a* **bijection** *if it is both monic and epic. Other popular terminology for these last three properties of functions are* **injective, surjective** *and* **bijective.** *Finally, a* **binary operation** *on a set A is a function $\circ : A \times A \to A$. We denote the image $\circ[(a, a')]$ by $a \circ a'$.*

We begin at the beginning, defining our first object of study.

Definition 1.1.1 *The pair $(G, \circ)$ is a* **group** *if the following axioms hold:*

1. *G is a set and $\circ$ is a binary operation on G.*

2. *There is an element $e \in G$ such that $e \circ x = x \circ e = x$ for all $x \in G$.*

3. For all $x, y, z \in G$, $x \circ (y \circ z) = (x \circ y) \circ z$. (The operation $\circ$ is associative.)

4. For all $x \in G$, there exists an element $y \in G$ such that $x \circ y = y \circ x = e$.

Axiom 3 allows us to write the product of any three elements $x, y, z \in G$ as $x \circ y \circ z$, without parentheses. Using induction (which we'll discuss later), one can show that the product of any finite number of elements can be computed by inserting parentheses in any manner. For example $x \circ y \circ z \circ w \circ u$ could be computed as $x \circ [y \circ (z \circ w)] \circ u$. This result is eminently believable and the proof is fairly cumbersome, so we won't present it.

The element e in Axiom 2 is called the **identity** of G; the element y in Axiom 4 is called the **inverse** of x, we write $y = x^{-1}$. If $(G, \circ)$ satisfies the additional axiom that $x \circ y = y \circ x$ for all $x, y \in G$, we call $(G, \circ)$ an **abelian** (or commutative) group. For ease of notation, we denote $x \circ y$ by xy; that is we represent the binary operation in an abstract group by multiplication. Of course, if the binary operation $\circ$ should naturally be written as addition we do so, and, adopting additive notation, denote the identity element by 0, the inverse of x by $-x$.

Exercise 1.1.1 *Prove that the element e in Axiom 2 is unique and that, for each $x \in G$, the element y in Axiom 4 is also unique. This justifies the terminology "the identity" and "the inverse of x".*

Exercise 1.1.2 *Prove that a group G satisfies the right and left cancellation laws: For $a, b, c \in G$, $ba = ca \implies b = c$ and $ab = ac \implies b = c$.*

(Throughout, we will use $\implies$ for logical implication and $\iff$ for logical equivalence.)

Exercise 1.1.3 *Show that in a group $(x_1 x_2 ... x_n)^{-1} = x_n^{-1} ... x_2^{-1} x_1^{-1}$.*

The notion of a group has been used in mathematics for approximately 150 years; the first formal definition was given near the close of the nineteenth century. Group theory originated in the study of polynomial equations. Currently, it is not only an active research area in its own right, but is fundamental in algebraic geometry, topology, analysis and physics.

Let $N = \{1, 2, 3, \cdots\}$ be the set of natural numbers, $Z = \{\cdots, -1, 0, 1, 2, \cdots\}$ be the set of integers and $Q = \{a/b : a, b \in Z, b \neq 0\}$ be the set of rational numbers. We've all been told since grade school that $(Z, +)$ and $(Q, +)$ are abelian groups. (Although, since the demise of the "new math", we probably weren't told using this exact terminology.) Of course, $(N, +)$ is not a group, lacking both an identity and additive inverses. In fact, these deficiencies are what motivated the extension of N to Z. The sets Q^+ of positive rationals and Q^* of nonzero rationals, each together with multiplication, provide two more familiar examples of abelian groups. The pair $(Z^*, \cdot)$ is not a group, since $2^{-1} \notin Z^*$. The set Q^* is the natural extension of the set Z^* to repair its lack of multiplicative inverses. Here is a different and more interesting sort of example.

Example 1.1.1 *Suppose we have a plastic square with top face black and bottom face white. Say that the vertices on each face are numbered clockwise from the upper left. The square sits on a table, black face up, its bottom edge parallel to the bottom edge of the table, in the following initial position:*
$$\begin{bmatrix} 1 & 2 \\ & \blacksquare & \\ 4 & 3 \end{bmatrix}$$
. We pick up the square, move it around, and replace it on the table with one edge of the square still parallel to the table's bottom edge. A moment's thought shows that we can place our square in 8 possible ways:

$$(M1)\ \begin{bmatrix} 1 & 2 \\ & \blacksquare & \\ 4 & 3 \end{bmatrix} \quad (M2)\ \begin{bmatrix} 4 & 1 \\ & \blacksquare & \\ 3 & 2 \end{bmatrix} \quad (M3)\ \begin{bmatrix} 3 & 4 \\ & \blacksquare & \\ 2 & 1 \end{bmatrix}$$

$$(M4)\ \begin{bmatrix} 2 & 3 \\ & \blacksquare & \\ 1 & 4 \end{bmatrix} \quad (M5)\ \begin{bmatrix} 1 & 4 \\ & \square & \\ 2 & 3 \end{bmatrix} \quad (M6)\ \begin{bmatrix} 2 & 1 \\ & \square & \\ 3 & 4 \end{bmatrix}$$

$$(M7)\ \begin{bmatrix} 3 & 2 \\ & \square & \\ 4 & 1 \end{bmatrix} \quad (M8)\ \begin{bmatrix} 4 & 3 \\ & \square & \\ 1 & 2 \end{bmatrix} .$$

   Let D_4 be the set of these eight motions, labeled $M_1, \cdots, M_8$ displayed above. Define a binary operation on D_4 by composition: that is $M_i M_j$ is the motion that acts by doing M_j then doing M_i. For example $M_5 M_2 = M_8$. (Check this so that you see what we're doing.) It is geometrically clear that the composition of elements of D_4 produces an element of D_4 and that composition is associative. The identity element $e = M_1$ is the motion that keeps the square in its initial position. Any motion M_i can be "undone" to produce the motion M_i^{-1}. For example $M_2^{-1} = M_4$. Thus D_4 is a group, called the **group of symmetries of the square**. (Here we are using the common practice of referring only to the set G as a group, rather than to the pair $(G, \circ)$-as long as the binary operation is understood.)

   We can handle these motions more easily with a little algebraic notation. Let $x = M_2$, the clockwise rotation of the square through $\pi/2$ radians from its initial position. Then we can represent $M_3 = xx = x^2$ and $M_4 =$

$xx^2 = x^3$. Let $y = M_5$, the reflection of the square around the top left-bottom right diagonal. Then the other symmetries can be represented as: $M_6 = xy$, $M_7 = x^2y$, $M_8 = x^3y$. Hence, we can present the elements of our symmetry group as $\{e, x, x^2, x^3, y, xy, x^2y, x^3y\}$. Note that $x^4 = y^2 = e$. Another useful relation is that $yx = x^3y$, both being representations of M_8. (Check this.) So D_4 is nonabelian. We can use this last relation to slide y past x in order to compute products of symmetries algebraically, rather than compose them geometrically. For example $(xy)x = x(yx) = x(x^3y) = x^4y = y$.

Exercise 1.1.4 *Write out a **group table** for D_4, an 8×8 array with rows and columns labeled by the elements of D_4 such that M_iM_j is the element in row i, column j. Note that, by the cancellation laws, each element will appear exactly once in any given row or column.*

In a similar manner, we can define D_n, the **group of all symmetries of the regular n-sided polygon**. The group D_n can be presented as $D_n = \{e, x, x^2, ..., x^{n-1}, y, xy, ..., x^{n-1}y\}$. Here x is a clockwise rotation through $2\pi/n$ radians and y is reflection around a principal diagonal if n is even or around a perpendicular bisection from a vertex to the opposite edge if n is odd.

Exercise 1.1.5 *What are the relations for $x, y \in D_n$ analogous to those in D_4?*

A finite group is a group with finitely many elements. The **order** of a finite group G, denoted $|G|$, is the number of elements in the set G.

Let $x \in G$. (The letter G always denotes a group.) We put $x^0 = e$, $x^1 = x$ and, if $\{x^1, ..., x^{n-1}\}$ have been defined, we can inductively define $x^n = xx^{n-1}$. (See the remark directly below.) The **order** of x, denoted $|x|$, is the least natural number n, if one exists, such that $x^n = e$. If $x^n \neq e$ for all $n > 0$, we say $|x| = \infty$. We have already used the notation x^{-1} for the inverse of x. If $n < -1$ we put $x^n = (x^{-1})^{|n|}$. The usual rules of exponents, $x^nx^m = x^{n+m}$, $(x^n)^m = x^{nm}$, hold. If G is an additive group, these ideas have the following notational expression. We define $0x = 0$ (the first zero is in Z, the second one is in G), $1x = x$ and, inductively, define $nx = x + (n-1)x$ for $n > 1$. As mentioned previously, $-x$ is the additive inverse of x. If $n < -1$ we put $nx = |n|(-x)$. If $n, m \in Z$ and $x \in G$, we have additive versions of the rules of exponents: $(n+m)x = nx + mx$ and $n(mx) = (nm)x$. If G is abelian, we also have $n(x+y) = nx + ny$.

Exercise 1.1.6 *Explain where we use the fact that $(G+)$ is abelian to conclude that $n(x+y) = nx + ny$.*

Remark 1.1.2 *In the above discussion we've used two important and equivalent axioms employed in the construction of the set of natural numbers N. The first allowed us to conclude that x^n is defined for all $n \in N$.*

Axiom 1.1.1 *(Induction) If $S \subseteq N$ contains 1 and if, for $n > 1$, $\{1, ..., n-1\} \subseteq S$ implies $n \in S$, then $S = N$.*

Our set S was the integers $j \in N$ such that x^j is defined.

We used the following axiom to choose the least element from the non-empty set $S = \{t \in N : x^t = e\}$.

Axiom 1.1.2 *(Well ordering) Every nonempty subset of N contains a least element.*

Exercise 1.1.7 *Prove that these axioms are equivalent. Hint: Consider complements.*

Before proceeding, we give examples of three more classes of groups. The first one requires a preliminary remark.

Remark 1.1.3 *We prove the **division algorithm** for Z. Let $a, b \in Z$ with $0 < a < b$. Then there exist unique $q, r \in Z$ with $0 \leq r < a$ such that $b = qa + r$. The existence part of the proof is by induction on b. For the inductive step replace b by $b - a$. If you haven't done a proof by induction for awhile, it would be a good idea to write out this one. To show that q, r are unique, suppose $b = qa + r = q'a + r'$ with $0 \leq r, r' < a$. Then $(q - q')a = r' - r$. Say $q \neq q'$. Without loss, suppose $q > q'$. Then $(q - q')a \geq a$ so that $(r' - r) \geq a$, contradicting the assumptions on r, r'. Hence, $q = q'$ and, therefore, $r = r'$. We can trivially extend the division algorithm to the case $a > 0$, $0 \leq b \leq a$; in this case $q = 0$ or $q = 1$. Finally, we extend the algorithm to the case $a > 0$, $b \in Z$ by noting, if $b < 0$, then $-b = qa + r$, $0 \leq r < a$, so that $b = -(q + 1)a + (a - r)$, $0 < a - r \leq a$. (If $a - r = a$ then $r = 0$ and $b = -qa$.)*

Example 1.1.2 *Let $n > 0$. (Here and hereafter this means let $n \in N$.) The group $Z(n)$ consists of the set $\{0, 1, ..., n - 1\}$ together with the binary operation $+_n$ of addition mod n. More precisely, if $0 \leq x, y \leq n - 1$, $x +_n y = r$, where $x + y = qn + r$ as in Remark 1.1.3. Plainly, $+_n$ is a binary operation on $\{0, 1, ..., n - 1\}$ with identity 0. If $x \in Z(n)$, its additive inverse, $-x$, is $n - x$. The only thing that needs a little proof is the associative law.*

Exercise 1.1.8 *Check the associative law for $Z(n)$.*

In the definition of the group $Z(n)$, we used $+_n$ to distinguish between addition mod n and ordinary addition of integers. In practice the subscript n is usually omitted.

Example 1.1.3 *Let S be a nonempty set. A **permutation** of S is a bijection $f : S \to S$. The group $P(S)$ is the set of all permutations of S, together with the binary operation $\circ$ of composition of functions; that is $(f \circ g)(x) = f[g(x)]$. It is well-known that $f, g \in P(S) \implies (f \circ g) \in P(S)$; 1_S, the identity function on S, is the identity element under function composition; composition of functions is associative and $f \in P(S) \implies f^{-1} \in P(S)$. Thus, $(P(S), \circ)$ is a group.*

If $S = \{1, ..., n\}$ then $P(S)$ is called the **symmetric group on** n **letters** and is denoted S_n. We will examine the group S_n in detail in section three.

Exercise 1.1.9 *Consider the group S_3 of all bijections of $\{1, 2, 3\}$. The elements of S_3 can be written $\{e, (12), (13), (23), (123), (132)\}$. Here (ab) is a **transposition**, the map that interchanges a and b and (abc) is a **three cycle**, the map that sends a to b, b to c, and c to a. Write out the multiplication table for S_3 . Use juxtaposition for function composition, e.g. write $(12)(13) = (132)$.*

Exercise 1.1.10 *For each $n > 0$, let $GL(n, \mathcal{R})$ be the set of invertible $n \times n$ real matrices with binary operation matrix multiplication. Assuming that matrix multiplication is associative, prove that $GL(n, \mathcal{R})$ is a group (called the **general linear group**).*

To prove that matrix multiplication is associative, one has start with the definition of matrix multiplication: the i-j entry in AB, where $A = (a_{kl}), B = (b_{rs})$ is $\sum_{s=1}^{n} a_{is}b_{sj}$. Then we put $C = (c_{uv})$ and work through the double summations corresponding to $(AB)C$ and $A(BC)$. This is tedious but not mathematically deep.

1.2   Subgroups, cyclic groups

Definition 1.2.1 *Let $(G, \circ)$ be a group. A subset $H \subseteq G$ is a **subgroup** of G if $(H, \circ)$, the set H together with the binary operation on G restricted to H, is a group.*

For $H \subseteq G$, we write $H \leq G$ to mean that H is a subgroup of G. The subsets $\{e\}$ and $\{G\}$ are subgroups of any group G, called the **trivial subgroups**. If $H \leq G$ and $\{e\} \subset H \subset G$, H is called a **proper subgroup**.

Some simple examples of subgroups are: (1) $(3Z, +) \leq (Z, +)$ (Here $3Z$ is the set of all integral multiples of 3.); (2) $(Z, +) \leq (Q, +)$ and (3) $(Q^+, \cdot) \leq (Q^*, \cdot)$.

The definition above is a little hard to employ to check whether or not a given subset is a subgroup. The following theorem provides a more convenient criterion.

Theorem 1.2.1 *(subgroup test theorem) Let $H \subseteq G$. Then $H \leq G$ if and only if $e \in H$ and $x, y \in H \Rightarrow xy^{-1} \in H$. Here e is the identity of G and y^{-1} denotes the inverse of y computed in the group G.*

Proof. Suppose that H is a subgroup of G. The group H contains an identity element e'. We have $e'e' = e' = ee'$. By right cancellation, $e = e'$. So $e \in H$ and e is the identity for both G and H. It follows that, for $y \in H, y^{-1}$ is unambiguously defined, whether computed in the group G or in the group H. Hence, if $x, y \in H$, then $xy^{-1} \in H$. Conversely, suppose $e \in H$ and $x, y \in H \Longrightarrow xy^{-1} \in H$. Let $y \in H$. Then $y^{-1} = ey^{-1} \in H$. If $x \in H$, then $xy = x(y^{-1})^{-1} \in H$. Since H inherits the associative law from G, H is a group. ∎

Exercise 1.2.1 *Let $\{H_i : i \in I\}$ be any collection of subgroups of G. Prove $\cap_{i \in I} H_i$ is a subgroup.*

For $\{x_1, ..., x_k\} \subseteq G$, let $< x_1, ..., x_k >$ be the intersection of all subgroups $H \leq G$ such that $\{x_1, ..., x_k\} \subseteq H$. By the above exercise, $< x_1, ..., x_k >$ will be a subgroup of G, plainly the smallest subgroup of G containing the set $\{x_1, ..., x_k\}$. The subgroup $< x_1, ..., x_k >$ is called the **subgroup of G generated by** $\{x_1, ..., x_k\}$. If $G = < x_1, ..., x_k >$, we say that $\{x_1, ..., x_k\}$ is a **set of generators** for G. Here's a more concrete description of the subgroup $< x_1, ..., x_k >$.

Theorem 1.2.2 *For $\{x_1, ..., x_k\} \subseteq G$, the subgroup $< x_1, ..., x_k >$ coincides with the set $S = \{x_{i_1}^{e_1} \cdots x_{i_t}^{e_t} : t \in N, \{e_1, ..., e_t\} \subseteq Z, \{i_1, ..., i_t\} \subseteq \{1, ..., k\}\}$.*

Proof. If H is any subgroup of G containing $\{x_1, ..., x_k\}$, then H must contain all elements of the form $x_{i_1}^{e_1} \cdots x_{i_t}^{e_t}$ as above. Hence, $S \subseteq < x_1, ..., x_k >$. Plainly, $\{x_1, ..., x_k\} \subseteq S$. To complete the proof we only need show that $S \leq G$. We use the subgroup test theorem above. First, $e = x_1^0 \in S$. If $a = x_{i_1}^{e_1} \cdots x_{i_t}^{e_t}$ and $b = x_{j_1}^{f_1} \cdots x_{j_u}^{f_u}$ are in S, then $ab^{-1} = x_{i_1}^{e_1} \cdots x_{i_t}^{e_t} x_{j_u}^{-f_u} \cdots x_{j_1}^{-f_1} \in S$, and the proof is complete. ∎

If $x \in G$, then $< x >$ is called the **cyclic subgroup of G generated by** x. We next examine the possibilities for the structure of $< x >$.

Theorem 1.2.3 *For $x \in G$ there are two possibilities: (1) If $\mid x \mid = n < \infty$, then $< x >$ is the set of n distinct elements $\{e, x, ..., x^{n-1}\}$. (2) If $\mid x \mid = \infty$, then $< x >$ is the countable set of distinct elements $\{x^k : k \in Z\}$.*

Proof. (1) By the previous theorem $< x > = \{x^k : k \in Z\}$. If $\mid x \mid = n$, let $H = \{e, x, ..., x^{n-1}\}$. For $k \in Z$, $k = qn + r$ with $0 \leq r \leq n - 1$. Hence, $x^k = x^{qn+r} = (x^n)^q x^r = e^q x^r = x^r \in H$. Thus, $< x > = H$. Moreover, if $0 \leq i < j \leq n - 1$ and $x^i = x^j$, then $0 < j - i < n$ with $x^{j-i} = e$, contradicting $\mid x \mid = n$. Hence, H contains n distinct elements. (2) If $\mid x \mid = \infty$, it is easy to check that all the elements of the set $\{x^k : k \in Z\}$ must be distinct. Thus, $< x >$ is this countable set of distinct elements. ∎

Incidentally, we've shown that our two notions of order are consistent; for $x \in G$, $\mid x \mid = \mid < x > \mid$.

A group G is called **cyclic** if there exists $x \in G$ such that $G = < x >$. A cyclic group will automatically be abelian. The groups $(Z, +), (Z(n), +_n)$ are cyclic. We'll soon see that, in a reasonable sense, these are the only cyclic groups. The groups D_4 and S_3 are non-cyclic since no element in either group will serve as a generator for the entire group. Using the division algorithm in Z it is easy to prove the following theorem. We assign the proof to you in Problem 1.3.8.

Theorem 1.2.4 *Let $G = < x >$ be a cyclic group and $< e > \neq H \leq G$. Then $H = < x^d >$, where d is the least natural number such that $x^d \in H$.*

At this point, we need to recall (or, perhaps, introduce) the notion of an equivalence relation.

Remark 1.2.1 *Let S be a nonempty set. A **relation** on S is any subset of $S \times S$. An **equivalence relation** is a subset $E \subseteq S \times S$ with the following properties: (1) The pair $(x, x) \in E$ for all $x \in S$. (2) If $(x, y) \in E$, then $(y, x) \in E$. (3) If $(x, y) \in E$ and $(y, z) \in E$, then $(x, z) \in E$. Following standard usage, we write $x \sim y$ and say that x is equivalent to y to mean that $(x, y) \in E$. The three axioms then read: (1) For all $x \in S$, $x \sim x$ (the reflexive law); (2) If $x \sim y$ then $y \sim x$ (the symmetric law); (3) If $x \sim y$ and $y \sim z$ then $x \sim z$ (the transitive law). For $x \in S$, define the **equivalence class** of x, denoted $[x]$, by $[x] = \{y \in S : y \sim x\}$. Using properties (1)-(3) it's not difficult to prove the following facts: (i) Any two equivalence classes are either disjoint or identical. (ii) The set S can be written as a disjoint union of equivalence classes. The following exercise appears as a theorem in most undergraduate abstract algebra texts.*

Exercise 1.2.2 *Prove (i) and (ii).*

For us here are two of the most important examples of equivalence relations and equivalence classes.

Example 1.2.1 *(Congruence mod n) Let $n > 1$. For $x, y \in Z$ define $y \equiv x$ (n) if $y - x$ is (evenly) divisible by n. It is easy to check that congruence mod n is an equivalence relation on Z. Using the division algorithm for Z, we see that the distinct equivalence classes for this equivalence relation are in 1-1 correspondence with the remainders after division by n : $[0], [1], ..., [n - 1]$.*

Example 1.2.2 *(Congruence mod a subgroup) Let $H \leq G$. For $x, y \in G$, define $y \sim x$ if $y = hx$ for some $h \in H$. Then $[x] = \{hx : h \in H\}$ This set is naturally denoted Hx and is called the* **right coset** *of x mod H.*

Exercise 1.2.3 *Prove that congruence mod H is an equivalence relation on G.*

The number of distinct right cosets Hx of G mod H is called the **index** of H in G, denoted $[G : H]$.

The following extremely useful fact was discovered by Lagrange in the late 18-th century. It shows that, for finite groups, the order of a subgroup must divide the order of the group. More specifically:

Theorem 1.2.5 *Let G be finite and $H \leq G$. Then $\mid G \mid = [G : H] \mid H \mid .$*

Proof. We can write G as a disjoint union of right cosets $G = Hx_1 \cup Hx_2 \cdots \cup Hx_k$, where $k = [G : H]$. Without loss, take $x_1 = e$, so that $Hx_1 = H$. For each fixed i, $2 \leq i \leq k$, the map $h \to hx_i$ is a bijection from the coset $Hx_1 = H$ to the coset Hx_i. Hence, each coset has the same number of elements as the subgroup H, and so $[G : H] \mid H \mid = \mid G \mid .$ ∎

Since, for $x \in G, \mid x \mid = \mid < x > \mid$ we have:

Corollary 1.2.1 *Let G be finite and $x \in G$. Then $\mid x \mid$ is finite and divides $\mid G \mid .$*

Recall that a natural number $p > 1$ is **prime** if p has no proper divisors, that is $p = ab$, $a, b \in N$ implies $a = 1$ or $b = 1$. Henceforth, the symbol p will always stand for a prime.

Corollary 1.2.2 *Let $\mid G \mid = p$ and $e \neq x \in G$. Then $G = < x > .$*

1.3   Factorization in Z

We provide a brief discussion of factorization in Z. For $a, b \in Z$ write $a \mid b$ to mean that a divides b and $a \nmid b$ to mean that a doesn't divide b.

Definition 1.3.1 *The natural number d is called a **greatest common divisor** (gcd) of $a, b \in Z$ if $d \mid a$, $d \mid b$ and, if $x \in Z$ is such that x divides both a and b, it follows that $x \mid d$. If d is a gcd of a, b we write $d = (a, b)$.*

Theorem 1.3.1 *Let $a, b \in Z$ with at least one of a, b nonzero. Then there exists a unique $d = (a, b)$. Moreover, $d = ua + vb$ for $u, v \in Z$.*

   Proof. First we prove unicity. For a, b as above, let $d, d' \in N$ satisfy the definition $d = (a, b), d' = (a, b)$. Then $d \mid a$, $d \mid b$, so, since d' is a gcd, $d \mid d'$. Similarly, $d' \mid d$. Hence, since d, d' are positive, $d = d'$. We've shown that a gcd, if it exists, is unique. For the existence part of the proof, let $H = \{ua + vb : u, v \in Z\}$. Then $\{0\} \neq H \leq Z$ (check this). By the additive version of Theorem 1.2.4 above, $H = <d>$, where $d = d1$ is the smallest natural number in H. By definition of H, $d = ua + vb$ for some $u, v \in Z$. We claim $d = (a, b)$. First, $a, b \in H = <d>$, so $a = td, b = sd$ for $s, t \in Z$. Thus, d is a common divisor of a and b. Now suppose $x \in Z$ is such that $x \mid a$, $x \mid b$. Then $x \mid (ua + vb) = d$. Hence, $x \mid d$ and $d = (a, b)$. ∎

Corollary 1.3.1 *Let p be a prime and $a, b \in Z$ with $p \mid ab$. Then $p \mid a$ or $p \mid b$.*

   Proof. Suppose $p \mid ab$ and $p \nmid a$. Then $(a, p) = 1$, so that $1 = ua + vp$ for $u, v \in Z$. Therefore, $b = 1b = uab + vpb$ is divisible by p. ∎

   We call $a, b \in Z$ **relatively prime** if $(a, b) = 1$.

Exercise 1.3.1 *Let $a, b, c \in Z$. (1) Then a, b are relatively prime $\iff$ there exist $u, v \in Z$ with $1 = ua + vb$. (2) Suppose $a \mid bc$ and $(a, b) = 1$. Then $a \mid c$.*

Example 1.3.1 *For $a, b \in Z(n)$, we define multiplication mod n in an analogous manner to addition mod n : $a \cdot b = r \in Z(n)$, where $ab = qn + r$, $0 \leq r \leq n$. It is immediate that $\cdot$ is a commutative binary operation on the set $Z(n)$ and that 1 is an identity element.*

Exercise 1.3.2 *Check that multiplication mod n is an associative operation on $Z(n)$.*

Exercise 1.3.3 *Is $Z(n)$ a group?*

Theorem 1.3.2 *(fundamental theorem of arithmetic for Z) Each nonzero integer n can be written as ± 1 times a product of primes, $n = (\pm 1)p_1 \cdots p_k$. The number k is unique, as are the primes p_i, up to arrangement.*

Proof. It suffices to prove each $n > 1$ is a product of primes, the number being unique and the prime factors themselves unique up to arrangement. For the existence of a prime factorization we induct on the set $\{n \in N : n \geq 2\}$. The number 2 is prime. Suppose we've established the existence of a prime factorization for all $m < n$. If n is prime, we're done. Otherwise, $n = ab$ for $1 < a, b < n$. By the inductive assumption, a and b are each a product of primes. Hence, so is $n = ab$. (That was easy enough.)

Now suppose $n = p_1 \cdots p_k = q_1 \cdots q_s$ with the p_i, q_j prime for $1 \leq i \leq k, 1 \leq j \leq s$. We have $p_1 \mid n = q_1(q_2 \cdots q_s)$. By Corollary 1.3.1, $p_1 \mid q_1$ or $p_1 \mid q_2 \cdots q_s$. If $p_1 \mid q_1$ then $p_1 = q_1$ since both are prime. If $p_1 \mid q_2 \cdots q_s = q_2(q_3 \cdots q_s)$, then $p_1 \mid q_2$ or $p_1 \mid q_3 \cdots q_s$. Hence, $p_1 = q_2$ or $p_1 \mid q_3 \cdots q_s$. Continue is this manner, until we obtain $p_1 = q_j$ for some j. Cancel $p_1 = q_j$ from the left and right sides of the equality $p_1 \cdots p_k = q_1 \cdots q_s$ to obtain $p_2 \cdots p_k = q_1 \cdots q_{j-1}q_{j+1} \cdots q_s$. Now argue with p_2 as we did with p_1 to match p_2 with some q_t. Eventually, we must match up each p_i with a q_j, $j = j(i)$- otherwise we would be left with a product of primes on one side of an equality equal to 1 on the other side. ∎

Normally, one gathers together all equal primes and writes the factorization as $n = \pm p_1^{e_1} \cdots p_r^{e_r}$ where the e_i are positive integers and $p_1 < p_2 < \cdots < p_r$.

The following was proved by Euclid, using the same proof given here.

Theorem 1.3.3 *There are an infinite number of primes.*

Proof. The proof is by contradiction. Assume the contrary, that there exists $k \in N$ such that $\{p_1, ..., p_k\}$ is the set of all primes. The integer $n = (p_1 \cdots p_k) + 1$ is a product of primes. But no p_i can divide n, a contradiction. Hence, there must be infinitely many primes. ∎

First Problem Set

Problem 1.3.1 *If $g \in G$ and $m > 0$ is such that $g^m = e$, then $\mid g \mid$ divides m.*

Problem 1.3.2 *Find all the subgroups of D_4 (not all are cyclic). For each subgroup $H \leq D_4$, list the distinct right cosets of D_4 mod H.*

Problem 1.3.3 *Let $Q' = Q \backslash \{-1\}$, the set of all rational numbers excluding -1. For $x, y \in Q'$, define $x \circ y = x + y + xy$. Prove that $(Q', \circ)$ is a group.*

Problem 1.3.4 *Let G be a set with an associative binary operation such that: (1) there is an element e such that $ex = x$ for all $x \in G$ and (2) for each $x \in G$ there exists an element $x' \in G$ such that $x'x = e$. Prove that G is a group.*

Problem 1.3.5 *Let C be a finite group and $A \leq B \leq C$. Show that $[C : A] = [C : B] [B : A]$.*

Problem 1.3.6 *Let $H \leq G$. Starting with the appropriate equivalence relation, develop the theory of left cosets xH. Prove the number of left cosets of G mod H is equal to the number of right cosets of G mod H. Hence $[G : H]$ can be defined as either the number of left or the number of right cosets of G mod H.*

Problem 1.3.7 *A word of caution: Despite the result of the previous problem, not every left coset of G mod H is a right coset of G mod H. Find an example to show that this is so.*

Problem 1.3.8 *Let $G = \langle x \rangle$ be a cyclic group and $\langle e \rangle \neq H \leq G$. Use the division algorithm for Z to prove that $H = \langle x^d \rangle$, where d is the least natural number such that $x^d \in H$.*

Problem 1.3.9 *Prove that $(Q^*, \cdot)$ is not generated by any finite subset of Q^*.*

Problem 1.3.10 *Generalize the statement and proof of Theorem 1.3.1 to the case of a finite set of integers $\{m_1, ... m_k\}$, not all $m_i = 0$.*

Problem 1.3.11 *The **least common multiple** (lcm) of a set of nonzero integers $\{l_1, ..., l_k\}$ is the natural number m such that: (1) $l_i \mid m, 1 \leq i \leq k$, and (2) if $x \in Z$ is such that $l_i \mid x, 1 \leq i \leq k$, then $m \mid x$. Use the fundamental theorem of arithmetic to prove the existence and unicity of the lcm for any finite set of nonzero integers.*

Problem 1.3.12 *Let $U(n) = \{a \in Z(n) : (a, n) = 1\}$. Prove that, for each $n > 1$, $(U(n), \cdot)$ is a group.*

Problem 1.3.13 *Prove or disprove: For all $n > 1, U(n)$ is cyclic.*

1.4   Isomorphism

Example 1.4.1 *Let $C(n) = \{[0], [1], ..., [n-1]\}$ be the set of distinct equivalence classes in Z for the relation congruence mod n. We can define a binary operation $\oplus_n$ on $C(n)$ by $[i] \oplus_n [j] = [i+j]$. Since we're defining the sum of two equivalence classes as the equivalence class of the sum of two representatives, we must prove that our operation is well defined, that is $i \equiv i_1 \ (n), j \equiv j_1 \ (n) \implies (i+j) \equiv (i_1 + j_1) \ (n)$. (See the exercise immediately below.) This having been proved, a moment's thought shows that $(C(n), \oplus_n)$ is a group. Another moment's thought reveals that $(C(n), \oplus_n)$ is just the group $(Z(n), +_n)$ written in a different form. More precisely, the correspondence $\theta : i \rightarrow [i]$ is a bijection from $Z(n)$ to $C(n)$ that preserves the group tables. To see this, note that $\theta(i +_n j) = \theta(r) = [r]$, where $i + j = qn + r$. Furthermore, $[r] = [i+j] = [i] \oplus_n [j] = \theta(i) \oplus_n \theta(j)$. Hence, for $i, j \in Z(n)$, $\theta(i +_n j) = \theta(i) \oplus_n \theta(j)$.*

Exercise 1.4.1 *Verify that addition mod n is a well-defined binary operation on $C(n)$.*

Definition 1.4.1 *Let $(G, \circ)$ and $(G', *)$ be groups. We say G is **isomorphic to** G', and write $G \cong G'$, if there is a bijection $\theta : G \rightarrow G'$ such that $\theta(x \circ y) = \theta(x) * \theta(y)$ for all $x, y \in G$. The map θ is called an **isomorphism from** G **to** G'.*

If $G \cong G'$, they may be isomorphic via more than one isomorphism. To indicate that G, G' are isomorphic via a particular isomorphism θ we write $\theta : G \cong G'$.

The proof of the following theorem is an immediate application of Theorem 1.2.3. Its proof is left as an exercise.

Theorem 1.4.1 *(1) Any cyclic group of order n is isomorphic to $Z(n)$. (2) Any infinite cyclic group is isomorphic to Z. (3) Any group of order p is isomorphic to $Z(p)$.*

Exercise 1.4.2 *Prove that $(Z, +) \cong (3Z, +)$.*

Exercise 1.4.3 *Find an isomorphism from $(\mathcal{R}^+, \cdot)$, the group of positive real numbers and multiplication, to $(\mathcal{R}, +)$, the additive group of reals. Think about calculus.*

Theorem 1.4.2 *Isomorphism satisfies the reflexive, symmetric and transitive properties.*

Proof. We verify these three properties in turn. (reflexive) For any group G, the identity map 1_G: $G \cong G$.

(symmetric) Suppose $\theta : G \cong G'$. Since θ is a bijection there is an inverse bijection $\theta^{-1} : G' \to G$. To claim that $\theta^{-1} : G' \cong G$, we just need check that θ^{-1} gets along with the group operations. Let $x, y \in G'$. We want the equality $\theta^{-1}(xy) = \theta^{-1}(x)\theta^{-1}(y)$. If we apply θ to both sides of our desired equality, we obtain the identity $xy = xy$. Since θ is one to one, our desired equality holds.

(transitive) Say $\theta : G \cong H$ and $\phi : H \cong K$. Then $\phi\theta : G \to K$ is a bijection. If $x, y \in G$, then $(\phi\theta)(xy) = \phi[\theta(xy)] = \phi[\theta(x)\theta(y)] = \phi[\theta(x)]\phi[\theta(y)] = (\phi\theta)(x)(\phi\theta)(y)$. Hence $\phi\theta : G \cong K$. ∎

The reason we didn't state the above theorem as "Isomorphism is an equivalence relation." is that an equivalence relation is defined on a set. To avoid the same sort of immediate paradoxes that can be constructed if we allow the collection of all sets to be a set, we cannot construct a set containing all groups. Look ahead to Section 4.2 for further discussion on sets and cardinality.

To prove two groups G, G' are isomorphic, we must construct an explicit isomorphism $\theta : G \cong G'$. To prove two groups aren't isomorphic, it's usually best to find a property of one, that would be carried over by an isomorphism, but that is not a property of the other. Trivially, for example, $Z(8)$ is not isomorphic to D_4, since $Z(8)$ is cyclic while D_4 is not even abelian.

Second Problem Set

Problem 1.4.1 *Let $Z(4) \oplus Z(2)$ be the group consisting of the set $Z(4) \times Z(2)$ together with componentwise addition. Prove that $Z(4) \oplus Z(2)$ is not isomorphic to $Z(8)$.*

Problem 1.4.2 *Prove that $(\mathcal{R}, +)$ is not isomorphic to $(\mathcal{R}^*, \cdot)$, the multiplicative group of nonzero reals.*

Problem 1.4.3 *Let $\theta : G \cong G'$ and $x \in G$. Prove $\mid x \mid = \mid \theta(x) \mid$.*

Problem 1.4.4 *Let V (the **Klein 4-group**) be the group consisting of the set $\{e, a, b, c\}$ with multiplication table $a^2 = b^2 = c^2 = e$, $ab = ba = c$, $bc = cb = a$, $ac = ca = b$. Show that V is isomorphic to $Z(2) \oplus Z(2)$. Here $Z(2) \oplus Z(2)$ is the additive group on the Cartesian product $Z(2) \times Z(2)$ determined by componentwise addition.*

Problem 1.4.5 *Prove that every group of order four is isomorphic to either $Z(4)$ or V.*

1.5 Homomorphisms

We now consider a map from one group to another that need not be bijective, but that still respects the group operations.

Definition 1.5.1 *A **homomorphism** from a group $(G, \circ)$ to a group $(G', *)$ is a map $\theta : G \to G'$ such that $\theta(x \circ y) = \theta(x) * \theta(y)$ for all $x, y \in G$.*

An onto homomorphism is called an **epimorphism**; a monic homomorphism is called a **monomorphism**. An isomorphism is a map which is both an epimorphism and a monomorphism, that is a bijective homomorphism. The following simple theorem will come in handy later, especially when dealing with objects defined by universal mapping properties. See Problem 1.9.4 for the first example of this phenomena.

Theorem 1.5.1 *Two groups G, G' are **isomorphic** if and only if there are homomorphisms $\alpha : G \to G', \beta : G' \to G$ such that $\beta\alpha = 1_G, \alpha\beta = 1_{G'}$. Here $1_G, 1_{G'}$ are the identity maps on G, G' respectively.*

Proof. If $f : G \cong G'$, put $\alpha = f, \beta = f^{-1}$. Conversely, suppose α, β are as in the statement of the theorem. If $\alpha(x) = \alpha(y)$, then $x = \beta\alpha(x) = \beta\alpha(y) = y$, so that α is monic. Let $z \in G'$. Then $z = \alpha\beta(z) = \alpha[\beta(z)]$, so that α is epic. Hence, $\alpha : G \cong G'$. ∎

Two natural subsets are associated with each homomorphism $\theta : G \to G'$.

Definition 1.5.2 *Let $\theta : G \to G'$ be a homomorphism and let e' be the identity element of G'. Then **kernel** $\theta = \{g \in G : \theta(g) = e'\}$ and **image** $\theta = \{g' \in G' : g' = \theta(g) \text{ for some } g \in G\}$.*

We abbreviate image θ by Im θ or $\theta(G)$ and kernel θ by $Ker\ \theta$. We collect a few basic facts concerning homomorphisms in the following theorem.

Theorem 1.5.2 *Let $\theta : G \to G'$ be a homomorphism and let e, e' be the identities of G, G'. Then:*

1. $\theta(e) = e'$;

2. $\theta(x^{-1}) = [\theta(x)]^{-1}$ for all $x \in G$;

3. θ is monic if and only if $Ker\ \theta = \{e\}$;

4. if $x \in G$ has finite order, then $|\theta(x)|$ divides $|x|$;

5. $Ker\ \theta \leq G$;

6. $Im\ \theta \leq G'$.

Proof. (1) Following our usual custom, denote both the operation in G and the operation in G' simply by multiplication. Then in the group G' we have the chain of equalities: $\theta(e)e' = \theta(e) = \theta(ee) = \theta(e)\theta(e)$. By left cancellation, we obtain $e' = \theta(e)$. (2) For $x \in G, \theta(x^{-1})\theta(x) = \theta(x^{-1}x) = \theta(e) = e'$. Since G' is a group, the latter equation is enough to conclude that $\theta(x^{-1}) = [\theta(x)]^{-1}$. (3) Since $\theta(e) = e'$, if θ is monic then $Ker\ \theta$ coincides with the singleton set $\{e\}$. Suppose that $Ker\ \theta = \{e\}$ and that $a, b \in G$ with $\theta(a) = \theta(b)$. Then $[\theta(b)]^{-1}\theta(a) = e'$. By (2) it follows that $e' = \theta(b^{-1})\theta(a) = \theta(b^{-1}a)$. Hence, $b^{-1}a \in Ker\ \theta = \{e\}$ so that $b^{-1}a = e$ and $b = a$. (4) Say $\mid x \mid = n$. Then $[\theta(x)]^n = \theta(x^n) = \theta(e) = e'$. Hence, $\mid \theta(x) \mid = m \leq n$. By Problem 1.3.1, $m \mid n$ and the proof of (4) is complete. The proofs of 5 and 6 are Problem 1.5.1 below. ∎

Here are some examples of homomorphisms.

Example 1.5.1 *For any groups G, G' the* **trivial homomorphism** *sending each element of G to the identity element $e' \in G'$ is an uninteresting example of a homomorphism. Here the kernel is G and the image is $\{e'\}$.*

Example 1.5.2 *Let $\pi : Z \to Z(n)$ be the epic map that sends $x \in Z$ to $x \in Z(n)$. Addition in $Z(n)$ has been defined so that π will be a epimorphism. The kernel of π is nZ.*

Example 1.5.3 *Let $GL(2, \mathcal{R})$ be the group of all invertible 2×2 real matrices with binary operation matrix multiplication. For a matrix $A \in GL(2, \mathcal{R})$, let $d(A)$ be the determinant of A. Since $A \in GL(2, \mathcal{R}) \iff d(A) \neq 0$ and $d(AB) = d(A)d(B)$, the map $d : A \to d(A)$ is an homomorphism from $GL(2, \mathcal{R})$ to $(\mathcal{R}^*, \cdot)$. Furthermore, if $r \in \mathcal{R}^*$, then $r = d\begin{bmatrix} r & 0 \\ 0 & 1 \end{bmatrix}$ so that $Im\ \theta = \mathcal{R}^*$. The kernel of d is the subgroup of $GL(2, \mathcal{R})$ consisting of all matrices of determinant one.*

Example 1.5.4 *Let $\theta : Z \to Z$ be the map sending x to $5x$. Since $5(x + y) = 5x + 5y$ for all $x, y \in Z, \theta$ is a homomorphism. Here $Ker\ \theta = \{0\}$ and $Im\ \theta = 5Z$.*

A slight generalization of the example above shows that $Z \cong nZ$ for all $n > 0$.

Third Problem Set

Problem 1.5.1 *Complete the proof of Theorem 1.5.2 by proving parts 5 and 6.*

Problem 1.5.2 *(a) Let D_3 be the group of all rotations of an equilateral triangle. Write a multiplication table for D_3 in terms of x, a $120°$ clockwise rotation of the triangle from its initial position, and y, a rotation around the perpendicular bisector drawn from one vertex of the triangle to its opposite edge. (b) Prove that $D_3 \cong S_3$.*

Problem 1.5.3 *Let $\theta : G \to H$ be a homomorphism of groups. (a) If $g \in G$ is an element of finite order prove that $\mid \theta(g) \mid$ divides $\mid g \mid$. (b) If θ is an isomorphism prove that $\mid \theta(g) \mid = \mid g \mid$. (c) If all elements $g \in G$ are of finite order and θ is such that $\mid \theta(g) \mid = \mid g \mid$ for all $g \in G$, what can you say about the homomorphism θ?*

Problem 1.5.4 *Prove that there are no homomorphisms from $(Q, +)$ to $(Z, +)$ except for the map sending every element of Q to $0 \in Z$.*

1.6 Normal subgroups and factor groups

We turn to the construction of factor groups. For $H \leq G$ denote the set of right cosets of G mod H by G/H. We consider the following problem: Try to define a binary operation on G/H such that G/H will be a group and such that the natural map $\pi : G \to G/H$, given by $\pi(g) = Hg$, will be a homomorphism. A little thought reveals that, if π is to be a homomorphism, we have no choice for the binary operation we put on G/H. We must define $HaHb$ so that $HaHb = \pi(a)\pi(b) = \pi(ab) = Hab$. Assume, for the moment, that the stipulation $HaHb = Hab$ produces a well defined coset multiplication on the set G/H . Then the coset He will play the role of the identity element. The coset Hx^{-1} will be the inverse for a given coset Hx. The associative law for coset multiplication follows trivially from the associative law in G.

Hence, in solving our problem, the only unclear point is whether or not our supposed binary operation is actually well defined. We gave the product of cosets in terms of the product of coset representatives. So we need to check that the product is independent of the choice of these representatives, equivalently that $Ha = Ha', Hb = Hb' \implies Hab = Ha'b'$. Since $a'b' = ea'b' \in Ha'b', Ha'b' = Hab \iff a'b' \in Hab$. (Recall that right cosets are equivalence classes, so are either disjoint or identical.) Thus,

$Hab = Ha'b' \iff a'b' = hab$ for some $h \in H$. Since $Ha = Ha', Hb = Hb'$, we have $a' = h_1 a, b' = h_2 b$ for some $h_1, h_2 \in H$. Hence, $a'b' = (h_1 a)(h_2 b) = h_1[a(h_2 b)] = h_1[(ah_2)b]$. If we could rewrite the element ah_2 in the form $h_3 a$ for some $h_3 \in H$, we could continue the chain of equalities in the previous sentence as follows: $h_1[(ah_2)b] = h_1[(h_3 a)b] = (h_1 h_3)(ab)$. Since $h_1 h_3 = h \in H$, we could conclude that $Hab = Ha'b'$, as desired. This line of reasoning leads to the following definition.

Definition 1.6.1 *A subgroup $H \leq G$ is called* **normal** *in G if $Ha = aH$ for all $a \in G$. In other words, the right and left coset mod H of any element $a \in G$ coincide as subsets of G.*

We write $H \lhd G$ to indicate that H is a normal subgroup of G. The two trivial subgroups, $\{e\}$ and G are normal in any group. If G is abelian every subgroup is normal. In the discussion prior to the definition, we have essentially proved the following theorem.

Theorem 1.6.1 *Let $H \lhd G$. Then the set G/H together with the multiplication $HaHb = Hab$ is a group, called the* **factor group** *of the group G mod its normal subgroup H. The* **natural factor map** $\pi : g \to Hg$ *is a epimorphism from G onto G/H.*

The order of the factor group above is $[G : H]$. It is high time for some examples. We ask the reader to wait one more minute, while we complete our circle of ideas. First, we prove a useful lemma. If $K \leq G$, let $g^{-1}Kg = \{g^{-1}kg : k \in K\}$.

Exercise 1.6.1 *Let $K \leq G$ and $g \in G$. Prove that $g^{-1}Kg \leq G$ and that $K \cong g^{-1}Kg$.*

Lemma 1.6.1 *A subgroup $K \leq G$ is normal if and only if $g^{-1}Kg \subseteq K$ for all $g \in G$.*

Proof. By direct calculation, $Kg = gK$ for all $g \iff g^{-1}Kg = K$ for all g. The latter set equality holds if and only if $g^{-1}Kg \subseteq K$ for all g, since $g^{-1}Kg \subseteq K$ for all g implies $g(g^{-1}Kg)g^{-1} = K \subseteq gKg^{-1}$ for all g. (Note that the final set containment is logically equivalent to $K \subseteq g^{-1}Kg$ for all g.) ∎

Here is an important characterization for normal subgroups.

Theorem 1.6.2 *A subset $H \subseteq G$ is a normal subgroup of G if and only if $H = Ker\,\theta$ for some homomorphism θ mapping G into some group G'.*

Proof. Suppose $H \lhd G$. Then $\pi : G \to G/H$ is a homomorphism. By definition, $Ker\ \pi = \{g \in G : Hg = He\} = \{g \in G : g \in He\} = H$. Conversely, let $\theta : G \to G'$ be a homomorphism and let $K = Ker\ \theta$. We show $g^{-1}Kg \subseteq K$ for all $g \in G$. If we apply θ to an element of the form $g^{-1}kg$ we obtain $\theta(g^{-1}kg) = \theta(g^{-1})\theta(k)\theta(g) = [\theta(g)]^{-1}e'\theta(g) = e'$. Hence $g^{-1}Kg \subseteq K$ for all g and the proof is complete. ∎

A group G' is called a **epimomorphic image** of a group G if there is an epimorphism mapping G onto G'. We've already noted that, if $H \lhd G$, then the factor group G/H is a epimorphic image of G via the natural factor map π. The next theorem, called the first isomorphism theorem, shows that the factor groups G/H are, up to isomorphism, the only epimorphic images of G.

Theorem 1.6.3 *(first isomorphism theorem) Let θ be a homomorphism from G onto G'. Then $G/Ker\ \theta \cong G'$.*

Proof. We know that $K = Ker\ \theta \lhd G$, so the factor group G/K is defined. Let $\bar\theta : G/K \to G'$ be "defined" by $\bar\theta(Kg) = \theta(g)$. We claim that $\bar\theta$ is our desired isomorphism. First of all, we must check that $\bar\theta$ is well defined. We have a chain of logical equivalences: $Kb = Ka \Longleftrightarrow b = ka$ for some $k \in K \Longleftrightarrow ba^{-1} \in K \Longleftrightarrow \theta(ba^{-1}) = e' \Longleftrightarrow \theta(b)[\theta(a)]^{-1} = e' \Longleftrightarrow \theta(b) = \theta(a) \Longleftrightarrow \bar\theta(Kb) = \bar\theta(Ka)$. This chain, when read from left to right, shows that $\bar\theta$ is well defined. When read from right to left it proves that $\bar\theta$ is 1-1. Since θ is onto so is $\bar\theta$. Finally, $\bar\theta$ is a homomorphism since $\bar\theta(KaKb) = \bar\theta(Kab) = \theta(ab) = \theta(a)\theta(b) = \bar\theta(Ka)\bar\theta(Kb)$. ∎

At last, some examples!

Example 1.6.1 *Since $(Z+)$ is abelian, every subgroup is normal. We already know that the set of subgroups of Z (including the trivial ones) are of the form dZ, where $d \geq 0$. The trivial factor groups are $Z/0Z = Z/(0) = \{(0)+m : m \in Z\} \cong Z$ and $Z/1Z = Z/Z = \{Z+0\} \cong <0>$. The nontrivial ones, occurring when $d > 1$, are $Z/dZ = \{dZ+0, ..., dZ+(d-1)\} \cong Z(d)$. These factor groups coincide with the epimorphic images of Z.*

Example 1.6.2 *Consider $S_3 = \{e, (12), (1,3), (2,3), (123), (132)\}$. (Refer back to Exercise 1.1.8.) Since S_3 has order 6, its only proper subgroups must have orders 2 or 3, hence be cyclic groups generated by an element of order 2 or 3. So the proper subgroups of S_3 are $< (12) >= \{e, (12)\}$, $< (13) >= \{e, (13)\}, < (23) >= \{e, (23)\}$ and $< (123) >=< (132) >= \{e, (123), (132)\}$. The subgroup $\{e, (12)\}$ is not a normal subgroup of S_3*

since $(13)\{e,(12)\} = \{(13),(13)(12)\} = \{(13),(123)\}$, *while* $\{e,(12)\}(13) =$
$\{(13),(12)(13)\} = \{(13),(132)\}$. *Similarly, neither of the other subgroups
of order 2 are normal. Let* $A_3 = \{e,(123),(132)\}$. *By the cancellation laws,
left and right multiplication by any* $h \in A_3$ *induces a bijection on the set*
A_3. *Hence,* $hA_3 = A_3 = A_3 h$ *for all* $h \in A_3$. *Thus,* $A_3 \lhd S_3 \iff$
$\sigma A_3 = A_3 \sigma$ *for each* $\sigma \notin A_3$. *(An analogous observation allows us to save
some work when trying to check whether or not any subgroup* H *of any
group* G *is normal.) It is easy to check that* $\tau A_3 = A_3 \tau$ *for each of the three
transpositions* τ. *So* $A_3 \lhd S_3$. *It follows that the only proper epimomorphic
image of* S_3 *is (up to isomorphism)* $S_3 / A_3 \cong Z(2)$.

Example 1.6.3 *For any group* G, *the* **center** *of* G, $C(G) = \{g \in G :
gx = xg$ *for all* $x \in G\}$. *Clearly,* $e \in C(G)$. *Let* $x, y \in C(G)$ *and* $g \in G$.
Since $y \in C(G)$, $yg = gy$ *so that* $g = y^{-1}gy$ *and* $gy^{-1} = y^{-1}g$. *Since* g
was arbitrary, $y^{-1} \in C(G)$. *Thus, since* $x \in C(G)$, $(xy^{-1})g = x(y^{-1}g) =
x(gy^{-1}) = (xg)y^{-1} = (gx)y^{-1} = g(xy^{-1})$. *We have shown that* $xy^{-1} \in
C(G)$. *By the subgroup test theorem,* $C(G) \le G$. *By definition of* $C(G)$,
$xC(G) = C(G)x$ *for all* $x \in G$ *and, thus,* $C(G) \lhd G$.

Exercise 1.6.2 *Find* $C(D_4)$. *Identify the factor group* $D_4 / C(D_4)$.

Definition 1.6.2 *Let* G *be a group. A* **commutator** *in* G *is an element
of the form* $c = xyx^{-1}y^{-1}$ *for some* $x, y \in G$. *Let* $G^1 \subseteq G$ *be the set of all
finite products of commutators,* $G^1 = \{c_1 \cdots c_n : 1 \le n < \infty\}$.

Theorem 1.6.4 *(a) The set* G^1 *is a subgroup of* G. *(b) Further,* $G^1 \lhd G$.
(c) G/G^1 *is abelian. (d) If* $H \lhd G$ *is such that* G/H *is abelian, then*
$G^1 \le H$.

Proof. (a) First, $e = xx^{-1}x^{-1}x$ is a commutator. Since the inverse
of a commutator is a commutator, the same holds for a finite product of
commutators. Thus, we can apply the subgroup test theorem to prove
$G^1 \le G$. For (b), first note that conjugation by any fixed $z \in G$ is a
homomorphism. Thus, it suffices to show that $z^{-1}xyx^{-1}y^{-1}z \in G^1$ for
any single commutator $xyx^{-1}y^{-1}$. But $z^{-1}xyx^{-1}y^{-1}z$ can be rewritten
as the commutator $(z^{-1}xz)(z^{-1}yz)(z^{-1}x^{-1}z)(z^{-1}y^{-1}z)$ and (b) is proved.
(c) For $x, y \in G$, $(G^1x)(G^1y)(G^1x^{-1})(G^1y^{-1}) = G^1xyx^{-1}y^{-1} = G^1e$. It
follows that $G^1xG^1y = G^1yG^1x$ for all $x, y \in G$. (d) Suppose $H \lhd G$ is
such that G/H is abelian. Then $HxHyHx^{-1}Hy^{-1} = He$ for all $x, y \in G$.
Hence, $xyx^{-1}y^{-1} \in H$ for all x, y. It follows that $G^1 \le H$ and the proof is
complete. ∎

The subgroup G^1 is called the **commutator subgroup** of G

We conclude this section with some useful results on subgroups and related isomorphisms.

For $N \lhd G$, there's a nice correspondence between subgroups of the factor group G/N and subgroups of G. If $N \subseteq H \leq G$, let $H/N = \{Nh : h \in H\} \subseteq G/N$. Since $H \leq G$, it follows immediately that $H/N \leq G/N$.

Theorem 1.6.5 *(subgroup correspondence) Let $N \lhd G$. Every subgroup X of G/N is of the form $X = H/N$ for some $N \leq H \leq G$. Furthermore, $X \lhd G/N$ if and only if $H \lhd G$.*

Proof. Let $X \leq G/N$. Put $H = \{y \in G : Ny \in X\}$. For $n \in N, Nn = Ne \in X$. Hence, $N \subseteq H$. If $y, z \in H$, then $Ny, Nz \in X$. Thus, $Nyz^{-1} = (Ny)(Nz)^{-1} \in X$ so that $yz^{-1} \in H$. Therefore, by the subgroup test theorem, $H \leq G$. By definition, $H/N = X$.

If $H \lhd G$, let $\pi : G/N \to G/H$ be the factor map, $\pi(Ng) = Hg$. Since $N \subseteq H$ it is easy to check that π is well defined and, if well defined, π is plainly a homomorphism. We have $Ker \ \pi = H/N = X$, so $X \lhd G/N$.

Suppose $X = H/N \lhd G/N$ and let $g \in G$. Then $(Ng)^{-1}(H/N)(Ng) \subseteq H/N$. Thus, $(g^{-1}Hg)/N \subseteq H/N$. This latter containment holds if and only if $(g^{-1}Hg) \subseteq H$. Hence, by Lemma 1.5.1, $H \lhd G$. ∎

Theorem 1.6.6 *(second isomorphism theorem) Suppose that $N \lhd G$ and $H \lhd G$ are arbitrary normal subgroups with $N \subseteq H$. Then $(G/N)/(H/N) \cong G/H$.*

Proof. Let $\pi : G/N \to G/H$ be the natural factor map as defined above. By the first isomorphism theorem, $G/(Ker \ \pi) \cong G/H$. Since $Ker \ \pi = H/N$, the proof is complete. ∎

Theorem 1.6.7 *(third isomorphism theorem) Let $H \leq G$ and $N \lhd G$. Then: (a) $H \cap N \lhd H$; (b) $NH = \{nh : n \in N, h \in H\} \leq G$ and (c) $H/H \cap N \cong NH/N$.*

Proof. (a) If $h \in H$, then $h^{-1}(H \cap N)h \subseteq H \cap (h^{-1}Nh) \subseteq H \cap N$, so that $H \cap N \lhd H$. (b) Plainly, $e \in NH$. Let $x = nh, y = n_1h_1 \in NH$. Then $xy^{-1} = nh(h_1)^{-1}(n_1)^{-1}$. Since $N \lhd G, [h(h_1)^{-1}](n_1)^{-1} = n_2[h(h_1)^{-1}]$ for some $n_2 \in N$. Thus, $xy^{-1} = (nn_2)[h(h_1)^{-1}] \in NH$ and $NH \leq G$. (c) First, note that $N \lhd NH$ so the factor group NH/N makes sense. Let $\theta : H \to NH/N$ be defined by $\theta(h) = Nh$. It's easy to check that θ is an epimorphism and that $Ker \ \theta = H \cap N$. By the first isomorphism theorem, the proof is complete. ∎

Fourth Problem Set

Problem 1.6.1 (a) Consider the proper subgroups of D_4, the group of symmetries of the square. Which of these subgroups are normal? (b) For each normal subgroup that you found, construct the group table for the corresponding factor group. (c) What are the proper homomorphic images of D_4 (in terms of groups we've already seen)?

Problem 1.6.2 If $G/C(G)$ is cyclic prove that $G = C(G)$, in other words G is abelian. If $G/C(G)$ is abelian, need G be abelian?

Problem 1.6.3 Let N, K be normal subgroups of G with $N \cap K = <e>$. Show that, for any $n \in N, k \in K$, we have $nk = kn$.

Problem 1.6.4 Prove that any homomorphic image of an abelian group is abelian and that any homomorphic image of a cyclic group is cyclic.

Problem 1.6.5 Suppose $C \leq B \leq A$ and that $C \lhd A$. Prove that $B/C \lhd A/C$ if and only if $B \lhd A$.

Problem 1.6.6 Prove that every subgroup H of index two in a group G is normal.

Problem 1.6.7 In proving the third isomorphism theorem, we showed that $H \leq G, N \lhd G \implies HN \leq G$. Prove that, if $N_1, N_2, ..., N_k$ are normal subgroups of G, then $N_1 N_2 \cdots N_k = \{n_1 n_2 \cdots n_k : n_i \in N_i\}$ is a normal subgroup of G.

Problem 1.6.8 For $d > 0$, prove the all subgroups $X \leq Z/dZ$ are of the form $X = mZ/dZ$, where $m \mid d$.

Problem 1.6.9 Let $\theta : G \cong G'$ and suppose $N \lhd G, N' \lhd G'$ are normal subgroups such that $\theta(N) = N'$. Prove that θ induces an isomorphism from G/N to G'/N'.

1.7   Simple groups and composition series

Definition 1.7.1 A group $G \neq <e>$ is called **simple** if G and $<e>$ are the only normal subgroups of G.

For p a prime, $Z(p)$ is a simple abelian group. In fact, cyclic groups of order p are the only simple abelian groups. (See Problem 1.7.2 at the end of this section.) For the last hundred years, the problem of classifying all finite simple groups has been a major mathematical undertaking. It has only recently come to completion (see [GLS]). The reason for the intense interest in finite simple groups can be understood as follows. Let G be a finite group. If G is not simple, we can choose a **maximal normal subgroup** G_1 of G; that is G_1 is a proper normal subgroup of G such that, if $N \lhd G$ with $G_1 \subseteq N \subseteq G$, then $N = G_1$ or $N = G$. If G_1 is not simple, take a maximal normal subgroup $G_2 \lhd G_1$. Since G is finite, eventually some subgroup G_k must be simple. We then will have a chain of subgroups: $< e > = G_{k+1} \lhd G_k \lhd \cdots \lhd G_2 \lhd G_1 \lhd G_0 = G$ such that each G_{i+1} is maximal normal in G_i, $0 \le i \le k$. By the subgroup correspondence theorem, each factor group G_i/G_{i+1} is simple. Thus, G can be thought of as being built up in "layers", each layer a simple group.

Definition 1.7.2 *Let G be a group. A chain of subgroups $< e > = G_{k+1} \lhd G_k \lhd \cdots \lhd G_1 \lhd G_0 = G$, each G_{i+1} normal in its predecessor G_i, $0 \le i \le k$, is called a **normal series** for G. The natural number k, which is the number of proper normal subgroups, is called the **length** of the series.*

Example 1.7.1 *The series $< e > \lhd < y > \lhd < x^2, y > \lhd D_4$ and $< e > \lhd < x > \lhd D_4$ are normal series for D_4 of lengths 2 and 1, respectively. Note that $< y >$ is not normal in D_4.*

Definition 1.7.3 *A normal series $< e > = G_{k+1} \lhd G_k \lhd \cdots \lhd G_1 \lhd G_0 = G$ such that each factor group G_i/G_{i+1} is simple, is a **composition series** for G. The set $\{G_i/G_{i+1} : 0 \le i \le k\}$ is called the **set of successive factor groups** of the series.*

Every finite group and some infinite ones will have a composition series. Given a finite group G, it is likely that there are many ways in that we could choose a descending chain of successive maximal normal subgroups to form a composition series.

Example 1.7.2 *Let $G = Z(20)$. Here all subgroups are normal. There are three composition series: (1) $< 0 > \lhd < 10 > \lhd < 5 > \lhd Z(20)$, (2) $< 0 > \lhd < 10 > \lhd < 2 > \lhd Z(20)$ and (3) $< 0 > \lhd < 4 > \lhd < 2 > \lhd Z(20)$. To verify that series (1) is a composition series, we compute the set of successive factor groups. Since $< 5 > = \{0, 5, 10, 15\}$ and $< 10 > = \{0, 10\}$, it follows that $| Z(20)/ < 5 > | = 20/4 = 5$, $| < 5 > / < 10 > | = 4/2 = 2$ and $| < 10 > / < 0 > | = 2/1 = 2$. Thus, the set of factor groups for the first composition series is $\{Z(5), Z(2), Z(2)\}$. Each factor group is simple so that (1) is a composition series.*

Exercise 1.7.1 *Compute the set of successive factor groups for series (2) and (3) above to verify that each is a composition series.*

Having just done the exercise, the following definition should seem natural to you.

Definition 1.7.4 *Two composition series for a group G, (1) $< e >= G_{k+1} \lhd G_k \lhd \cdots \lhd G_2 \lhd G_1 \lhd G_0 = G$ and (2) $< e >= H_{l+1} \lhd H_l \lhd \cdots \lhd H_2 \lhd H_1 \lhd H_0 = G$ are* **equivalent** *if $k = l$ and if the sets of factor groups of each series, counting repetitions, are equal. (When comparing the sets of factor groups, we identify isomorphic groups.)*

It is routine to check that this relation is an equivalence relation on the set of all composition series of a given group G. The next theorem, called the Jordan-Holder Theorem, is fairly amazing. The proof we give is a somewhat expanded version of the one in [W]. The argument is not deep, but is fairly intricate, so it takes careful reading. The best thing to do is first get an overview of the plan of the proof, then check the details.

Theorem 1.7.1 *(Jordan-Holder Theorem) Any two composition series for a group are equivalent.*

Proof. If G has a composition series of length 0, $< e > \lhd G$, then G is simple, so that the only composition series for G is $< e > \lhd G$. Inductively, suppose $k \geq 1$ and that any group with a composition series of length less than k has the property that all its composition series are equivalent. Say that G has a composition series:

$$(1) < e >= G_{k+1} \lhd G_k \lhd \cdots \lhd G_2 \lhd G_1 \lhd G.$$

Let

$$(2) < e >= H_{l+1} \lhd H_l \lhd \cdots \lhd H_2 \lhd H_1 \lhd G$$

be another composition series for G. Consider the normal series:

$$(3) < e >= G_{k+1} \cap H_1 \lhd G_k \cap H_1 \lhd \cdots \lhd G_2 \cap H_1 \lhd G_1 \cap H_1 \lhd G_1 \lhd G$$

and

$$(4) < e >= H_{l+1} \cap G_1 \lhd H_l \cap G_1 \lhd \cdots \lhd H_2 \cap G_1 \lhd H_1 \cap G_1 \lhd H_1 \lhd G.$$

Since $G_{i+1} \cap H_1 \lhd G_i \cap H_1$, the third isomorphism theorem yields $(G_i \cap H_1)/(G_{i+1} \cap H_1) = (G_i \cap H_1)/[G_{i+1} \cap (G_i \cap H_1)] \cong [G_{i+1}(G_i \cap$

$H_1)]/G_{i+1}$. Now $G_{i+1}(G_i \cap H_1) \lhd G_i$, since it is the product of two normal subgroups of G_i. Since G_i/G_{i+1} is simple, either $G_{i+1}(G_i \cap H_1) = G_{i+1}$ or $G_{i+1}(G_i \cap H_1) = G_i$. This means that $(G_i \cap H_1)/(G_{i+1} \cap H_1)$ is either isomorphic to G_i/G_{i+1} or to $G_{i+1}/G_{i+1} = <e>$.

Therefore, if we remove repetitions from $<e> = G_{k+1} \cap H_1 \lhd G_k \lhd \cdots \lhd G_2 \cap H_1 \lhd G_1 \cap H_1 \lhd G_1$, we get a composition series for G_1. By induction, this series is equivalent to the series $<e> = G_{k+1} \lhd G_k \lhd \cdots \lhd G_2 \lhd G_1$; hence (1) and (3′) [(3) (with repetitions removed] are equivalent. If $G_1 = H_1$, we can use the induction hypothesis to conclude that (1) $\sim$ (2).

If $G_1 \neq H_1$, then $G_1 H_1, H_1 G_1$ are normal subgroups of G properly containing G_1, so $G_1 H_1 = H_1 G_1 = G$. Thus, $G_1/(G_1 \cap H_1) \cong (G_1 H_1)/H_1 = G/H_1$ and $H_1/(H_1 \cap G_1) \cong (H_1 G_1)/G_1 = G/G_1$. We have matched up the first two factor groups of series (3) and (4). But now, in view of the inductive hypothesis, the pieces of (3′) and (4′) [(4) with repetitions removed] from the group $G_1 \cap H_1 = H_1 \cap G_1$ on down are equivalent. Hence, (1) $\sim$ (4′). If we delete H_0 from (2) and from (4′), we get composition series (2′) and (4″) for the group H_1. Since length (4″) $<$ length (4′) = length (1), we can use induction to claim that (2′) $\sim$ (4″). So (2) $\sim$ (4′) and, hence, (1) $\sim$ (2). The proof is complete. ■

Fifth Problem Set

Problem 1.7.1 *Use the Jordan Holder Theorem to prove the fundamental theorem of arithmetic.*

Problem 1.7.2 *Prove that every simple abelian group $(G, +)$ is isomorphic to $Z(p)$ for some prime p.*

Problem 1.7.3 *Prove that any normal series for a group G can be refined, by adding terms, to produce a composition series for G.*

1.8 Symmetric groups

We start our study of symmetric groups with an abstract theorem, proved by Cayley in the 19-th century. What the theorem really says is that symmetric groups are extremely complicated. However, the theorem does have some useful applications in the study of finite groups.

Theorem 1.8.1 *Every group is isomorphic to a subgroup of a group of permutations.*

Proof. Let G be a group. Recall that $S(G)$ is the group whose objects are bijections from G to itself and whose binary operation is function composition. For each fixed $g \in G$, let $L_g : G \to G$ be the function induced by left multiplication by g, that is $L_g(x) = gx$ for all $x \in G$. By left cancellation, L_g is monic. If $y \in G$, then $y = L_g(g^{-1}y)$. Thus, $L_g \in S(G)$. Now consider the map $L : G \to S(G)$ given by $L(g) = L_g$. We have $L(gg') = L_{gg'}$. Applying the map $L_{gg'}$ to a fixed $x \in G$ gives $L_{gg'}(x) = (gg')x = g(g'x) = L_g[L_{g'}(x)]$. Hence, $L(gg') = L_g \circ L_{g'}$ and L is a homomorphism from G into $S(G)$. Suppose that $L(g) = L(g')$ for $g, g' \in G$. Then $L_g = L_{g'}$, so that $g = ge = L_g(e) = L_{g'}(e) = g'e = g'$. Hence L is 1-1 and $G \cong L(G) \le S(G)$. ∎

Corollary 1.8.1 *Every group of order n is isomorphic to a subgroup of S_n.*

Proof. Order the elements of G in any way and write $G = \{g_1, ..., g_n\}$. Now each L_g can be regarded as an element of S_n. ∎

Exercise 1.8.1 *Number the elements of D_4 according to the following list $\{e, x, x^2, x^3, y, xy, x^2y, x^3y\}$. In the associated embedding $D_4 \to S_8$, what permutations correspond to x, y, x^3y?*

We next look more closely at the group S_n, first noting that $\mid S_n \mid = n!$. If $\sigma \in S_n$ we could display σ by writing the numbers from 1 to n in a row and, below each $1 \le i \le n$, putting the number $\sigma(i)$. However this notation is very cumbersome; we hasten to find a better method.

Definition 1.8.1 *Let $\{k_1, ..., k_j\} \subseteq \{1, ..., n\}$ with $k_i \ne k_{i'}$ if $i \ne i'$ and $j > 1$. A j-cycle $(k_1, ..., k_j)$ in S_n is the map that sends k_i to k_{i+1} for $1 \le i \le j - 1$, sends k_j back to k_1, and fixes each integer $m \notin \{k_1, ..., k_j\}$. A transposition is a two cycle (a, b). Two cycles $(k_1, ..., k_j)$ and $(l_1, ..., l_r)$ are called disjoint if $\{k_1, ..., k_j\} \cap \{l_1, ..., l_r\} = \phi$.*

If the integers $k_1, ..., k_j$ are imagined to be arranged in a circle, the reason for the terminology is clear. A j-cycle can be written in j different ways. For example, in S_8, $(2, 5, 3, 7) = (5, 3, 7, 2) = (3, 7, 2, 5) = (7, 2, 5, 3)$.

Exercise 1.8.2 *Let $\lambda, \sigma \in S_n$ be disjoint cycles. Prove in at most three lines that $\lambda\sigma = \sigma\lambda$.*

Here is an extremely useful result.

Theorem 1.8.2 *Every $e \ne \sigma \in S_n$ can be written as the product of disjoint cycles. This representation is unique up to the order of the factors.*

Proof. For convenience, put $\bar{n} = \{1, ..., n\}$ and fix $\sigma \neq e$. For $i, j \in \bar{n}$, say $j \sim i$ if $j = \sigma^t(i)$ for some $t \in Z$. We show that $\sim$ is an equivalence relation. Since $\sigma^0 = e$, $i = \sigma^0(i)$ and $i \sim i$ for all i. Suppose $j \sim i$. Then $j = \sigma^t(i)$. Thus, $\sigma^{-t}(j) = i$ and $i \sim j$. If $j = \sigma^t(i)$ and $k = \sigma^{t'}(j)$, then $k = \sigma^{t'}[\sigma^t(i)] = \sigma^{t'+t}(i)$. This shows that $k \sim j, j \sim i \Longrightarrow k \sim i$. Hence, $\sim$ is an equivalence relation. The equivalence class of an element i is the set $\{\sigma^t(i) : t \in Z\}$.

Suppose that $\mid \sigma \mid = m$. For each $i \in \bar{n}$ let $m_i \leq m$ be the least natural number such that $\sigma^{m_i}(i) = i$. Define the **orbit** of i, $O(i)$, to be the set $\{i = \sigma^0(i), \sigma(i), ..., \sigma^{m_i-1}(i)\}$. It is easy to check that the elements listed in $O(i)$ are pairwise distinct. If $m_i = 0$ then $O(i) = \{i\} = \{\sigma^t(i) : t \in Z\}$. Suppose $m_i > 0$. by the division algorithm, for all $t \in Z, t = qm_i + r$ with

$$0 \leq r < m_i. \text{ Thus } \sigma^t(i) = \sigma^{qm_i+r}(i) = \sigma^r(i) \in O(i). \text{ Hence, for each } i \in \bar{n},$$

the equivalence class of i coincides with $O(i)$.

It follows that $\bar{n}$ is partitioned into a disjoint union $\bar{n} = \cup_{j=1}^s O(i_j)$, where the sets $O(i_j), 1 \leq j \leq s$, are the distinct orbits. By inspection, the action of σ on n is the same as the action of the product of the disjoint cycles $c_1 c_2 \cdots c_s$, where $c_j = (i_j, \sigma(i_j), ..., \sigma^{m_{i_j}-1}(i_j))$. Since disjoint cycles commute, the product $c_1 c_2 \cdots c_s$ can be written in any order. If σ were to be written as another product of disjoint cycles $\sigma = d_1 d_2 \cdots d_r$, then the numbers appearing in each d_j must again be the numbers in one of the equivalence classes of $\sim$. So $r = s$ and, after rearrangement, $d_j = c_j$ for all j. ∎

Writing a given σ as a product of disjoint cycles is simplicity itself. We take any $i \in \bar{n}$ and spin out its orbit by successively applying σ. This gives us a cycle. Then we take $j \notin O(i)$ and do it again. For example, let $\sigma = \begin{pmatrix} 1 & 2 & 3 & 4 & 5 & 6 & 7 & 8 & 9 \\ 9 & 3 & 6 & 4 & 1 & 8 & 7 & 2 & 5 \end{pmatrix}$. If we start with $i = 1$, we spin out the cycle $(1, 9, 5)$. Choosing an element not in $O(1)$, say 2, gives us the cycle $(2, 3, 6, 8)$. Hence, $\sigma = (1, 9, 5)(2, 3, 6, 8)$. By convention, any number not listed in a cycle decomposition of a permutation is assumed to be fixed by that permutation. Thus, we don't list singleton cycles; in this case (4) and (7).

Corollary 1.8.2 *Every $\sigma \in S_n, n \geq 2$, can be written as a product of transpositions.*

Corollary 1.8.3 *The identity element $e = (1, 2)(1, 2)$. If $\sigma \neq e$, write $\sigma = c_1 c_2 \cdots c_s$ as a product of disjoint cycles (by convention, of length greater than one). By direct computation, the cycle $(k_1, ..., k_j)$ is equal to the product $(k_1, k_j)(k_1, k_{j-1}) \cdots (k_1, k_4)(k_1, k_3)(k_1, k_2)$.*

For the permutation σ directly after the theorem above, we have $\sigma = (1, 9, 5)(2, 3, 6, 8) = [(1, 5)(1, 2)][(2, 8)(2, 6)(2, 3)]$. Note that no claim of unicity is made in the corollary. For example, in S_3, $(1, 3)(1, 2)(2, 3)(1, 3) = (1, 3, 2) = (1, 2)(1, 3)$. However, our next theorem shows that there is one sense in which the factorization of a permutation into a product of transpositions is unique. Two natural numbers k and l are said to have the same **parity** if either both are even or both are odd.

Theorem 1.8.3 *Let $\sigma \in S_n$ be written as a product of transpositions in two ways, $\sigma = \tau_1 \cdots \tau_k = \upsilon_1 \cdots \upsilon_l$. Then k and l have the same parity.*

Proof. There are a number of different proofs of this theorem, some easier than others. The proof here comes from [Fr]. Let $\sigma \in S_n$ and let $\tau = (i, j)$ be a transposition. To prove the theorem is suffices to prove the following claim: the number of orbits of σ and of $\tau\sigma$ differ by one. To see that the claim suffices, suppose $\sigma = \tau_1 \cdots \tau_k = \upsilon_1 \cdots \upsilon_l$. Since the identity permutation e has n singleton orbits and $\sigma = \sigma e = (\tau_1 \cdots \tau_k)e = (\upsilon_1 \cdots \upsilon_l)e$, by repeated applications of the claim we see that the number of orbits of differs from n by a number whose parity is both that of k and of l. Hence, k and l have the same parity.

Now to prove the claim, we consider two cases. I) Suppose that i and j lie in different orbits of σ. (a) If both $O(i)$ and $O(j)$ are singletons then multiplying σ on the left by τ just adds the orbit $\{i, j\}$. (b) If exactly one of these orbits, say $O(i)$, is a singleton, write $\sigma = (j, \times, \times, \times, k)c_2 \cdots c_s$ as a product of disjoint cycles, with i not involved in any cycle. Here the $\times$'s denote possible other numbers in the first orbit. It's easy to check in this case that $(i, j)\sigma = (i, j)(j, \times, \times, \times, k)c_2 \cdots c_s = (j, \times, \times, \times k, i)c_2 \cdots c_s$. Here we've decreased the number of orbits by one. (c) Say i, j lie in different orbits, neither of which is a singleton. Then $\sigma = (b, j, \times, \times, \times)(a, i, \times, \times, \times)c_3 \cdots c_s$. In this case $(i, j)\sigma = (a, j, \times, \times, \times, b, i, \times, \times, \times)c_3 \cdots c_s$ and the first two cycles have been joined into one.

II) Suppose σ has an orbit of the form $(a, i, \times, \times, \times, b, j, \times, \times, \times)$. Then, as a product of disjoint cycles, $\sigma = (a, i, \times, \times, \times, b, j, \times, \times, \times)c_2 \cdots c_s$ and $(i, j)\sigma = (a, j, \times, \times, \times)(b, i, \times, \times, \times)c_3 \cdots c_s$. The first cycle has been split into two. The proof is now complete. ∎

Definition 1.8.2 *A permutation is called **even** if it can be written as a product of an even number of transpositions. Otherwise, it's called **odd**. Define $\text{sgn}(\sigma) = 1$ if σ is even, $\text{sgn}(\sigma) = -1$ if σ is odd. Let $A_n = \{\sigma \in S_n : \sigma \text{ is even}\}$.*

Theorem 1.8.4 *For $n \geq 2$, A_n is a normal subgroup of S_n of index 2.*

Proof. Let s mapping S_n onto the multiplicative group $\{\pm 1\}$ be given by $s(\sigma) = sgn(\sigma)$. Plainly, for $\sigma, \tau \in S_n$, we have $\text{sgn}(\sigma\tau) = \text{sgn}(\sigma)\text{sgn}(\tau)$. Thus, s is an epimorphism Since $Ker\ s = A_n$, $A_n \lhd S_n$. Finally $[S_n : A_n] = |S_n/A_n| = |\{\pm 1\}| = 2$. ∎

We close this section with a theorem giving a new infinite set of simple groups, in addition to our first example $\{Z(p) : p$ a prime$\}$. Our proof is a slight revision of the one in [W]. As in the proof of the Jordan Holder Theorem, the argument is not deep but more technical than we would like.

Theorem 1.8.5 *For* $n \geq 5, A_n$ *is simple.*

Proof. Let $< e > \neq N \lhd A_n$ for $n \geq 5$. We'll show $N = A_n$. Note that $(a, b, c) = (b, c)(a, c)$ so that A_n contains all 3-cycles. The proof will involve 3 steps: (1) Show N contains a 3-cycle. (2) Show N contains all 3-cycles. (3) Show A_n is generated by 3-cycles.

The proof uses the following technique repeatedly: Suppose $\alpha \in N, \beta \in A_n$. Then $\alpha^{-1}\beta^{-1}\alpha\beta = \alpha^{-1}(\beta^{-1}\alpha\beta) \in N$. We select $\beta \in A_n$ so that $\alpha^{-1}\beta^{-1}\alpha\beta$ has a special form.

For example, suppose the 5-cycle $\alpha = (i_1, ..., i_5) \in N$. Let $\beta = (i_1, i_2, i_3)$. Then $\alpha^{-1}\beta^{-1}\alpha\beta = (i_2, i_3, i_5) \in N$.

Let $e \neq \alpha \in N$ and write α as a product of disjoint cycles $\alpha = \alpha_1 \cdots \alpha_k$. Say $\alpha_1 = (i_1, ..., i_r)$. First, suppose $r > 3$ and let $\beta = (i_1, i_2, i_r)$. Then $\alpha^{-1}\beta^{-1}\alpha\beta = (i_1, i_2, i_{r-1}) \in N$.

If $r = 3$ and $k = 1$, then $(i_1, i_2, i_3) \in N$. If $r = 3$ and $k > 1$, say $\alpha_2 = (j_1, ..., j_s)$. Put $\beta = (i_1, i_2, j_1)$. Then $\alpha^{-1}\beta^{-1}\alpha\beta = (i_1, i_2, j_1, i_3, j_s) \in N$. Since N contains a 5-cycle, we employ the argument above to conclude that there is a 3-cycle in N.

Suppose $r = 2$. Then $k > 1$, since α is even. Let $\beta = (i_1, i_2, j_1)$ with j_1 the first listed element of α_2. Then $\alpha^{-1}\beta^{-1}\alpha\beta = (i_1, j_s)(i_2, j_1)$ is an element of N. Choose an integer a with $1 \leq a \leq n$ such that a is different from i_1, i_2, j_1, j_2. (This is the one and only place that we use the hypothesis $n \geq 5$.) Then, if we put $\beta = (i_1, i_2, a), \alpha = (i_1, j_s)(i_2, j_1)$, we have $\alpha^{-1}\beta^{-1}\alpha\beta = (i_1, i_2, j_1, j_s, a) \in N$. Again, a 5-cycle is in N so a 3-cycle is in N.

We have completed step (1), showing that N contains a 3-cycle.

On to step (2)! Let $\alpha = (i_1, i_2, i_3) \in N$, and choose a such that $1 \leq a \leq n, a \neq i_1, i_2, i_3$. If $\beta = (i_1, a, i_2,)$, then $\beta^{-1}\alpha\beta = (a, i_3, i_2) \in N$, whence $(a, i_3, i_2)^2 = (a, i_2, i_3) \in N$. We have shown: (*) $(x, y, z,) \in N, a \neq x, y, z \implies (a, y, z) \in N$.

Let (a, b, c) be an arbitrary 3-cycle. If $a \neq i_1, i_2, i_3$, by (*), $(a, i_2, i_3) \in N$. Say that a is one of i_1, i_2, i_3 and, without loss, suppose $a = i_1$.

Then $(i_1, i_2, i_3) = (a, i_2, i_3) \in N$. So, in any event, we can conclude that $(a, i_2, i_3) = (i_2, i_3, a) \in N$.

Suppose $b \neq i_3$. If $b = i_2$ then $(a, i_2, i_3) = (a, b, i_3) = (i_3, a, b) \in N$. If $b \neq i_2$, then, by $(*)$, $(b, i_3, a) = (i_3, a, b) \in N$. We have shown $b \neq i_3$ implies $(i_3, a, b) \in N$.

Now suppose $b = i_3$. Then, from above, $(i_2, b, a) \in N$. So, in view of the previous paragraph, either (i_3, a, b) or (i_2, b, a) is in N. By $(*)$, since $c \neq i_3$ or $c \neq i_2$, either (c, a, b) or (c, b, a) is in N. Since $(a, b, c) = (c, a, b) = (c, b, a)^2$, we have shown that $(a, b, c) \in N$. This completes step (2).

Finally, noting that $(i_1, i_3)(i_1, i_2) = (i_1, i_2, i_3)$ and that $(i_1, i_2)(i_3, i_4) = (i_1, i_3, i_2)(i_1, i_3, i_4)$, and remembering that every element of A_n is a product of an even number of transpositions, we conclude that every element of A_n is a product of 3-cycles. The proof is complete. ∎

Sixth Problem Set

Problem 1.8.1 *Consider* $\sigma, \tau \in S_7$ *given by*

$$\sigma = \begin{pmatrix} 1 & 2 & 3 & 4 & 5 & 6 & 7 \\ 7 & 5 & 3 & 2 & 6 & 1 & 4 \end{pmatrix}, \tau = \begin{pmatrix} 1 & 2 & 3 & 4 & 5 & 6 & 7 \\ 2 & 3 & 5 & 1 & 7 & 4 & 6 \end{pmatrix}.$$

Write $\sigma\tau^{-1}$ as a product of disjoint cycles, then as a product of transpositions.

Problem 1.8.2 *Prove that the order of the product of disjoint cycles is the least common multiple of the lengths of the cycles.*

Problem 1.8.3 *Prove that if* $\sigma \in S_n$ *and* $r \leq n$, *then* $\sigma(k_1, k_2, ..., k_r)\sigma^{-1} = (\sigma(k_1), \sigma(k_2), ..., \sigma(k_r))$.

Problem 1.8.4 *Let* $\alpha, \beta \in S_n$. *Prove that* $\beta = \sigma\alpha\sigma^{-1}$ *for some* $\sigma \in S_n$ *if and only if* α, β *have the* **same cycle structure**, *that is they factor into products of the same number of disjoint cycles of the same lengths.*

Problem 1.8.5 *Prove that the commutator subgroup of* S_n *is* A_n.

Problem 1.8.6 *Prove that* $< (1, 2), (1, 3), ..., (1, n) > = S_n$.

Problem 1.8.7 *If* $n \geq 5$, *prove* A_n *is the only proper normal subgroup of* S_n.

Problem 1.8.8 *Regard* $S_3 \leq S_4$ *in the natural way,* S_3 *being identified as the subgroup* $\{\sigma \in S_4 : \sigma(4) = 4\}$. *Use the right coset decomposition of* S_4 *mod* S_3 *to get a list of the elements of* S_4. *Is* $S_3 \triangleleft S_4$?

Problem 1.8.9 *Show, by filling in the details in the steps below, that the group A_4, a group of order 12, has no subgroup of order 6. (a) Suppose $N \leq A_4$ and N has order 6. Then $N \lhd A_4$. (b) Let $\sigma \in A_4, \sigma \notin N$. Then $\sigma^2 \in N$. Thus, the square of every element of A_4 lies in N. (c) Every 3-cycle is a square of an element of A_4, so all 3-cycles are in N. (d) This contradicts $\mid N \mid = 6$.*

1.9 Conjugacy classes, p-groups, solvable groups

Let G be a group. We define a new equivalence relation on G as follows.

Definition 1.9.1 *For $x, y \in G$ define $y \sim x$ if $y = g^{-1}xg$ for some $g \in G$.*

It is easy to check that the above relation is an equivalence relation. If $y \sim x$ we say that y is a **conjugate** of x or that x and y are conjugate. We use $Cl(x)$, called the **conjugacy class** of x, to denote the equivalence class of a fixed element x. Recall that $C(G)$, the center of G, is the normal subgroup of G consisting of all elements in G which commute with every element of G.

Exercise 1.9.1 *Prove that $x \in C(G)$ if and only if $Cl(x) = \{x\}$.*

For a fixed element $x \in G$, let $C(x) = \{g \in G : gx = xg\}$.

Exercise 1.9.2 *Check that $C(x) \leq G$ and that $C(G) = \cap_{x \in G} C(x)$.*

The next theorem gives us a way of counting the number of elements in a given conjugacy class.

Theorem 1.9.1 *Let $x \in G$. Then $\mid Cl(x) \mid = [G : C(x)]$.*

Proof. Let $y, z \in G$. We have a chain of logical equivalences: $y^{-1}xy = z^{-1}xz \iff zy^{-1}x = xzy^{-1} \iff zy^{-1} \in C(x) \iff C(x)z = C(x)y$. Thus, $y^{-1}xy \neq z^{-1}xz \iff C(x)z \neq C(x)y$. The result follows. ∎

Corollary 1.9.1 *Let G be finite and $x \in G$. Then $\mid Cl(x) \mid$ divides $\mid G \mid$.*

Example 1.9.1 *We exhibit the disjoint conjugacy classes for $G = D_4$. By direct calculation $C(G) = \{e, x^2\}$. Hence, there are two singleton classes, $Cl(e) = \{e\}$ and $Cl(x^2) = \{x^2\}$. Since $\{e, x, x^2, x^3\} \subseteq C(x)$ and $D_4 \neq C(x)$, $C(x) = \{e, x, x^2, x^3\}$. Thus $\mid Cl(x) \mid = [G : C(x)] = 2$. Plainly, $x \in Cl(x)$, so in D_4 there's only one more conjugate of x. To get a different conjugate, let's try conjugating by y: $y^{-1}xy = (yx)y = (x^3y)y = x^3y^2 =$*

x^3. Hence, $Cl(x) = \{x, x^3\}$. Since $x^2, y \in C(y)$ and $C(y) \neq D_4$, $C(y) = \{e, y, x^2, x^2y\}$. Reasoning as before, $Cl(y)$ contains y and one additional conjugate. We try $x^{-1}yx = x^3(yx) = x^3(x^3y) = x^2y$. We have $Cl(y) = \{y, x^2y\}$. Since $xy \notin C(G)$, we know $Cl(xy)$ properly contains $\{xy\}$, hence it must be that $Cl(xy) = \{xy, x^3y\}$.

A finite group G is called a p-**group** if $| G | = p^n$ for some $n \in N$. The next theorem has powerful consequences in the study of finite groups.

Theorem 1.9.2 *(the class equation) Let G be a finite group. We have:* $| G | = | C(G) | + \sum_{i=1}^{k}[G : C(x_i)]$, *where $\{x_1, ..., x_k\}$ is a listing of one representative from each conjugacy class of G containing more than one element.*

Proof. Let $Cl(z_1), ..., Cl(z_t), Cl(x_1), ..., Cl(x_k)$ be a listing of the distinct conjugacy classes of G, where the singleton classes, corresponding to elements of $C(G)$, are listed first. Since G is a disjoint union of its conjugacy classes, $| G | = | C(G) | + \sum_{i=1}^{k} | C(x_i) | = | C(G) | + \sum_{i=1}^{k}[G : C(x_i)]$. ∎

Corollary 1.9.2 *If $| G | = p^n$ then $C(G) \neq\, <e>$.*

Proof. Let $| G | = p^n$ and suppose $C(G) = <e>$. Then, by the class equation, $p^n = 1 + \sum_{i=1}^{k}[G : C(x_i)]$, with the x_i as above. Since, for each i, $| Cl(x_i) | > 1$ and $| Cl(x_i) |$ divides p^n, then $| Cl(x_i) | = p^{e_i}, 1 \leq e_i \leq n$. Thus, $p^n = 1 + \sum_{i=1}^{k} p^{e_i}$, a contradiction. ∎

The corollary allows us to prove theorems about p-groups by induction on $| G |$. We do this by going from the group G to the group $G/C(G)$. We'll have an example of this procedure in a moment. First, we want to define one more kind of group that will play an important role in Chapter 5.

Definition 1.9.2 *A group G is **solvable** if there is a normal series S : $<e> = G_{k+1} \lhd G_k \lhd \cdots \lhd G_1 \lhd G_0 = G$ with the successive factor groups abelian. If S is a normal series that demonstrates the solvability of G, we say that G is **solvable with solvable series** S.*

Example 1.9.2 *Any abelian group G is solvable with series $<e> \lhd G$. The group S_3 is solvable with series $<e> \lhd A_3 \lhd S_3$. The group D_4 is solvable with series $<e> \lhd <x> \lhd D_4$. If $n \geq 5, S_n$ is not solvable, since A_n is the only proper normal subgroup of S_n and A_n is simple. In Problem 1.9.6 below, you're asked to show that S_4 is solvable.*

It will be convenient to have an alternate description of solvability. Recall that G^1 denotes the commutator subgroup of a group G, the smallest normal subgroup of G such that G/G^1 is abelian.

Definition 1.9.3 *The **derived series** for a group G is the normal series $G = G^0 \rhd G^1 \rhd G^2 = (G^1)^1 \rhd \cdots \rhd G^i \rhd G^{i+1} = (G^i)^1 \rhd \cdots$. The group $G^i, i \geq 0$, is called the i-th **derived subgroup**.*

What we do is start with G and repeatedly take commutator subgroups. The derived series may or may not terminate at the identity subgroup.

Example 1.9.3 *(a) The derived series for S_3 is $S_3 \rhd A_3 \rhd < e >$. (b) The derived series for S_5 is $S_5 \rhd A_5 \rhd A_5 \rhd \cdots$. (The group A_5 is nonabelian and simple.)*

Here is the alternate description of solvability, followed by an application.

Theorem 1.9.3 *A group G is solvable if and only if $G^n = < e >$ for some $n \geq 0$.*

Proof. Suppose first that G is solvable with a solvable series $< e > = G_{k+1} \lhd G_k \lhd \cdots \lhd G_1 \lhd G_0 = G$. Since G_0/G_1 is abelian, the commutator subgroup G^1 is contained in G_1. Suppose inductively that $G^i \leq G_i$ for $i \geq 1$. Then, since G_i/G_{i+1} is abelian, the commutator subgroup $(G_i)^1$ of the group G_i is contained in G_{i+1}. By the inductive hypothesis $G^i \leq G_i$ so that $G^{i+1} = (G^i)^1 \leq (G_i)^1 \leq G_{i+1}$. We have shown $G^i \leq G_i$ for all $i \geq 1$. Hence, $G^{k+1} = < e >$.

Conversely, if $G^n = < e >$ for some $n \geq 0$, then the derived series itself is a solvable series for G. ∎

The next lemma will have several uses.

Lemma 1.9.1 *Let $H \lhd G$. Then G is solvable if and only if H and G/H are solvable.*

Proof. Suppose G is solvable and $H \lhd G$. Since $H^k \leq G^k$ for all k, we can apply the previous theorem to conclude that H is solvable. Let $\pi : G \to G/H$ be the natural factor map. It is not hard to prove (see Problem 1.9.5) that $\pi(G^i) = \pi(G)^i$ for all $i \geq 0$. Again, Theorem 1.9.3 applies to show that G/H is solvable.

Conversely, suppose $H \lhd G$ with H and G/H solvable groups. We have, for some l, $(G/H)^l = H/H$. Directly from the definitions, for any $H \lhd G$, we have $(G/H)^1 = G^1H/H$. Thus, inductively $(G/H)^i = G^iH/H$

for all $i \geq 1$. It follows that $G^l \leq H$. But then, since $H^j = < e >$ for some j, $G^{l+j} = (G^l)^j \leq H^j = < e >$ and the proof is complete. ∎

We close this section with a theorem, giving a large class of solvable groups.

Theorem 1.9.4 *Every p-group is solvable.*

Proof. The proof is by induction on n, where $\mid G \mid = p^n$. If $\mid G \mid = p$ then $G \cong Z(p)$ so that G is abelian, hence solvable. Suppose that $n > 1$ and that each group of order p^m is solvable for $m < n$. By Corollary 1.8.2, $C = C(G) \neq < e >$. If $C = G$, we're done. Otherwise, C is a proper normal subgroup of G. By the induction hypothesis, both C and G/C are solvable. We apply Lemma 1.8.1 to conclude that G is solvable. ∎

Seventh Problem Set

Problem 1.9.1 *Prove if $\mid G \mid = p^2$ then G is abelian.*

Problem 1.9.2 *Prove if $\mid G \mid = p^n$ then G has a composition series with factors $\{\overbrace{Z(p), ..., Z(p)}^{n \ copies}\}$.*

Problem 1.9.3 *Use Problem 1.7.4 to describe the conjugacy class of the permutation $\sigma \in S_7$ given by $\sigma = \begin{pmatrix} 1 & 2 & 3 & 4 & 5 & 6 & 7 \\ 5 & 1 & 7 & 2 & 4 & 3 & 6 \end{pmatrix}$.*

Problem 1.9.4 *List the conjugacy classes of S_3.*

Problem 1.9.5 *Complete the proof of Theorem 1.9.3 by showing that $\pi(G^i) = \pi(G)^i$ for all $i \geq 0$.*

Problem 1.9.6 *Let $N = \{e, (1,2)(3,4), (1,3)(2,4), (1,4)(2,3)\} \subseteq S_4$. (a) Prove that $N \leq S_4$ and that $N \cong V$ the Klein 4-group. (b) Prove $N \lhd A_4$. (c) Prove that S_4 is solvable.*

1.10   Direct products

In this section we discuss a way of constructing new groups from old ones and of decomposing a given group into a product of subgroups.

Definition 1.10.1 *Let $G_1, ..., G_n$ be groups. The **(external) direct product** $\prod_{i=1}^{n} G_i$ is the group $(G_1 \times \cdots \times G_n, \circ)$, where $G_1 \times \cdots \times G_n$ is the Cartesian product of the sets $G_1, ..., G_n$ and $\circ$ is coordinatewise multiplication; that is, $(g_1, ..., g_n) \circ (h_1, ..., h_n) = (g_1 h_1, ..., g_n h_n)$. The groups G_i are called the **factors** of the direct product.*

If all of the groups G_i are additive, we call the Cartesian product with componentwise addition the (**external**) **direct sum** and adopt the notation $\bigoplus_{i=1}^{n} G_i$. The groups G_i are then called the **summands** of the direct sum. We caution the reader that, when dealing with a collection of groups $\{G_i : i \in I\}$, where the index set I is infinite, the notations $\prod_{i \in I} G_i$ and $\bigoplus_{i \in I} G_i$ are defined differently. This situation will arise in several of the latter chapters.

Exercise 1.10.1 *Verify that $\prod_{i=1}^{n} G_i$ is a group.*

For all i, let $e_i \in G_i$ be the identity element. For each fixed $1 \le j \le n$, let $\mu_j : G_j \to G'_j$ be given by $\mu_j(g) = (e_1, e_2, ..., e_{j-1}, g, e_{j+1}, ..., e_n)$. Then put $G'_j = \operatorname{Im} v_j = \{(e_1, e_2, ..., e_{j-1}, g, e_{j+1}, ..., e_n) : g \in G_j\}$. Plainly, each $G'_j \le \prod_{i=1}^{n} G_i$ and $\mu_j : G_j \cong G'_j$. We call μ_j the **natural embedding** of G_j into $\prod_{i=1}^{n} G_i$.

We collect some facts on the subgroups G'_j.

Lemma 1.10.1 *Let $G = \prod_{i=1}^{n} G_i$. Then each $G'_j \lhd G$ and $G = G'_1 G'_2 \cdots G'_n$. Further, for each j, $G'_j \cap [G'_1 G'_2 \cdots \bar{G}'_j \cdots G'_n] = \; <e> \; .$ Here the superscript – indicates deletion of the factor below which it lies.*

Proof. First, for all $x \in G$, we have $x^{-1}(e_1, e_2, ..., e_{j-1}, g, e_{j+1}, ..., e_n)x = (e_1, e_2, ..., e_{j-1}, x^{-1}gx, e_{j+1}, ..., e_n) \in G'_j$. Thus, each $G'_j \lhd G$. Next, we recall (Problem 1.6.7) that, since each $G'_j \lhd G$, the set $G'_1 G'_2 \cdots G'_n = \{g'_1 g'_2 \cdots g'_n : g'_j \in G'_j\}$ is a subgroup of G. If, for some fixed j, $y = (e_1, e_2, ..., e_{j-1}, g, e_{j+1}, ..., e_n)$ and $y = g'_1 g'_2 \cdots g'_{j-1} g'_{j+1} \cdots g'_n$, $g'_i \in G'_i$ it follows directly that $y = (e_1, e_2, ..., e_{j-1}, e_j, e_{j+1}, ..., e_n) = e$. Finally, an arbitrary element of the direct product $(g_1, g_2, ..., g_j, ..., g_n)$ can be written as a product $g'_1 g'_2 \cdots g'_n$ with $g'_j = (e_1, e_2, ..., e_{j-1}, g_j, e_{j+1}, ..., e_n)$. Thus, $G = G'_1 G'_2 \cdots G'_n$. ∎

Exercise 1.10.2 *Prove that $Z(2) \oplus Z(2) \cong V$ (Klein 4-group, p. 14) and that $Z(3) \oplus Z(5) \cong Z(15)$.*

We now take the opposite point of view, that of decomposing a group into a product of factors.

Definition 1.10.2 *A group G is the **internal direct product** of its normal subgroups $G'_1, G'_2, ..., G'_n$ if $G = G'_1 G'_2 \cdots G'_n$ and if, for each $j, G'_j \cap [G'_1 G'_2 \cdots \bar{G}'_j \cdots G'_n] = \; <e> \; .$ We write $G = \operatorname{int} \prod_{i=1}^{n} G'_i$.*

The next lemma will help us show that internal and external directs products are just different ways of describing the same thing.

Lemma 1.10.2 Let $G = int \prod_{i=1}^n G_i'$. Then (1) each element $g \in G$ has a unique representation as a product $g = g_1'g_2' \cdots g_n'$ with $g_i' \in G_i'$ and (2) if $h \in G$ has representation $h = h_1'h_2' \cdots h_n'$, then $gh = (g_1'h_1')(g_2'h_2') \cdots (g_n'h_n')$.

Proof. (1) The proof is by induction on n, the case $n = 1$ being trivial. Suppose $G = int \prod_{i=1}^n G_i'$ for $n > 1$ and that unicity of product representation holds for groups that are internal direct products of $m < n$ normal subgroups. Let $g \in G$ and say $g = g_1'g_2' \cdots g_n' = g_1^*g_2^* \cdots g_n^*$ with $g_i', g_i^* \in G_i', 1 \le i \le n$. Then $g_n^*(g_n')^{-1} = (g_1^*g_2^* \cdots g_{n-1}^*)^{-1}g_1'g_2' \cdots g_{n-1}' \in G_n' \cap G_1'G_2' \cdots G_{n-1}' = <e>$. Hence $g_n' = g_n^*$ and right cancellation yields the equation $g_1'g_2' \cdots g_{n-1}' = g_1^*g_2^* \cdots g_{n-1}^*$. These products can be regarded as product representations of an element in the subgroup $H = int \prod_{i=1}^{n-1} G_i'$. By induction, $g_i' = g_i^*$ for $1 \le i \le n$. This proves (1).

(2) Again, the proof is by induction on n. For the inductive step, note that, by the internal direct product hypothesis, the elements $(g_1'g_2' \cdots g_{n-1}')$ and g_n' lie in normal subgroups with intersection $<e>$. Hence, these elements commute (Problem 1.6.2). So, if $h = h_1'h_2' \cdots h_n'$, then $gh = (g_1'g_2' \cdots g_{n-1}')g_n'(h_1'h_2' \cdots h_{n-1}')h_n' = g_n'[(g_1'g_2' \cdots g_{n-1}')(h_1'h_2' \cdots h_{n-1}')]h_n'$. Working in the group H defined above, and invoking the inductive hypothesis to compute the bracketed product, we obtain $gh = g_n'[(g_1'h_1')(g_2'h_2') \cdots (g_{n-1}'h_{n-1}')]h_n'$. Since the bracketed element in the last equation is in H, it commutes with g_n'. Therefore $gh = (g_1'h_1')(g_2'h_2') \cdots (g_n'h_n')$, proving (2). ∎

We close this section by showing that the internal and external direct products are essentially the same thing. So, hereafter, we will ignore the distinction and call either one the direct product, written as $\prod_{i=1}^n G_i$. We will regard $\prod_{i=1}^n G_i$ as either internal or external, whichever is more convenient for our particular situation. Of course, the same remarks hold, with a change to additive notation, for internal and external direct sums.

Theorem 1.10.1 (1) Let $G = \prod_{i=1}^n G_i$. Then $G = int \prod_{i=1}^n G_i'$, where G_j is naturally isomorphic to G_j' for each fixed j. (2) Let $G = int \prod_{i=1}^n G_i'$. Then $G \cong \prod_{i=1}^n G_i'$.

Proof. Claim (1) follows immediately by Lemma 1.10.1. For the proof of (2), let $\theta : int \prod_{i=1}^n G_i' \to \prod_{i=1}^n G_i'$ be given by $\theta(g_1'g_2' \cdots g_n') = (g_1', g_2', ..., g_n')$. Part one of Lemma 1.10.2 shows that θ is a well defined bijection. Part two of the same lemma shows that θ is a homomorphism. ∎

Eighth Problem Set

Problem 1.10.1 *Let $n > 1$ have prime factorization $n = p_1^{e_1} \cdots p_k^{e_k}$, where each $e_i > 0$ and the primes $p_1, ..., p_k$ are pairwise distinct. Prove that $Z(n) \cong Z(p_1^{e_1}) \oplus \cdots \oplus Z(p_k^{e_k})$.*

Problem 1.10.2 *Let $\theta_i : H \to G_i$ be homomorphisms from a fixed group H into the groups $G_i, 1 \leq i \leq n$. For each fixed $1 \leq j \leq n$, let $\rho_j : \prod_{i=1}^n G_i \to G_j$ be the "natural projection map": $\rho_j(g_1, ..., g_j, ..., g_n) = g_j$. Prove that there exists a unique map $\theta : H \to \prod_{i=1}^n G_i$ such that $\rho_j \theta = \theta_j$ for all j.*

Problem 1.10.3 *Let $\phi_i : G_i \to H$ be homomorphisms from the abelian groups $G_i, 1 \leq i \leq n$ into a fixed abelian group H. Prove that there exists a unique homomorphism $\phi : \prod_{i=1}^n G_i \to H$ such that $\phi \mu_i = \phi_i, 1 \leq i \leq n$. Here $\mu_i : G_i \to G_i'$ is the natural embedding map discussed earlier.*

Problem 1.10.4 *Show that the claim of the previous problem is false if we omit the word "abelian".*

Problem 1.10.5 *Draw two diagrams illustrating the situations in Problems 1.10.2 and 1.10.3. Draw solid arrows for the given maps and a dashed arrow for the map claimed to exist. What is the relationship between the diagrams? Why is the claim of Problem 1.10.4 a little surprising?*

1.11 Sylow Theorems

We present the some of the most powerful results in finite group theory, the Sylow Theorems. The basic idea is this: Lagrange's Theorem says that if $H \leq G$ then $| H |$ divides $| G |$. However Problem 1.8.9 outlines the construction of an example to show that the converse of Lagrange's Theorem is false; if m divides $| G |$, G will not necessarily have a subgroup of order m. However, the Sylow Theorems do tell us that if p^k divides $| G |$ then G will have a subgroup of order p^k. They also provide a good bit of information about these subgroups. The treatment here of the Sylow Theorems is a mild revision of that of [W].

The first step toward the Sylow Theorems is a theorem by Cauchy. The claim is not nearly as obvious as it might seem at first glance.

Theorem 1.11.1 *Suppose p is a prime that divides the order of a group G. Then G has an element of order p.*

Proof. The proof is by induction on $\mid G \mid$. If $\mid G \mid = 1$, the theorem holds vacuously. Suppose $\mid G \mid = n > 1$ and that every group of order less than n has an element of order p whenever p divides its order. Let p be a prime dividing n. Consider the class equation of G : $\mid G \mid = \mid C(G) \mid + \sum_{i=1}^{k} [G : C(x_i)]$. Suppose that p divides $\mid C(x_i) \mid$ for at least one i. Since $x_i \notin C(G), C(x_i)$ is a proper subgroup of G. By the inductive hypothesis, $C(x_i)$, hence G, has an element of order p.

Assume now that $\mid C(x_i) \mid$ is prime to p for all i. Since, for all i, $\mid G \mid = \mid C(x_i) \mid [G : C(x_i)]$, we can conclude that p divides the index of each $C(x_i)$ in G. Thus, p divides $\mid G \mid - \sum_{i=1}^{k} [G : C(x_i)] = \mid C(G) \mid$. Let x be a nonidentity element of $C = C(G)$. If $p \mid \mid < x > \mid$, the cyclic group $< x >$, hence G, trivially has an element of order p. If $p \nmid \mid < x > \mid$, then p divides $\mid C / < x > \mid = \mid C \mid / \mid < x > \mid$. By the inductive hypothesis, the factor group $C / < x >$ has an element of order p. Hence, there exists $z \in C$ such that $z^p \in < x >$. We have $z^p = x^k$ for some k. If $\mid x^k \mid = m$, then $\mid z \mid = pm$ so that $\mid z^m \mid = p$, and the proof is complete. ∎

For finite groups there are two sensible definitions for a p-group. We have said that G is a p-group if $\mid G \mid = p^n$ for some $n > 0$. Another possible definition would be that G is a p-group if and only if $g \in G \Longrightarrow \mid g \mid = p^{k(g)}$, where $k(g)$ is a nonnegative integer depending on g. Using what we've done so far, it's easy to prove that these two definitions are equivalent.

Exercise 1.11.1 *Let G be a finite group. Prove that G is a p-group if and only if $g \in G \Longrightarrow \mid g \mid = p^{k(g)}$.*

We are ready for the first Sylow Theorem . In the statement, p^n refers to an arbitrary prime power with $n \geq 1$.

Theorem 1.11.2 *(First Sylow Theorem) Let p^n divide $\mid G \mid$. Then G has a subgroup of order p^n.*

Proof. Again, the proof is by induction on $\mid G \mid$, the case $\mid G \mid = 1$ being trivial. Suppose $\mid G \mid = m > 1$ and that the theorem holds for all groups of order less than m. If G has a proper subgroup H such that $[G : H]$ is prime to p, then p^n divides $\mid H \mid$ and, by induction, we're done. Assume that $p \mid [G : H]$ for all proper subgroups $H \leq G$. Arguing as in the proof of the previous theorem, we can conclude that p divides the order of $C = C(G)$. Thus, C has an element c of order p. Let $S = < c >$. Since $S \subseteq C, S \lhd G$. We have p^{n-1} divides $\mid G/S \mid < \mid G \mid$, so, by induction, G/S has a subgroup H/S (subgroup correspondence theorem) of order p^{n-1}. Hence, $\mid H \mid = p^n$, so, in this case too, we're done. ∎

Definition 1.11.1 *Fix a prime p dividing the order of a finite group G. Let p^n be the highest power of p that divides $\mid G \mid$. A subgroup S of order p^n is called a **Sylow p-subgroup** of G.*

Theorem 1.11.2 guarantees us the existence of at least one Sylow p-subgroup for each p dividing $\mid G \mid$. We want to find the number of Sylow p-subgroups for a given p and G. We also will show that any two Sylow p-subgroups S and S' are conjugate, that is there exists $g \in G$ such that $S' = g^{-1}Sg$. We need a preliminary definition and lemma.

Definition 1.11.2 *Let $A, B \leq G$. An **A-conjugate** of B is a subgroup of the form $a^{-1}Ba$, where $a \in A$.*

A G-conjugate of a subgroup B is simply called a **conjugate** of B.

Exercise 1.11.2 *Show that, if $B \leq G$ and $g \in G$, then $g^{-1}Bg \leq G$. Furthermore, $B \cong g^{-1}Bg$. Thus, if S is a Sylow p-subgroup of G, so is $g^{-1}Sg$.*

Exercise 1.11.3 *For U a subgroup of a group G, let $N(U) = \{g \in G : g^{-1}Ug = U\}$. Show that $N(U)$ is a subgroup of G with $U \lhd N(U)$.*

If $U =< x >$, $N(U)$ coincides with what we've called $C(x)$. If $U \leq G, N(U)$ will be the largest subgroup of G in which U is a normal subgroup. Our next lemma is a generalization of Theorem 1.9.1. The proof is similar to that of Theorem 1.9.1 and is left as an exercise (Problem 1.11.1).

Lemma 1.11.1 *Let $A, B \leq G$. Then the number of A-conjugates of B is the index $[A : N(B) \cap A]$.*

Theorem 1.11.3 *(Second Sylow Theorem) Let G be a finite group and p a prime dividing $\mid G \mid$. Then any $H \leq G$ such that $\mid H \mid$ is a power of p is contained in a Sylow p-subgroup of G. Furthermore, any two Sylow p-subgroups are conjugate.*

Proof. Let $H \leq G$ and suppose $\mid H \mid = p^k$. Say $\mid G \mid = p^n m$, with $k \leq n$ and m prime to p. Let S be a Sylow p-subgroup of G. We have that $\mid S \mid = p^n$ and, hence, that $[G : S] = m$. By Lemma 1.11.1 with $A = G, B = S$, the number of conjugates of S is $[G : N(S)]$. By Problem 1.3.5, since $S \leq N(S) \leq G, [G : S] = [G : N(S)][N(S) : S]$. Since $[G : S] = m$, it follows that $[G : N(S)]$ is prime to p. Hence, the number of conjugates of S is prime to p.

Now H induces an equivalence relation $\sim$ on the set of G-conjugates of S. Two conjugates of S are equivalent if they are conjugate by an element of H. (Check that $\sim$ is an equivalence relation.)

Let S_1 be a conjugate of S. By Lemma 1.11.1, the number of H-conjugates of S_1 is $[H : N(S_1) \cap H]$. This number is a power of p since $|H| = p^k$. Thus, in the partition of the set of conjugates of S corresponding to $\sim$, the number of elements in each equivalence class is a power of p. But the number of conjugates of S is prime to p. So at least one equivalence class has size $p^0 = 1$. Say this class is $\{T\}$. Then $[H : N(T) \cap H] = 1$, so $H \subseteq N(T)$. Therefore, $TH \leq N(T)$ so that $T \lhd TH$. By the second isomorphism theorem $TH/T \cong H/(H \cap T)$ is a p-group. Since T is a p-group, it follows that TH is a p-group. But, $T \cong S$ is a Sylow p-subgroup. Thus, $TH = T$ so that $H \subseteq T$. We have proven the first claim of our theorem.

If H is already a Sylow p-subgroup then $H = T$. Since T is conjugate to S and S was arbitrary, the second claim follows as well. ∎

Corollary 1.11.1 *Let S be a Sylow p-subgroup of G. Then S is the only Sylow p-subgroup in $N(S)$.*

Corollary 1.11.2 *The group G has a unique Sylow p-subgroup S_p for some prime p if and only if $S_p \lhd G$.*

Corollary 1.11.3 *Let S be a Sylow p-subgroup of G. Then $N(N(S)) = N(S)$.*

Proof. Plainly, $N(S) \subseteq N(N(S))$. Let $x \in N(N(S))$. Then $x^{-1}N(S)x = N(S)$ so that conjugation by x is an isomorphism from $N(S)$ to itself. Since S is the unique Sylow p-subgroup of $N(S)$ it must be carried to itself by this isomorphism. Hence, $x^{-1}Sx = S$ and $x \in N(S)$. ∎

Theorem 1.11.4 *(Third Sylow Theorem) The number of Sylow p-subgroups of G is congruent to 1 mod p and divides $[G : S]$.*

Proof. Let S be a fixed Sylow p-subgroup of G. We know by Theorem 1.11.3 that the set of all Sylow p-subgroups of G coincides with the set of all G-conjugates of S. The number of conjugates of S is equal to $[G : N(S)]$, a divisor of $|G : S|$.

The equivalence relation of being S-conjugate partitions the set of all Sylow p-subgroups of G into equivalence classes. By the proof of Theorem 1.11.3, each of these classes has a power of p elements and at least one class has one element. But if $\{T\}$ is a singleton equivalence class, then $s^{-1}Ts = T$ for all $s \in S$. This latter condition means that $S \subseteq N(T)$. As in the proof of Theorem 1.11.3, ST is a p-group. Since S, T are maximal p-subgroups of G, $S = T = ST$. Thus, the only singleton equivalence class under S-conjugation is $\{S\}$. We conclude that the total number of Sylow p-subgroups of G is congruent to 1 mod p. ∎

We present an interesting result, which will have applications in our next section. First, we need a preparatory lemma.

Lemma 1.11.2 *Let $N_1, ..., N_k$ be normal subgroups of a group G such that, for each fixed j, $N_j \cap [N_1 \cdots N_{j-1} N_{j+1} \cdots N_k] =< e >$. Then $\mid N_1 \cdots N_k \mid = \mid N_1 \mid \cdots \mid N_k \mid$.*

Proof. Recall that, by Problem 1.5.10, the product of any finite collection of normal subgroups is a normal subgroup.

The proof of our lemma is by induction on k, the case $k = 1$ being trivial. Suppose that $k > 1$ and that the result holds for any appropriate collection of m normal subgroups, whenever $m < k$. By the third isomorphism theorem $(N_1 \cdots N_k)/N_k = [(N_1 \cdots N_{k-1})N_k]/N_k \cong (N_1 \cdots N_{k-1})/[(N_1 \cdots N_{k-1}) \cap N_k] = (N_1 \cdots N_{k-1})/ < e > \cong N_1 \cdots N_{k-1}$. By the inductive assumption, $\mid N_1 \cdots N_{k-1} \mid = \mid N_1 \mid \cdots \mid N_{k-1} \mid$. Thus, $\mid N_1 \cdots N_k \mid = \mid (N_1 \cdots N_k)/N_k \mid \mid N_k \mid = \mid N_1 \mid \cdots \mid N_k \mid$. ∎

Henceforth, for $n > 1$, when we write $n = p_1^{e_1} \cdots p_k^{e_k}$, we assume that all the e_i are all greater than zero and that the p_i are distinct primes.

Theorem 1.11.5 *Let $\mid G \mid = n = p_1^{e_1} \cdots p_k^{e_k}$ and suppose that G has a unique Sylow p_i-subgroup S_i for $1 \leq i \leq k$. Then $G = \prod_{i=1}^{k} S_i$.*

Proof. By Corollary 1.11.2, each $S_i \lhd G$. (Here, we're taking the point of view of decomposing G into an internal direct product of normal subgroups). We need to verify the two defining properties of the internal direct product. We'll verify the second one first. Fix an index j and let $z \in S_j \cap [S_1 \cdots S_{j-1} S_{j+1} \cdots S_k]$. Since $z \in S_j$, $\mid z \mid = p_j^t$ for some $t \geq 0$. We also have $z = g_1 \cdots g_{j-1} g_{j+1} \cdots g_k$. Note that, if $i \neq j$, then $S_i \cap S_j =< e >$. This follows since every nonidentity element of S_i (resp. S_j) has order a power of p_i (resp. p_j). By Problem 1.6.2, if $i \neq j$ and $x \in S_i, y \in S_j$, then $xy = yx$. Let $m = n/p_j^{e_j}$. Then $z^m = (g_1 \cdots g_{j-1} g_{j+1} \cdots g_k)^m = g_1^m \cdots g_{j-1}^m g_{j+1}^m \cdots g_k^m = e$. (Each individual $g_r^m = e$, since $g_r \in S_r$ and $\mid S_r \mid = p^{e_r} \mid m$.) Hence $\mid z \mid$ divides m. But $(m, p_j) = 1$. We must have $\mid z \mid = 1$ so that $z = e$.

It remains to prove that G coincides with the subgroup $S_1 S_2 \cdots S_k$. But, by Lemma 1.11.2, $\mid S_1 S_2 \cdots S_k \mid = p_1^{e_1} \cdots p_k^{e_k} = \mid G \mid$, and the proof is complete. ∎

We close the section with some examples of applications of the Sylow Theorems. The examples are presented in increasing order of difficulty.

Example 1.11.1 *We show that there is no simple group of order $1000 = 2^3 5^3$. Let $\mid G \mid = 1000$. The number of Sylow 5-subgroups of G must be of the form $1 + k5$ and also divide 8. One is the only such number. Hence, G has a unique normal Sylow 5-subgroup.*

Example 1.11.2 *We show that there is no simple group of order 30. Suppose $| G | = 30 = 2 \cdot 3 \cdot 5$. First, we look at s_5, the number of Sylow 5-subgroups of G. By the Sylow Theorems, $s_5 = 1 + k5$ and $s_5 \mid 6$. Thus, $s_5 = 1$ or $s_5 = 6$. If $s_5 = 1$, then G has a unique-hence normal-Sylow 5-subgroup, and we're done. Suppose that $s_5 = 6$. Since each Sylow 5-subgroup is isomorphic to $Z(5)$, the pairwise intersection of any two distinct ones is $< e >$. Thus, the Sylow 5-subgroups account for $25 = 6 \cdot 4 + 1$ distinct elements. Next, look at s_3. If $s_3 = 1$, we're done. So, suppose that $s_3 = 10$, the only other possibility. In this case, reasoning as above, we account for $20 = 10 \cdot 2$ additional elements. But we've already used more elements than are in the group! Thus, the conditions $s_5 = 6$ and $s_3 = 10$ cannot hold simultaneously. It follows that G has a normal subgroup of order 5 or of order 3.*

Example 1.11.3 *We show that there is no simple group of order 108. Continue with the notation above and let $| G | = 108 = 2^2 3^3$. Then s_3 is either 1 or 4. If $s_3 = 1$, we're done. Suppose that $s_3 \doteq 4$. Conjugation by any $g \in G$ produces a permutation on $S(3)$, the set of Sylow 3-subgroups of G. If we number the elements of $S(3)$, we obtain a corresponding homomorphism $c : G \rightarrow S_4$. The homomorphism c sends each $g \in G$ to the permutation induced by conjugation by g. Since $| G | = 108 < 24 = | S_4 |$, the kernel of the map c will be a proper normal subgroup of G. Note that c cannot map each element $g \in G$ to the identity of S_4, since any two 3-Sylow subgroups are conjugate.*

Example 1.11.4 *Our final example has a different flavor. We construct a nonabelian group of G of order 21. Consider a group G with $| G | = 3 \cdot 7$. Then $s_7 = 1$, that is there is a unique normal Sylow 7-subgroup $S \le G$. Since $| S | = 7$, S is cyclic, say $S = < y >$. Let $T = < x >$ be any Sylow 3-subgroup of G. The distinct right cosets of $G / < y >$ can be written as $< y > e, < y > x, < y > x^2$. Hence, the elements of G can be listed: $G = \{e, y, ..., y^6, x, yx, ..., y^6x, x^2, yx^2, ..., y^6x^2\}$. Just as in the construction of the group D_4 at the beginning of the chapter, once we determine the product xy, all the other products in the group table for G can be calculated by sliding x's past y's. Since $< y > \lhd G$, $xyx^{-1} = y^j$ for $1 \le j \le 6 (j = 0$ yields a contradiction). If $j = 1$, then $xy = yx$ and G will be abelian. So we want to choose $2 \le j \le 6$. Then, the stipulation $xy = y^j x$ will produce our nonabelian group. However, our choice of j is not free, due to the following considerations. Let $c_x : < y > \rightarrow < y >$ be defined by the condition $c_x(y) = xyx^{-1}$. (Then c_x automatically extends to powers of y.) It's easy to check that c_x is an isomorphism. Since $x^3 = e$, it follows that the map $(c_x)^3$ coincides with the identity map from $< y >$ to itself. But $c_x(y) = y^j$, so that $y = (c_x)^3(y) = y^{j^3}$ (check this last equality). We have $y = y^{j^3}$, or,* equivalently, $j^3 \equiv 1$ *(mod 7)*. *So for j we need a cube root of 1, other than 1 itself, in the group $U(7)$. The element 2 works nicely. We set $xy = y^2 x$ to produce our example. (The element $4 \in U(7)$ is another cube root of 1, so we could have put $xy = y^4 x$ to produce a different example.)*

Ninth Problem Set

Problem 1.11.1 *Prove Lemma 1.10.1.*

Problem 1.11.2 *Prove that any group of order 15 is isomorphic to $Z(15)$.*

Problem 1.11.3 *Let G be a finite group and $N \lhd G$. Let S_p be a Sylow p-subgroup of G. Prove that $S_p \cap N$ is a Sylow p-subgroup of N and that $S_p N / N$ is a Sylow p-subgroup of G/N.*

Problem 1.11.4 *Let $\mid G \mid = pq$ with p, q primes and $p < q$. If $p \nmid (q - 1)$ discuss the possibilities (up to isomorphism) for G.*

Problem 1.11.5 *Using the results we've developed so far, prove that any group of order n with $1 < n < 60$ is either of prime order or has a proper normal subgroup. Hint: Often when you've found a group with a proper normal subgroup, many others will have a proper normal subgroup for the same reason. Organize your proof that way.*

Problem 1.11.6 *Let G be a finite abelian group. For each p dividing $\mid G \mid$, let $T_p(G) = \{g \in G : g^{p^k} = e$ for some $k \geq 0\}$. Prove that $T_p(G)$ is the unique Sylow p-subgroup of G. (So, if we stick to finite abelian groups, the Sylow Theorems are unnecessary.)*

1.12 The structure of finite abelian groups

Throughout this section, G will be a finite abelian group written additively. We prove the well known decomposition theorem for finite abelian groups: A finite abelian group decomposes into a direct sum of cyclic summands of prime power order. This theorem was first proved by Frobenius and Stickelberger in 1878. Our proof of the decomposition theorem is exactly that in [F].

Definition 1.12.1 *For an additive abelian group G and a prime p, the **p-torsion subgroup of** G is the subgroup $T_p = T_p(G) = \{g \in G : p^k g = 0$ for some $k \geq 0\}$. (See Problem 1.10.6 in the previous section.)*

We start with a fact of independent interest.

Theorem 1.12.1 *Let* $\mid G \mid = p_1^{e_1} \cdots p_k^{e_k}$. *Then* $G = \bigoplus_{i=1}^{k} T_{p_i}$.

Proof. The proof we give is cheap (some might say elegant) but, from the point of view of abelian group theory, highly unnatural. Problem 1.12.1 asks you to use greatest common divisors to come up with a more direct proof. In Chapter 3 we present a generalized version of this theorem.

On with the proof. Since, for each fixed p, the subgroup T_p contains all elements in G having order a power of p, T_p is obviously the unique maximal p-subgroup of G. By Theorem 1.11.5, written additively, $G = \bigoplus_{i=1}^{k} T_{p_i}$. ∎

If we work with finite abelian groups, Theorem 1.12.1 allows us to concentrate on abelian groups G with $\mid G \mid = p^e$, that is finite abelian p-groups. Therefore, until the statement of the general structure theorem for finite abelian groups, we will add without fanfare the additional assumption that $\mid G \mid = p^e$.

Definition 1.12.2 *Let* $K \leq G$. *A subgroup* $H \leq G$ *maximal with respect to the property that* $H \cap K = <0>$ *is called a* K-**high** *subgroup of* G.

Since G is finite, K-high subgroups exist for any $K \leq G$.

Example 1.12.1 *Let* $G = Z(2) \oplus Z(2)$, $K = Z(2) \oplus <0> = \{(0,0), (1,0)\}$. *Then* $H_1 = <0> \oplus Z(2)$ *and* $H_2 = \{(0,0), (1,1)\}$ *are each* K-*high subgroups of* G.

We next prove two highly technical lemmas. We'll then be rewarded with a very interesting result.

Lemma 1.12.1 *Let* $K \leq G$ *and let* H *be a* K-*high subgroup. Form the (internal) direct sum* $K \oplus H = \{k + h : h \in H, k \in K\} \leq G$. *If* $g \in G$ *is such that* $pg \in H$, *then* $g \in K \oplus H$.

Proof. If $g \in H$, we're done. Otherwise $< H, g >$, the subgroup of G generated by H and the new element g, properly contains H. It's easy to check (check it) that $< H, g > = \{h + mg : h \in H, 0 \leq m \leq p - 1\}$. Since H is K-high $< H, g > \cap K$ contains a nonzero element b. We have $b = h + mg$ as above and $b \in K$. Since $H \cap K = <0>$, then $0 < m \leq p-1$. Hence, $(m, p) = 1$ and there exist integers r, s with $rm + sp = 1$. So $g = (rm + sp)g = r(mg) + s(pg) = r[b + (-h)] + s(pg)$. Since both $[b + (-h)]$ and pg are in $K \oplus H$, the result follows. ∎

Lemma 1.12.2 *Let G, K, H be as above. Suppose, for all $g \in G$, that $pg = (k + h) \in K \oplus H$ implies $k = pk'$ for some $k' \in K$. Then $G = K \oplus H$.*

Proof. Suppose the implication stated in the hypothesis holds, and that $G \neq K \bigoplus H$. We'll derive a contradiction. Let $g \in G$ be such that $g \notin K \bigoplus H$. Without loss, we can assume that $pg \in K \oplus H$. Say $pg = (k+h)$. By assumption, $k = pk'$, so that $p(g - k') = pg - k = h$. Thus, the element $g - k'$ satisfies the hypothesis of the previous lemma. We can conclude that $(g - k') \in K \oplus H$. But, since $k' \in K, g \in K \oplus H$, a contradiction. ∎

Here is our reward.

Lemma 1.12.3 *Let $\mid G \mid = p^e$ and let $x \in G$ be an element of maximal order, say p^k. Then $G = <x> \oplus H$ for some complementary direct summand $H \leq G$.*

Proof. Let H be a $<x>$-high subgroup of G. If something like this doesn't work, nothing will! Form the (internal) direct sum $<x> \oplus H = \{mx \oplus h : 0 \leq m < p^k, h \in H\} \leq G$. We apply Lemma 1.12.2 with $K = <x>$. Suppose (1) : $pg = mx \oplus h \in <x> \oplus H$. Multiplying (1) by p^{k-1} yields the equation: $0 = p^{k-1}mx \oplus p^{k-1}h \in <x> \oplus H$. Hence, $p^{k-1}mx = 0$. Since $\mid x \mid = p^k$, it follows that $m = pm'$. Thus, $mx = pm'x$, and the hypothesis of Lemma 1.12.2 holds. By Lemma 1.12.2, $G = <x> \oplus H$. ∎

The decomposition theorem now follows easily.

Theorem 1.12.2 *(decomposition theorem for finite abelian groups) Let G be a finite abelian group. Then G is a direct sum of cyclic groups of prime power order.*

Proof. Let $\mid G \mid = p_1^{e_1} \cdots p_k^{e_k}$. By Theorem 1.12.1, $G = \bigoplus_{i=1}^k T_{p_i}$. It remains to show that each p_i-torsion subgroup T_{p_i} decomposes into a direct sum of cyclic subgroups (necessarily of p_i-power order). The proof is an easy induction on $\mid T_{p_i} \mid$. For the inductive step use Lemma 1.12.3. ∎

This proof should remind the reader of the proof that every natural number is the product of primes.

There is a corresponding **unicity of decomposition for finite abelian groups**: For any prime p and nonnegative integer k, let $f_k^p(G)$ be the number of cyclic summands of order p^{k+1} in some decomposition of G into a direct sum of cyclic subgroups of prime power order. We show that the numbers $f_k^p(G)$ are independent of the particular direct sum decomposition used to count them. In other words the numbers $f_k^p(G)$ are **invariants** of the group G. (There is a good reason why it's better to have $f_k^p(G)$ be the number of $Z(p^{k+1})$-summands rather than $Z(p^k)$-summands, but we have to wait until the last section of the book to see it.)

Conversely, if G, H are finite abelian groups with invariants $f_k^p(G) = f_k^p(H)$ for all p, k, then, obviously, $G \cong H$. So, in this instance, we call the set of nonnegative integers: $\{f_k^p(G) : p \text{ a prime}, k \geq 0\}$ a **complete set of isomorphism invariants** for G in the class of all finite abelian groups.

We prove the hard part, that the numbers $f_k^p(G)$ are invariants of G. Firstly, if q is a prime not dividing $\mid G \mid$, then, by Corollary 1.2.1 and independent of any decomposition, $f_k^q(G) = 0$ for all $k \geq 0$. Now suppose that $G = < x_1 > \oplus < x_2 > \oplus \cdots \oplus < x_n >$ with $\mid x_i \mid = p_i^{e_i}$. Fix a prime q dividing $\mid G \mid$. As a first step, do the following exercise.

Exercise 1.12.1 *With G as above, explain why $T_q(G)$ is the direct sum of those summands $< x_i >$ such that $\mid x_i \mid = q^{e_i}$.*

In view of the exercise and that fact that $G = \bigoplus T_p$ with each p-torsion subgroup T_p uniquely determined by G, it is enough to prove that any finite abelian q-group T_q as above has a unique decomposition as a direct sum of q-power order cyclic summands. Suppose $\mid T_q \mid = q^e$ and that $T_q = < x_1 > \oplus < x_2 > \oplus \cdots \oplus < x_m >$ with $\mid x_i \mid = q^{e_i}$. Since T_q is abelian, we can arrange the cyclic summands such that $e_1 \leq e_2 \leq \cdots \leq e_m$. Let $\{e_1, ..., e_l\}$ be the set of distinct natural numbers in the set $\{e_1, ..., e_m\}$, arranged so that $e_1 < \cdots < e_l$. For each fixed j with $1 \leq j \leq l$, let T_{qj} be the direct sum of the $< x_i >$ such that $\mid x_i \mid = q^{e_j}$. Then $T_q = \bigoplus_{j=1}^l T_{qj}$. We want a way of counting the number t_{qj} of cyclic summands of T_{qj}, that is determining the number $f_{e_j-1}^q(T_q)$. **But our method must not refer to the given decomposition $T_q = < x_1 > \oplus < x_2 > \oplus \cdots \oplus < x_m >$**-only to the group T_q itself.

The following lemma does the job. It needs careful reading. Statement (b) requires a little knowledge of vector spaces. Statement (b) can be ignored if you want; for what follows we can rely on (a). First, here is an additional bit of notation: For any abelian group A, let $A[q] = \{a \in A : qa = 0\}$.

Exercise 1.12.2 *Prove, for abelian groups A, that $A[q] \leq A$.*

Lemma 1.12.4 *With notation as above, let $U(e_j - 1, q)$ denote the factor group $(q^{e_j-1}T_q)[q]/(q^{e_j}T_q)[q]$. Then (a) $q^{t_{qj}} = \mid U(e_j - 1, q) \mid$ and (b) $t_{qj} = \dim_{Z/qZ} U(e_j - 1, q)$.*

Proof. We have $T_q = T_{q1} \oplus \cdots \oplus T_{q(j-1)} \oplus T_{qj} \oplus T_{q(j+1)} \oplus \cdots \oplus T_{ql}$. Thus, $q^{e_j-1}T_q = q^{e_j-1}T_{qj} \oplus q^{e_j-1}T_{q(j+1)} \oplus \cdots \oplus q^{e_j-1}T_{ql}$, all of the earlier summands being wiped out by multiplication by q^{e_j-1}. Now we identify the subgroup of elements of order $\leq q$ in $q^{e_j-1}T_q$. A little thought reveals that $(q^{e_j-1}T_q)[q] = q^{e_j-1}T_{qj} \oplus q^{e_{j+1}-1}T_{q(j+1)} \oplus \cdots \oplus q^{e_l-1}T_{ql}$. (See Problem 1.12.2 below). Similarly, $(q^{e_j}T_q)[q] = q^{e_{j+1}-1}T_{q(j+1)} \oplus \cdots \oplus q^{e_l-1}T_{ql}$. Hence, $U(e_j - 1, q) = (q^{e_j-1}T_q)[q]/(q^{e_j}T_q)[q] \cong q^{e_j-1}T_{qj}$. But, if $T_{qj} = \bigoplus_{i=1}^{t_{qj}} <x_i>$, then $q^{e_j-1}T_{qj} = \bigoplus_{i=1}^{t_{qj}} < q^{e_j-1}x_i >$, a direct sum of t_{qj} many cyclic groups, each isomorphic to $Z(q)$. We can conclude that (a) $\mid U(e_j-1, q) \mid = \mid q^{e_j-1}T_{qj} \mid = q^{t_{qj}}$. Also, since $qU(e_j - 1, q) = 0$, this factor group can be naturally regarded as a vector space over the finite field Z/qZ. Assuming a little knowledge of vector spaces, we have (b) $t_{qj} = \dim_{Z/qZ} U(e_j - 1, q)$.

∎

It is important to note that, for each j, the factor group $U(e_j - 1, q)$ is independent of the chosen decomposition of T_q as a direct sum of cyclics. Thus, the number $f_{e_j-1}^q(G) = t_{qj}$ is an invariant of G. We should finally point out that, if k is a natural number with $f_k^q(T_q) = 0$ (equivalently $k \neq e_j$ for any $1 \leq j \leq l$), then $U(k - 1, q) =< 0 >$. These notions will appear again in Sections 3.5 and 8.4.

Exercise 1.12.3 *Check that our final claim immediately above is correct.*

Putting together everything we've done yields the following

Theorem 1.12.3 *(structure theorem for finite abelian groups) Let G be finite abelian group. Then G is a direct sum of cyclics. The set of numbers $S = \{f_k^p(G) : p$ a prime, $k \geq 0\}$, defined as the number of $Z(p^{k+1})$-summands in any direct sum decomposition of G, can be computed either by the formula (a) or formula (b) of Lemma 1.12.4. Thus, S is a **complete set of isomorphism invariants** for G in the class of all finite abelian groups.*

Example 1.12.2 *It's now easy to list all possible nonisomorphic abelian groups of a given (relatively small) order. Simply note that, if $G =< x_1 > \oplus < x_2 > \oplus \cdots \oplus < x_m >$ and $\mid x_i \mid = p_i^{e_i}$, then $\mid G \mid$ must equal $p_1^{e_1} \cdots p_m^{e_m}$ (here the primes are not necessarily distinct). For example $100 = 2^2 5^2$. The nonisomorphic abelian groups of order 100 are $Z(4) \oplus Z(25)$, $Z(2) \oplus Z(2) \oplus Z(25)$, $Z(4) \oplus Z(5) \oplus Z(5)$ and $Z(2) \oplus Z(2) \oplus Z(5) \oplus Z(5)$.*

Tenth Problem Set

Problem 1.12.1 *Suppose G is a finite abelian group written additively with $\mid G \mid = p_1^{e_1} \cdots p_n^{e_n}$. Use gcd's to prove: (1) any $g \in G$ can be written $g = g_1 + \cdots + g_n$ with $g_i \in T_{p_i}$ and (2) for each $1 \leq i \leq n$, $T_{p_i} \cap \sum_{j \neq i} T_{p_j} = 0$.*

Problem 1.12.2 *Suppose G is a torsion group, that is an additive abelian group such that every element $g \in G$ has finite order $m = m(g)$. Extend the theorem and proof of Problem 1.12.1 to this case.*

Problem 1.12.3 *Write $U(30)$ as a direct product of cyclic groups of prime power order. (It's more natural to use multiplicative notation here.)*

Problem 1.12.4 *Complete the proof of Lemma 1.12.4 by showing that $(q^{e_j-1}T_q)[q] = q^{e_j-1}T_{qj} \oplus q^{e_{j+1}-1}T_{q(j+1)} \oplus \cdots \oplus q^{e_l-1}T_{ql}$.*

Problem 1.12.5 *Find all the nonisomorphic abelian groups of order 24.*

Problem 1.12.6 *Suppose that G and H are finite abelian groups such that $G \oplus G \cong H \oplus H$. Prove $G \cong H$.*

Problem 1.12.7 *(Chapter Review Problem) Using the techniques developed in this chapter, find all the (not just abelian) nonisomorphic groups of order ≤ 10. (Do $|G| = 8$ last, since it's the hardest case.)*

Further reading: [R3] Rotman's old group theory book has a wide variety of nice material.

Chapter 2

Rings (mostly domains)

2.1 Definitions and elementary properties

Definition 2.1.1 *A **ring** consists of a set R together with two binary operations + and · , naturally called addition and multiplication, on R that satisfy some very familiar axioms.*

1. $(R, +)$ is an abelian group. (As usual, we denote the additive identity by 0 and the additive inverse of an element x by $-x$.)

2. Multiplication is an associative operation with an identity element denoted by 1. To avoid the trivial case $R = \{0\}$, we require that $1 \neq 0$.

3. Multiplication is left and right distributive over addition; that is, $x \cdot (y + z) = x \cdot y + x \cdot z$ and $(y + z) \cdot x = y \cdot x + z \cdot x$ for all $x, y, z \in R$.

If the ring R satisfies the additional axiom that $y \cdot z = z \cdot y$ for all $y, z \in R$, then R is called a **commutative ring**. Henceforth, we drop the symbol · and simply write products by juxtaposition. Also, for convenience, we write $a - b$ for $a + (-b)$.

We give a collection of more or less familiar examples of rings. The sets of numbers $Z, Q, \mathcal{R}$ and $\mathcal{C}$ (the complex numbers), together with their ordinary addition and multiplication, are commutative rings. For a fixed $n > 0$, the set of $n \times n$ real matrices with ordinary matrix addition and multiplication is a noncommutative ring. The set $Z(n)$ with addition and multiplication mod n is a commutative ring.

Here is a less familiar example.

Example 2.1.1 *Let* $\mathcal{H} = \{a+bi+cj+dk : a, b, c, d \in Q\}$. *Define addition in* $\mathcal{H}$ *by simply adding the coefficients of like terms:* $(a+bi+cj+dk)+(a'+b'i+c'j+d'k) = (a+a')+(b+b')i+(c+c')j+(d+d')k$. *Multiplication is defined by setting* $i^2 = j^2 = k^2 = -1, ij = k, jk = i, ki = j, ji = -k, kj = -i, ik = -j$ *and requiring the distributive laws to hold. It is not too hard to check that* $\mathcal{H}$ *is a ring (although checking the associate law is tedious). The ring* $\mathcal{H}$ *is called the ring of* **Hamilton quaternions**.

Problem 2.1.3 will guide you in proving that every nonzero $h \in \mathcal{H}$ has a multiplicative inverse. Thus, $\mathcal{H}$ is an example of a ring satisfying all the field axioms except that of commutative multiplication. Such a ring is called a **division ring**. These rings play an essential role in Chapter 6.

Our first theorem shows that arbitrary rings satisfy some familiar arithmetical properties.

Theorem 2.1.1 *Let R be a ring and $a, b, c \in R$. The following equalities hold:*

1. $0 = a0 = 0a$

2. $-(ab) = (-a)b = a(-b)$

3. $(-1)a = a(-1) = -a$

4. $(-a)(-b) = ab.$

Proof. 1. Since 0 is the additive identity, the left distributive law gives: $a0 = a(0 + 0) = a0 + a0$. Hence, by left cancellation in the group $(R, +)$, $0 = a0$. Similarly, $0 = 0a$. 2. By part 1 and the right distributive law: $ab + (-a)b = (a - a)b = 0b = 0$. Since the inverse of an element in a group is unique, $-(ab) = (-a)b$. Similarly, $-(ab) = a(-b)$. 3. By part 1 and the right distributive law, $a + (-1)a = 1a + (-1)a = (1 - 1)a = 0a = 0$. As in part 2, $(-1)a = -a$. Similarly, $a(-1) = -a$. 4. By part 2 $(-a)(-b) = a[-(-b)] = ab$. ∎

In contrast to our standard number systems, a general ring R may contain **zero divisors**, that is there can be nonzero elements $a, b \in R$ with $ab = 0$. For example $2 \cdot 3 = 0$ in the ring $Z(6)$ and $\begin{bmatrix} 1 & -1 \\ 3 & -3 \end{bmatrix} \begin{bmatrix} 2 & -4 \\ 2 & -4 \end{bmatrix} = \begin{bmatrix} 0 & 0 \\ 0 & 0 \end{bmatrix}$ in the ring of 2×2 matrices with real entries. (We will discuss matrix rings in more detail in Chapter 4.)

Definition 2.1.2 *An **integral domain** (or, more simply, **domain**) is a commutative ring with no zero divisors.*

Definition 2.1.3 *A field is a commutative ring F such that every nonzero $a \in F$ has a multiplicative inverse.*

If F is a field and $0 \neq a \in F$, we denote the multiplicative inverse of a by a^{-1}. An alternate definition of a field is a commutative ring $(F, +, \cdot)$ such that $(F^*, \cdot)$ is a group. Here F^* is the set of nonzero elements of F.

Exercise 2.1.1 *Prove that the multiplicative inverse of a nonzero field element is unique.*

The number systems $Q, \mathcal{R}$ and $\mathcal{C}$ provide familiar examples of fields. The ring of integers Z is a domain but not a field.

We have the following two simple results.

Theorem 2.1.2 *Every field is a domain.*

Exercise 2.1.2 *Prove Theorem 2.1.2 in no more than five lines.*

Theorem 2.1.3 *Every finite domain is a field.*

Proof. Let I be a finite domain and I^* be the distinct nonzero elements of I. We need to show each $j \in I^*$ has a multiplicative inverse. Since I has no zero divisors, j_l (left multiplication by j) maps I^* into itself. Suppose $ji = ji'$. Then $j(i - i') = 0$ and we must have $(i - i') = 0$, so that $i = i'$. Hence, j_l is a monomorphism from I^* into itself. Since I^* is finite, j_l must also be an epimorphism. Thus, there exists $i \in I^*$ such that $ji = 1 \in I^*$. The proof is complete. ∎

We now present a series of simple definitions and results for rings, rather analogous to those that we presented for groups.

Definition 2.1.4 *A subset S of a ring R is called a **subring** of R if the following properties hold:*

1. $(S, +, \cdot)$ is a ring.

2. The multiplicative identity of R, 1_R, is in S.

We adopt the same notation for subrings as for subgroups; we write $S \leq R$ to indicate that S is a subring of R.

Condition 2 of the above definition may look a little odd at first glance. We impose it since it will be convenient for a subring and its overlying ring to share the same identity. In general, we cannot always conclude from Condition 1 alone that $1_S = 1_R$. For example, the set S of 2×2 real matrices of the form $\begin{bmatrix} r & 0 \\ 0 & 0 \end{bmatrix}$ with the standard operations form a ring

with multiplicative identity $1_S = \begin{bmatrix} 1 & 0 \\ 0 & 0 \end{bmatrix}$. Note that $S \subseteq R$, the ring of

all 2×2 real matrices, but $1_S \neq \begin{bmatrix} 1 & 0 \\ 0 & 1 \end{bmatrix} = 1_R$. See Problem 2.1.2.

We have a **subring test theorem**, analogous to the subgroup test theorem. The simple proof is left for you to supply in Problem 2.1.6.

Theorem 2.1.4 *A subset S of a ring R is a subring of R if S satisfies the following properties: 1. $1_R \in S$ and 2. $a, b \in S \Longrightarrow a - b \in S$ and $ab \in S$.*

Definition 2.1.5 *Let R, S be rings. A function $\theta : R \to S$ is called a **ring homomorphism** if $\theta(a + b) = \theta(a) + \theta(b)$ and $\theta(ab) = \theta(a)\theta(b)$ for all $a, b \in R$. We also require that $\theta(1_R) = 1_S$.*

Again, the last requirement may look a little strange, but it is very useful and does not follow from the first two. (See Problem 2.1.7 below.) A bijective ring homomorphism is called a **ring isomorphism**. If a ring homomorphism $\theta : R \to S$ is monic, it's called a **ring embedding** R into S. Then $R \cong \theta(R) \leq S$.

For an arbitrary domain I, there is a method of constructing a field $q(I)$ such that $I \cong I' \leq q(I)$. The construction, outlined below, mimics the extension of the domain $(Z, +, \cdot)$ to the field $(Q, +, \cdot)$.

Theorem 2.1.5 *Let I be a domain. Then there is a field $q(I)$ and a ring embedding $v : I \to q(I)$. Furthermore, the embedding v satisfies the following **universal mapping property**: For every field F' and embedding $v' : I \to F'$, there is a unique ring homomorphism $\theta : q(I) \to F'$ such that $\theta v = v'$.*

Before giving the proof, we illustrate the situation in the following commutative diagram:

$$I \xrightarrow{\;v\;} q(I)$$
$$v' \searrow \quad \downarrow \theta$$
$$F'$$

Proof. Let $I \times I^* = \{(a, b) : a, b \in I, b \neq 0\}$. We define a relation $\sim$ on $I \times I^*$ by saying $(a, b) \sim (a', b')$ if $ab' = ba'$. It is routine to check that $\sim$ is an equivalence relation. Let a/b be the equivalence class of a pair (a, b) and let $q(I) = \{a/b : (a, b) \in I \times I^*\}$.

We want to define an addition and multiplication on $q(I)$ so that $q(I)$ becomes a field. Thinking about addition and multiplication in Q, we define $a/b + a'/b' = (ab' + a'b)/bb'$ and $(a/b)(a'/b') = aa'/bb'$. Since we've defined operations on equivalence classes in terms of prechosen representative elements, we must check that our operations are well defined. We'll do addition, the harder of the two, leaving the proof that multiplication is well defined as Problem 2.1.1. Say $a/b \sim a_1/b_1$ and $a'/b' \sim a_1'/b_1'$. We want:

$$(ab' + a'b)/bb' \sim (a_1 b_1' + a_1' b_1)/b_1 b_1' \text{ or }$$

$$(?) : (ab' + a'b) b_1 b_1' = bb'(a_1 b_1' + a_1' b_1).$$

We have (1) $ab_1 = ba_1$ and (2) $a'b_1' = b'a_1'$. Multiplying (1) by b_1' and (2) by b yields (1') $ab_1 b_1' = b_1' b a_1$ and (2') $ba'b_1' = bb'a_1'$. Multiplying out, our desired equation (?) becomes (??) : $ab'b_1 b_1' + a'bb_1 b_1' = bb'a_1 b_1' + bb'a_1' b_1$. Using (1') and (2'), we transform (??) by substituting for $ab_1 b_1'$ and $bb'a_1'$. These substitutions produce the identity $(b_1' ba_1)b' + a'bb_1 b_1' = bb'a_1 b_1' + (ba'b_1')b_1$. Since all steps reverse, our addition is well defined.

Now we need to show that $(q(I), +, \cdot)$ is a field. Direct computations using the definitions show that $+$ and $\cdot$ are commutative and associative binary operations on $q(I)$ and that multiplication distributes over addition. The additive and multiplicative identities are $0/1$ and $1/1$, respectively. The additive inverse of a/b is $(-a)/b$. Suppose that $a/b \neq 0 = 0/1$. Then $a \neq 0$ so that $b/a \in q(I)$. We have $(a/b)(b/a) = (ab)/(ba) = 1/1 = 1$. We have verified that $(q(I), +, \cdot)$ is a field.

Let $\upsilon : I \to q(I)$ be given by $\upsilon(x) = x/1$. It is routine to check that υ is a ring embedding. The proof of the universal mapping property is Problem 2.1.8 below. ∎

The field $q(I)$ is called a **quotient field** of I. The field $q(I)$ is unique up to isomorphism. See Problem 2.1.8 below.

Eleventh Problem Set

Problem 2.1.1 *Check that multiplication is well defined in $q(I)$.*

Problem 2.1.2 *Let $(R, +, \cdot)$ be a domain with identity 1_R. Let $S \subset R$ be such that $(S, +, \cdot)$ is a ring with identity 1_S. Prove that we must have $1_S = 1_R$.*

Problem 2.1.3 *Refer back the ring of Hamilton quaternions in the first example of the chapter. Let $h = (a + bi + cj + dk)$ and $h' = (a - bi - cj - dk)$. Compute hh' and $h'h$. Then prove that any nonzero quaternion has a two sided multiplicative inverse.*

Problem 2.1.4 *For a fixed prime p, let $Z_p = \{r/s \in Q : (r, s) = 1, p \nmid s\}$. Prove that Z_p is a subring of Q.*

Problem 2.1.5 *Prove that $Z(n)$ is a field if and only if $n = p$, a prime.*

Problem 2.1.6 *Prove Theorem 2.1.4.*

Problem 2.1.7 *Find an example of rings R, S and a map $\theta : R \to S$ such that $\theta(a + b) = \theta(a) + \theta(b)$ and $\theta(ab) = \theta(a)\theta(b)$ for all $a, b \in R$ but $\theta(1_R) \neq 1_S$. Can you add a condition to an additive and multiplicative map θ between rings such that we must have $\theta(1_R) = 1_S$? (Other than simply saying $\theta(1_R) = 1_S$!)*

Problem 2.1.8 *Verify the universal mapping property for the inclusion map $v : I \to q(I)$ in the statement of Theorem 2.1.5.*

Problem 2.1.9 *Suppose that F, F' satisfy the universal mapping property with respect to ring embeddings $\lambda : I \to F$ and $\lambda' : I \to F'$. Prove that there is a ring isomorphism $\theta : F \cong F'$ such that $\theta\lambda = \lambda'$. This proves that the quotient field $q(I)$ is unique up to an isomorphism respecting the inclusion map.*

Problem 2.1.10 *Show that the set of 2×2 real matrices of the form $F = \left\{ \begin{bmatrix} a & b \\ -b & a \end{bmatrix} : a, b \in \mathbf{R} \right\}$ is a field under the usual matrix operations. Is F isomorphic to any well-known field?*

Problem 2.1.11 *Suppose that $(A, +, \cdot)$ is a set with binary operations satisfying all the conditions in the definition of a ring except that A need not have a multiplicative identity. Let A_1 Be the Cartesian product $A \times Z$ with binary operations defined by $(a, m) + (b, n) = (a + b, m + n)$ and $(a, m) \cdot (b, n) = (a \cdot b + na + mb, mn)$. Prove that A_1 is a ring (with identity) and that there is an additive and multiplicative monomorphism from A into A_1. This shows that every "rng" (ring without identity) can be embedded in a ring with identity.*

2.2　Homomorphism, ideals and factor rings

As in the case of group homomorphisms, the kernel and image of a ring homomorphism are useful notions.

Definition 2.2.1 *Let R, R' be rings and $\theta : R \to R'$ be a ring homomorphism. Let $0'$ be the additive identity of R'. Then $Ker\theta = \{r \in R : \theta(r) = 0'\}$ and $Im\,\theta = \{r' \in R' : r' = \theta(r) \text{ for some } r \in R\}$.*

Thus, the kernel and image of ring homomorphisms have exactly the same definitions as for additive group homomorphisms. Moreover, since each ring homomorphism is an additive group homomorphism, our work in Chapter 1 tells us that $Ker\theta$ is a subgroup of $(R, +)$ (necessarily normal since $(R, +)$ is abelian) and that $\text{Im }\theta$ is a subgroup of $(R', +)$.

Exercise 2.2.1 *(a) Verify that* $\text{Im }\theta$ *is a subring of* R'. *(b) If* $Ker\theta$ *is a subring of* R, *what can you say about the map* θ?

Following the development in Chapter 1, we consider the problem of defining factor rings. If $(I, +)$ is a proper subgroup of $(R, +)$ we already know that the set of additive cosets $R/I = \{I + r : r \in R\}$ can be made into an additive abelian group in the natural way: $(I + r) + (I + r') = I + (r + r')$. We want to define coset multiplication equally naturally, that is by the formula $(I + r)(I + r') = I + rr'$. If we can find appropriate conditions on I such that coset multiplication is well defined, it is clear that R/I with coset addition and multiplication will satisfy the ring axioms. Additionally, the map $\pi : r \to I + r$ will be a ring homomorphism from R onto R/I. Let's see what additional conditions we need to impose on a subgroup I so that the natural coset multiplication is well defined.

Suppose that $I + r = I + r'$ and $I + s = I + s'$. It follows that $r - r'$ and $s - s'$ are both in I. We want $I + rs$, the product computed one way, to be equal to $I + r's'$, the product computed the other way. In other words, we need $rs - r's'$ to be in I.

First, assume that I is closed under products. Then we have: $(r - r')(s - s') = (rs - rs' - r's + r's') \in I$. It follows that $(rs - r's') \in I$ if and only if $(rs - r's') - (rs - rs' - r's + r's') = (-2r's' + rs' + r's) \in I$. This latter expression can be regrouped and factored as $r'(s - s') + (r - r')s'$. So what we want is for $r'(s - s') + (r - r')s'$ to be in I, given that $r - r'$ and $s - s'$ are both in I. Our somewhat indirect line of reasoning leads to the following definition.

Definition 2.2.2 *Let* R *be a ring. A subset* $I \subseteq R$ *is called an **ideal** of* R *if* $(I, +)$ *is a subgroup of* $(R, +)$ *and* $i \in I, r \in R$ *implies that both* ri *and* ir *are in* I.

Loosely speaking, an ideal is a subgroup that not only is closed under products but actually absorbs products. Note that (0) and R are ideals of any ring R and that an ideal $I = R$ if and only if $1 \in I$. Since subrings must contain 1, the only additive subgroup of R that is both a subring and an ideal is R itself. Some simple examples of ideals in rings are: (1) mZ is an ideal of Z and (2) $mZ(n)$ is an ideal of $Z(n)$. For convenience, we write $I \lhd R$ to mean that I is **proper** ideal of R. Since our factor rings R/I

must have $1_{R/I} = I + 1 \neq 0_{R/I} = I + 0$, dealing only with proper ideals will be essential.

Let $I \lhd R$ and refer back to the discussion prior to the definition. Since $(s - s') \in I$ then $r'(s - s') \in I$. Since $(r - r') \in I$ then $(r - r')s' \in I$. Hence, $[r'(s - s') + (r - r')s'] \in I$. Modulo the trivial verification of the ring axioms for R/I, we have proved the following theorem.

Theorem 2.2.1 *Let* $I \lhd R$. *Then* R/I *is a ring via the natural coset addition and multiplication. Further,* R/I *is a ring epimorphic image of* R *via the natural map* $r \rightarrow I + r$.

Exercise 2.2.2 *Take a few minutes to check out the ring axioms for* R/I.

We have a ring theoretic analogue of Theorem 1.5.4 with ideals playing the role of normal subgroups.

Theorem 2.2.2 *Let* R *be a ring and* $I \subset R$. *Then* $I \lhd R \iff I = Ker\theta$ *for some ring homomorphism* θ *mapping* R *into some ring* R',

Proof. If I is a proper ideal of R then I is the kernel of the natural map $\pi : R \rightarrow R/I$. Conversely, suppose $I = Ker\theta$ for $\theta : R \rightarrow R'$. Since any ring homomorphism is a homomorphism of the underlying additive groups, we already know that $(I, +)$ will be a subgroup of $(R, +)$. We need only to check that I absorbs products. Let $i \in I, r \in R$. Then $\theta(ri) = \theta(r)\theta(i) = \theta(r)0 = 0$. Hence, $ri \in I$. Similarly, $ir \in I$ so I is an ideal. Also note that, since $R' \neq \{0\}, I \neq R$. The proof is complete. ∎

As in the group case, we now are, at least theoretically, in a position to survey all ring epimorphic images of R. We find the ideals $I \lhd R$ and then compute the factor rings R/I. As a simple example where this procedure is possible, consider the case $R = Z$ (Problem 2.2.1 below).

There are ring theoretic analogues of the subgroup correspondence theorem and the three group isomorphism theorems. The proofs here are very similar to those in Chapter 1. The only additional thing needed is to check that the maps used in Chapter 1 are, in this setting, compatible with ring multiplication. Hence, we simply state the results.

Theorem 2.2.3 *(subring and ideal correspondence) Let* R *be a ring and* $I \lhd R$. *The subrings of* R/I *are all of the form* S/I, *where* S *is a subring of* R *with* $I \subsetneq S \subseteq R$. *The ideals of* R/I *are all of the form* J/I, *where* J *is an ideal of* R *with* $I \subseteq J \subseteq R$.

Theorem 2.2.4 *(first ring isomorphism theorem) Let* $\theta : R \rightarrow R'$ *be a ring homomorphism. Then* $R/(Ker\ \theta) \cong Im\ \theta$.

Theorem 2.2.5 *(second ring isomorphism theorem) Let I, J be ideals of a ring R with $I \subseteq J \subsetneq R$. Then $(R/I)/(J/I) \cong R/J$.*

Theorem 2.2.6 *(third ring isomorphism theorem) Let S be a subring of R and $I \lhd R$. Then $S + I = \{s + i : s \in S, i \in I\}$ is a subring of R and $S \cap I$ is an ideal of S. Furthermore, $(S + I)/I \cong S/(S \cap I)$.*

We close this section with a discussion of two special ideals in a commutative ring R.

Definition 2.2.3 *Let R be a commutative ring and $I \lhd R$. Then I is called* **prime** *if $a, b \in R, ab \in I \implies a \in I$ or $b \in I$.*

(See Problem 2.2.2 for the origin of this terminology.)

Definition 2.2.4 *Let R be a commutative ring and $I \lhd R$. Then I is called* **maximal** *if, whenever J is an ideal of R with $I \subseteq J \subseteq R$ then $J = I$ or $J = R$.*

The following theorem will be very useful in the rest of the chapter.

Theorem 2.2.7 *Let R be a commutative ring. Then R is a field if and only if the only ideals of R are (0) and R.*

Proof. Suppose R is a field and let $(0) \neq I$ be an ideal of R. Take $0 \neq i \in I$. Since R is a field, the multiplicative inverse $i^{-1} \in R$. Since I is an ideal $1 = i^{-1}i \in I$. Then, as remarked earlier, $I = R$.

Conversely, suppose R is a commutative ring having only (0) and R as ideals. To show that R is a field we need show that every $0 \neq a \in R$ has a multiplicative inverse. Let $0 \neq a \in R$ and consider $Ra = \{ra : r \in R\}$. Since R is commutative, it is easy to check that Ra is an ideal of R, in fact the smallest ideal of R containing the element a. Since $Ra \neq \{0\}$, we must have $Ra = R$. It follows that there exists $r \in R$ with $ra = 1$. The element r is the desired multiplicative inverse of a and the proof is complete. ∎

Theorem 2.2.8 *Let R be a commutative ring and $I \lhd R$. Then: (a) I is prime if and only if R/I is a domain. (b) I is maximal if and only if R/I is a field.*

Proof. Part (a) is immediate from the definition of multiplication in the ring R/I. For (b), suppose I is maximal. This means that there are no proper ideals between I and R. By the ideal correspondence theorem, the ring R/I has no proper ideals. Thus, by Theorem 2.2.7, R/I is a field.

Conversely, if R/I is a field, then R/I has no proper ideals. By the ideal correspondence theorem, I is maximal. ∎

Since every field is a domain and every finite domain is a field, we have the following two corollaries.

Corollary 2.2.1 *Every maximal ideal in a commutative ring is prime.*

Corollary 2.2.2 *Let R be a commutative ring and $I \lhd R$ a prime ideal of R. If R/I is finite then I is maximal.*

Twelfth Problem Set

Problem 2.2.1 *Find all ring homomorphic images of the ring Z.*

Problem 2.2.2 *(a) Prove that the prime ideals in the ring Z coincide with the ideals $Z0 = \{0\}$ and $Zp = \{np : n \in Z, p \text{ a prime}\}$. (b) Find all maximal ideals in Z.*

Problem 2.2.3 *Find all ideals in $\mathcal{H}$, the ring of Hamilton quaternions.*

Problem 2.2.4 *Let I be an ideal of a commutative ring R and a an element of R that is not in I. Find the ideal (I, a), the smallest ideal of R containing the given ideal I and the new element a.*

Problem 2.2.5 *Let $C[0, 1]$ be the set of all continuous real valued functions on the closed interval $[0, 1]$. Define addition and multiplication in $C[0, 1]$ pointwise; that is $(f + g)(x) = f(x) + g(x)$ and $(fg)(x) = f(x)g(x)$. It is easy to check that $C[0, 1]$ with these operations is a ring. For a fixed $r \in [0, 1]$, let $I_r = \{f \in C[0, 1] : f(r) = 0\}$. Prove that each ideal I_r is a maximal ideal in $C[0, 1]$.*

Problem 2.2.6 *Find all the subrings of Q that contain Z.*

Problem 2.2.7 *Find all maximal ideals in the ring $Z(n)$.*

Problem 2.2.8 *An element x of a ring R is called **nilpotent** if there exists a natural number $n(x)$ such that $x^{n(x)} = 0$. Prove that if R is a commutative ring the set N of all nilpotent elements is an ideal of R and that the factor ring R/N has no nonzero nilpotent elements.*

2.3 Principal ideal domains

In this section our goal is to introduce a special kind of domain in which we can generalize the fundamental theorem of arithmetic. We have already noted that, in a commutative ring, the ideal $Ra = \{ra : r \in R\}$ is the smallest ideal of R containing the element a. We call Ra the **principal ideal generated by** a and use the notations (a) and Ra interchangeably. We look at domains in which every ideal in principal.

Definition 2.3.1 *An domain R is called a **principal ideal domain** (pid) if every ideal $I \subseteq R$ is of the form $I = Ra$, for some $a \in I$.*

The ring of integers Z is a pid and is the source from which this notion was taken. Fields are pid's since they have so few ideals. In the next section, we show that $F[x]$, the ring of polynomials in one variable with coefficients in a field, is a pid. In Problem 2.3.5 is an example of a domain which is not a pid. The goal in this section is to show that we can do arithmetic in a pid in much the same way as we can in Z.

If $a, b \in R$, write $a \mid b$ to mean that $ax = b$ for some $x \in R$. Call an element $u \in R$ a **unit** if u has a multiplicative inverse in R. We consider two notions that reflect the divisibility and the factorization properties of primes in Z.

Definition 2.3.2 *An element $p \in R$ is **prime** if $p \neq 0$, p is not a unit, and $p \mid ab$ for $a, b \in R$ implies $p \mid a$ or $p \mid b$.*

Definition 2.3.3 *An element $q \in R$ is **irreducible** if $q \neq 0$, q is not a unit, and whenever $q = ab, a, b \in R$, either a or b is a unit.*

Thus, what we were originally calling primes in Z we now should call irreducibles. Theorem 2.3.3 below will show that no harm was done.

Theorem 2.3.1 *Every prime in a domain R is irreducible.*

Proof. Let $p \in R$ be prime and suppose $p = ab$. Since $p \mid p = ab$ and p is prime, then $p \mid a$ or $p \mid b$. Say $p \mid a$. Then $px = a$ so that $pxb = ab = p$. Hence, $p(xb - 1) = 0$. Since R is a domain, $bx - 1 = 0, bx = 1$, and b is a unit. ∎

We will show in Theorem 2.3.3 that, in a pid, the converse to Theorem 2.3.1 holds; in a pid every irreducible is prime. In Problem 2.3.4, at the end of this section, we will produce an example of an irreducible element in a domain (obviously not a pid) that is not prime.

For the remainder of this section, R is a pid.

The notion of a greatest common divisor in R is identical to the notion in Z.

Definition 2.3.4 Let $a, b \in R$. A **greatest common divisor** (gcd) of a and b is an element $d \in R$ such that $d \mid a, d \mid b$, and, whenever $x \in R$ is such that $x \mid a, x \mid b$, it follows that $x \mid d$.

Theorem 2.3.2 (*existence of gcd's in a pid*) Let $a, b \in R$, with at least one of a, b nonzero. Then a, b have a gcd $d \in R$ with $d = ra + sb$ for some $r, s \in R$. Furthermore, if $d' \in R$ is another gcd of a, b, then $d' = du$ with u a unit.

Proof. Consider (a, b), the smallest ideal of R containing the elements a, b. It's easy to check that $(a, b) = \{ra + sb : r, s \in R\}$. Since R is a pid, $(a, b) = (d)$ for an element $d \in R$. Plainly $d = ra + sb$ for some $r, s \in R$. We'll show that d is a gcd of a and b. First, since $a \in (a, b) = (d)$, then $a = cd$ for some $c \in R$, that is $d \mid a$. Similarly, $d \mid b$. Now let $x \in R$ be such that $x \mid a, x \mid b$. Then $xa' = a, xb' = b$. Hence, $x(ra' + sb') = ra + sb = d$ and $x \mid d$. Finally, suppose that d' is another gcd for a, b. Since $d \mid a$ and $d \mid b$ it follows that $d \mid d'$. Since $d' \mid a$ and $d' \mid b$ it follows that $d' \mid d$. We have $du = d'$ and $d'v = d$ for $u, v \in R$. Thus, $d = d'v = duv$ and $d(1 - uv) = 0$. We cannot have $d = 0$, since $(a, b) = (d)$ and one of a, b is nonzero. So $1 = uv$ and both u and v are units. The proof is complete. ∎

Thus, in a pid we can speak of "the gcd" of a, b, understanding that it is defined only up to a unit multiple. This convention can often be used to simplify arguments, as in that of the following proof.

Theorem 2.3.3 *In a pid the set of irreducibles coincides with the set of primes.*

Proof. By Theorem 2.3.1, it is enough to check that, in a pid, irreducibles are prime. Let $q \in R$ be irreducible and suppose $q \mid ab$ but $q \nmid a$. Since q is irreducible, the gcd of q and anything is either 1 or q. But $(q, a) \neq q$ since $q \nmid a$. Hence, $(q, a) = 1$ so that $1 = rq + sa$. Multiplication by b yields $b = rqb + sab$. Since $q \mid ab$ it follows that $q \mid b$, and the proof is complete. ∎

Theorem 2.3.4 *A pid R satisfies the **ascending chain condition for ideals** (acc). That is, if $I_1 \subseteq I_2 \subseteq \cdots \subseteq I_n \subseteq \cdots$ is an ascending chain of ideals of R, there exists a natural number n such that $I_n = I_{n+1} = I_{n+2} = \cdots = I_{n+j} = \cdots$.*

Proof. Let $I_1 \subseteq I_2 \subseteq \cdots \subseteq I_k \subseteq \cdots$ be an ascending chain of ideals of R. Put $I = \cup_{i=1}^{\infty} I_i$. Then I is an ideal of R (Problem 2.3.2 below) so that $I = Ra$. Since $a \in \cup_{i=1}^{\infty} I_i, a \in I_n$ for some n. Hence $a \in I_{n+j}$ for all $j \geq 0$.

It follows that, for $j \geq 0$, $I = Ra \subseteq I_{n+j} \subseteq I$. Therefore, $I_n = I_{n+j}$ for all $j \geq 0$. ∎

We are now ready for the main theorem of this section.

Theorem 2.3.5 *(fundamental theorem of arithmetic for pid's) Let R be a pid. Then every nonzero element $r \in R$ is a finite product $r = up_1 \cdots p_k$ where the p_i are irreducible and u is a unit. Further, if $r = vq_1 \cdots q_l$ where the q_i are irreducible and v is a unit, then $k = l$ and, after rearrangement, $q_i = e_i p_i$ with e_i a unit $1 \leq i \leq k$.*

Proof. Let $0 \neq r \in R$. We first prove the existence of a factorization such as that above. If r is a unit, we're done. (In this case there are no p_i. Let's agree to allow this as a degenerate case of the statement of our theorem. Otherwise, we must make the statement a little more cumbersome.) If r is irreducible, we put $u = 1, r = p_1$. Otherwise, $r = r_1 r_2$ with neither factor a unit. Thus, by Problem 2.3.3 below, $(r) \subsetneq (r_1)$. If r_1 is not irreducible, write $r_1 = r_{11} r_{12}$ with neither factor a unit. We then have a proper ascending chain of ideals $(r) \subsetneq (r_1) \subsetneq (r_{11})$. By the acc, at some stage we must have $r_{1\ldots1}$ an irreducible element, which we label p_1. Since $(r) \subseteq (p_1)$ we can write $r = p_1 r'$. If r' is a unit or is irreducible, we're done. Otherwise, we repeat the process with r' to obtain $r' = p_2 r''$ with p_2 irreducible. Then, $r = p_1 p_2 r''$. We now have a properly ascending chain $(r) \subsetneq (r') \subsetneq (r'')$. By the acc, at some stage we must have $r'^{\ldots}$ either irreducible or a unit. At this stage we have produced our factorization.

Let $r = up_1 \cdots p_k = vq_1 \cdots q_l$ be factorizations as in the statement of the theorem. We prove unicity by induction on k, starting at $k = 0$. Suppose $r = u = vq_1 \cdots q_l$. Multiplication by u^{-1} produces the equation $1 = u^{-1} v q_1 \cdots q_l$. No q_i can appear on the right hand side since irreducibles are not units. Hence, $r = u = v$ and, in this case, we're done.

Now suppose that $k > 0$, $r = up_1 \cdots p_k = vq_1 \cdots q_l$ and that our unicity statement holds whenever an element can be written as a unit times a product of fewer than k irreducibles. We have $p_1 \mid q_1(vq_2 \cdots q_l)$. Since irreducibles in R are prime, either $p_1 \mid q_1$ or $p_1 \mid (vq_2 \cdots q_l)$. If $p_1 \mid q_1$ then $q_1 = e_1 p_1$ with e_1 a unit. In this case, we substitute $e_1 p_1$ for q_1 in the right hand factorization, cancel p_1 from both factorizations, and we're done by induction. If $p_1 \mid (vq_2 \cdots q_l)$ we repeat the argument we have just outlined. Eventually, we must have $p_1 \mid q_i$ for some i (p_1 cannot divide the unit v). We substitute and cancel as above, and we're done by induction. The proof is complete. ∎

A domain that satisfies the conclusion of Theorem 2.3.5 is called a **unique factorization domain** (ufd). So, Theorem 2.3.5 says that every pid is a ufd. Later, we'll give examples of ufd's that are not pid's.

Thirteenth Problem Set

Problem 2.3.1 *Let R be a pid and $\{a_1, ..., a_n\} \subseteq R$ with not all the $a_i = 0$. Define a gcd d for $\{a_1, ..., a_n\}$ and prove that $\{a_1, ..., a_n\}$ has a gcd. If d, d' are gcd's for $\{a_1, ..., a_n\}$, what is the relationship between d and d'?*

Problem 2.3.2 *Let R be a commutative ring and $I_1 \subseteq I_2 \subseteq \cdots \subseteq I_k \subseteq \cdots$ be a countable ascending chain of ideals of R. Prove that $\cup_{i=1}^{\infty} I_i$ is an ideal of R.*

Problem 2.3.3 *Let R be a domain. If $a = bc$ is a factorization in R with neither b nor c a unit, we call b, c **proper divisors** of a. Prove that b is a proper divisor of a if and only if $Ra \subsetneq Rb$.*

Problem 2.3.4 *We'll construct, in a series of steps, a domain with an irreducible non-prime element.*
Let $R = \{a + b\sqrt{3}i : a, b \in Z\} \leq C$. It is easy to check that R is a subring of C. Thus, R is a domain.
a) Define $N : R \to Z$ by $N(a + b\sqrt{3}i) = a^2 + 3b^2$. Prove that, for $r, r' \in R$, $N(rr') = N(r)N(r')$.
b) Prove that $u \in R$ is a unit if and only if $N(u) = 1$. Find all units of R.
c) Prove that, in R, $4 = 2 \cdot 2 = (1 + \sqrt{3}i)(1 - \sqrt{3}i)$ but that 2 divides neither $1 + \sqrt{3}i$ nor $1 - \sqrt{3}i$. Hence, the element $2 \in R$ is not prime.
d) Use part (a) to prove that 2 is an irreducible element of R. Of course, by Theorem 2.3.5, this implies that R is not a pid.

Problem 2.3.5 *Let $R = \mathcal{R}[x, y]$ be the ring of polynomials with real coefficients in two variables x, y. (To do this problem, you need only a minimal knowledge about polynomials. We will make things precise in the next section.) Let $I = (x, y)$, the ideal of R generated by x, y. (1) Describe the polynomials in I. (2) Using standard facts about polynomial multiplication as given, prove that I cannot be of the form $I = Rp(x, y)$ for some fixed polynomial $p(x, y) \in R$.*

2.4 Polynomials

2.4.1 Unique Factorization

In this section we deal with $F[x]$ (defined precisely below), the familiar ring of polynomials in one variable x. Throughout the section, F denotes a field. In the first of our two subsections our goal is to prove that $F[x]$ is a pid. Hence, the unique factorization theorem of our last section will hold for $F[x]$.

Remark 2.4.1 *The polynomial ring $F[x]$ consists of the set of expressions $\{f(x) = c_0 + c_1 x + \cdots + c_n x^n : n \geq 0, c_i \in F\}$. Except in the case that $f(x) = 0$, the zero polynomial, when write a polynomial we assume that the leading coefficient c_n is nonzero. The number n is called the **degree** of $f(x)$ (abbreviated deg $f(x)$). We define the degree of the zero polynomial to be $-\infty$. Two polynomials $f(x) = c_0 + c_1 x + \cdots + c_n x^n$ and $g(x) = b_0 + b_1 x + \cdots + b_k x^k$ are equal if $n = k$ and $c_i = b_i$ for all i.*

We add and multiply polynomials exactly as we learned to do in high school, but here the coefficients come from the abstract field F. Specifically, for $f(x), g(x)$ as above, the coefficient of x^t in $f(x) + g(x)$ is $(c_t + b_t)$ and the coefficient of x^t in $f(x)g(x)$ is $d_t = \sum_{i=0}^{t} c_i b_{t-i}$. Terms may appear in the preceding addition and multiplication formulas with an index higher than their associated polynomial's degree. We will simply interpret such terms as 0.

Exercise 2.4.1 *Let $f(x), g(x) \in F[x]$. Prove that $\deg[f(x) + g(x)] \leq \max\{\deg f(x), \deg g(x)\}$ and $\deg f(x)g(x) = \deg f(x) + \deg g(x)$. (The last equality illustrates one reason why it's convenient to define the degree of the zero polynomial as $-\infty$.)*

Exercise 2.4.2 *Prove that $F[x]$ is a domain but not a field.*

There is a division algorithm for $F[x]$ much like that for Z.

Theorem 2.4.1 (division algorithm for $F[x]$) *Let $f(x), g(x) \in F[x]$ with $0 \leq \deg f(x)$. Then there exist polynomials $q(x), r(x) \in F[x]$ with $\deg r(x) < \deg f(x)$ such that $g(x) = q(x)f(x) + r(x)$.*

Proof. If $g(x) = 0$, put $q(x) = r(x) = 0$. (Here is another good reason to define $deg(0) = -\infty$.) From now on assume $0 \leq \deg g(x)$. Henceforth, the proof will be by induction on deg $g(x)$. Suppose that deg $g(x) = 0$. If deg $f(x) > 0$, put $q(x) = 0$ and $r(x) = g(x)$. If deg $f(x) =$ deg $g(x) = 0$, say that $g(x) = g_0 \neq 0$ and $f(x) = f_0 \neq 0$. Then set $q(x) = g_0/f_0 \in F$ and $r(x) = 0$.

Having disposed of the trivial cases, suppose deg $g(x) = m > 0$ and that the theorem is true for any appropriate $f(x), g(x)$, whenever deg $g(x) < m$. Write $f(x) = f_0 + f_1 x + \cdots + f_n x^n$ and $g(x) = g_0 + g_1 x + \cdots + g_m x^m$ with $f_n, g_m \neq 0$. If $n > m$ put $q(x) = 0, r(x) = g(x)$. If $n \leq m$, consider the polynomial $h(x) = g(x) - (g_m/f_n)x^{m-n}f(x)$. The polynomial $h(x)$ is a difference of two polynomials of degree m with equal x^m coefficients, hence

$h(x)$ has degree less than m. By the inductive hypothesis, with $h(x)$ playing the role of $g(x)$, there exists $q(x), r(x) \in F[x]$ with deg $r(x) <$ deg $f(x)$ such that $h(x) = q(x)f(x) + r(x)$. That this, $g(x) - (g_m/f_n)x^{m-n}f(x) = q(x)f(x) + r(x)$ so that $g(x) = [(g_m/f_n)x^{m-n} + q(x)]f(x) + r(x)$. The proof is complete. ∎

For a given $f(x), g(x)$ we can compute $q(x), r(x)$ by long division of polynomials, just as in high school-except here the coefficients come from an arbitrary field. Try Problem 2.4.1 below.

We now can achieve the goal of this subsection.

Theorem 2.4.2 $F[x]$ *is a pid.*

Proof. Let I be an ideal of $F[x]$. If $I = \{0\}$, then $I = F[x]0$. If $I \supset \{0\}$, choose a polynomial $f(x)$ of least degree among the nonzero polynomials in I. Since I is an ideal and $f(x) \in I$, then $F[x]f(x) \subseteq I$. We will prove that $I = F[x]f(x)$. That is, any nonzero ideal I in $F[x]$ will be principal and generated by any $0 \neq f(x) \in I$ of minimal degree among the set of nonzero polynomials in I. Let $g(x) \in I$. By the division algorithm, $g(x) = q(x)f(x) + r(x)$ for $q(x), r(x) \in F[x]$ with deg $r(x) <$ deg $f(x)$. But then $r(x) = [g(x) - q(x)f(x)] \in I$ and deg $r(x) <$ deg $f(x)$. By choice of $f(x)$, we must have $r(x) = 0$. Hence $g(x) = q(x)f(x) \in (f(x))$ and the proof is complete. ∎

We now can apply the unique factorization of the previous section to the pid $F[x]$. First, we should identify the units and irreducibles in $F[x]$. This is very easy to do.

Exercise 2.4.3 *(a) Prove that $u(x) \in F[x]$ is a unit if and only if deg $u(x) = 0$. (b) Prove that $p(x) \in F[x]$ is an irreducible element if and only if deg $p(x) = n \geq 1$ and $p(x)$ cannot factor in $F[x]$ as a product of two polynomials of degrees less than n. More commonly, we call such a $p(x) \in F[x]$ an* **irreducible polynomial** *in $F[x]$.*

Example 2.4.1 *Of course, irreducibility can depend on the ground field. The polynomial $p(x) = (x^2+1)(x^2-2)$ is a product of quadratic irreducibles in $Q[x]$, factors in $\mathcal{R}[x]$ as a product of irreducibles $p(x) = (x^2 + 1)(x - \sqrt{2})(x + \sqrt{2})$ and completely in $C[x]$ as a product of irreducibles $p(x) = (x - i)(x + i)(x - \sqrt{2})(x + \sqrt{2})$.*

Putting together all that we've done, we can state the following:

Theorem 2.4.3 *(unique factorization in $F[x]$) Let $0 \neq f(x) \in F[x]$. Then $f(x) = up_1(x) \cdots p_k(x)$, where the $p_i(x)$ are irreducible polynomials in $F[x]$ and $0 \neq u \in F$. Further, if $f(x) = vq_1(x) \cdots q_l(x)$ where the $q_i(x)$ are irreducible in $F[x]$ and $0 \neq v \in F$, then $k = l$ and, after rearrangement, for all i, $q_i(x) = e_i p_i(x)$ with $0 \neq e_i \in F$.*

When your Math 101 students are factoring $p(x) = x^6 - 1$ in $\mathcal{R}[x]$ they could proceed in one of two ways; factor considering $p(x)$ as a difference of two squares, $p(x) = (x^3 - 1)(x^3 + 1)$ and then factor further, or factor considering $p(x)$ as a difference of two cubes, $p(x) = (x^2-1)(x^4+x^2+1)$ and then factor further. You'll now be able to assure them, with authority, that no matter what method they use they'll end up with the same factorization into irreducibles.

2.4.2 Irreducibles in $F[x]$

In our second subsection, we wish to find criteria for a polynomial $p(x) \in F[x]$ to be irreducible in $F[x]$. Obviously, polynomials of degree 1 are irreducible. So, in this subsection all polynomials $p(x)$ are assumed to be of degree ≥ 2.

Definition 2.4.1 *Let $p(x) = p_0 + p_1 x + \cdots + p_n x^n \in F[x]$. The element $a \in F$ is a **root** of $p(x)$ if $p(a) = p_0 + p_1 a + \cdots + p_n a^n = 0 \in F$.*

Theorem 2.4.4 *(factor theorem) The element $a \in F$ is a root of $p(x)$ if and only if $x - a$ divides $p(x)$.*

Proof. Certainly, if $p(x) = q(x)(x-a)$ then $p(a) = 0$. For the converse, write $p(x) = q(x)(x - a) + r$, where $r \in F$. If $p(a) = 0$, then $0 = q(a)(a - a) + r = r$. Hence, $p(x) = q(x)(x - a)$ and the proof is complete. ∎

Corollary 2.4.1 *Let $p(x) \in F[x]$ be of degree 2 or 3. Then $p(x)$ is irreducible in $F[x]$ if and only if $p(x)$ has no root in F.*

The corollary allows us to test the irreducibility of $p(x)$ with $\deg p(x) \leq 3$ in the polynomial ring $(Z/qZ)[x]$, for q a not too large prime. We simply test all the possibilities for roots in the field Z/qZ.

Exercise 2.4.4 *Are the following polynomials irreducible in $(Z/7)[x]$? If not, factor them into a product of irreducibles. (a) $p(x) = 2x^2 + 3x + 3$ (b) $p(x) = x^2 + 1$.*

We can extend the technique provided us by the above corollary a good bit further. (See Problem 2.4.4 below.)

Our final topic here is factorization in $Q[x]$ versus factorization in $Z[x]$. The reason for considering this is that it's often possible to decide if a polynomial $z(x) \in Z[x]$ of positive degree cannot be factored as a product of polynomials in $Z[x]$ of lower degree. It turns out that this will imply $z(x)$ is actually irreducible in $Q[x]$. We need a series of lemmas to obtain our main result. The treatment here follows that of [R2].

Definition 2.4.2 *A polynomial* $z(x) = z_0 + z_1 x + \cdots + z_n x^n \in Z[x]$ *with* $n \geq 1$ *is called* **primitive** *if* $\gcd \{z_0, z_1, ..., z_n\} = 1$.

The definition above and the following lemma, originally proved by Gauss, generalize directly to polynomials in $I[x]$ for I a ufd.

Lemma 2.4.1 *The product of primitive polynomials is primitive.*

Proof. Let $z(x) = z_0 + z_1 x + \cdots + z_n x^n$ and $w(x) = w_0 + w_1 x + \cdots + w_k x^k$ be primitive polynomials and consider their product. If $z(x)w(x)$ is not primitive, there must be a prime p that divides all its coefficients. We'll show that this supposition leads to a contradiction. Say that this prime p divides z_i for $i < t$ but $p \nmid z_t$. (The prime p cannot divide all the coefficients of the primitive polynomial $z(x)$.) Similarly, say that p divides w_i for $i < l$ but $p \nmid w_l$. Note that, since $p \mid z_0 w_0$, we cannot have $t = l = 0$. Let y_{t+l} be the coefficient of x^{t+l} in $z(x)w(x)$. Consider equation $\triangle$:

$$y_{t+l} = [z_0 w_{t+l} + z_1 w_{t+l-1} + \cdots + z_{t-1} w_{l+1}] + z_t w_l + [z_{t+1} w_{l-1} + \cdots + z_{t+l} w_0].$$

(In $\triangle$ any z_j with $j > n$ or w_j with $j > k$ is set equal to zero.) Since p divides y_{t+l} and also divides both bracketed expressions on the right hand side of $\triangle$, p divides $z_t w_l$. But $p \nmid z_t$ and $p \nmid w_l$, a contradiction. ∎

Remark 2.4.2 *To avoid possible confusion, we call a polynomial* $p(x) \in I[x], I$ *a integral domain,* **non-factorable** *in* $I[x]$ *if* $\deg p(x) = n \geq 1$ *and* $p(x)$ *cannot factor in* $I[x]$ *as a product of two polynomials of degrees less than* n. *If* $I = F$, *a field, non-factorable is the same as irreducible. In* $Z[x]$, *the polynomial* $p(x) = 2x^2 + 4$ *is non-factorable but has the irreducible factorization* $p(x) = 2(x^2 + 2)$.

Lemma 2.4.2 *Every nonzero* $q(x) \in Q[x]$ *has a unique factorization of the form* $q(x) = c(q)z(x)$, *where* $c(q)$ *is a positive rational, called the* **content** *of* $q(x)$, *and* $z(x)$ *is a primitive polynomial in* $Z[x]$.

Proof. Let $D > 0$ be an integer such that $Dq(x) \in Z[x]$. Then let $C > 0$ be the (positive) gcd of the coefficients of the integral polynomial $Dq(x)$. We have $Dq(x) = Cz(x)$ with $z(x)$ a primitive integral polynomial. Thus, $q(x) = c(q)z(x)$ with $c(q) = C/D$.

Now to prove unicity, suppose $q(x) = cz(x) = c'z'(x)$ with $z(x), z'(x)$ primitive and c, c' positive rationals. Write $c = a/b, c' = a'/b'$ in lowest terms with all integers positive. Then $ab'z(x) = a'bz'(x)$. Since the content of the polynomial on the left equals that of the polynomial on the right $ab' = a'b$. It follows that $a = a', b = b'$ and therefore that $z(x) = z'(x)$. ∎

Exercise 2.4.5 *Fill in the details to prove $a = a', b = b'$.*

Now we can prove our desired result.

Theorem 2.4.5 *Let $p(x) \in Z[x] \leq Q[x]$ be a polynomial of degree $n > 1$. Then $p(x)$ is irreducible in $Q[x]$ if and only if $p(x)$ is non-factorable in $Z[x]$.*

Proof. Since $Z[x] \leq Q[x]$, certainly, if $p(x)$ is irreducible in $Q[x]$, then $p(x)$ is non-factorable in $Z[x]$. Conversely, suppose $p(x)$ is non-factorable in $Z[x]$ but $p(x) = q(x)r(x)$, where $q(x)$ and $r(x)$ are polynomials in $Q[x]$ each of degree less than n. We will obtain a contradiction. By the previous lemma, we have $p(x) = c(p)z(x), q(x) = c(q)z_1(x), r(x) = c(r)z_2(x)$ with the $c's$ positive rationals and the $z(x)'s$ primitive integral polynomials. Then $c(p)z(x) = c(q)c(r)[z_1(x)z_2(x)]$. Since $z(x)$ and $z_1(x)z_2(x)$ are primitive, it follows that $c(p) = c(q)c(r)$. Cancellation yields $z(x) = z_1(x)z_2(x)$ so that $p(x) = c(p)z(x) = [c(p)z_1(x)][z_2(x)]$. Since $\deg c(p)z_1(x) = \deg q(x)$ and $\deg z_2(x) = \deg r(x)$, we have contradicted the assumption that if $p(x)$ is non-factorable in $Z[x]$. The proof is complete. ∎

Armed with Theorem 2.4.5, we'd like to have some criteria to tell us when $z(x) \in Z[x]$ is non-factorable in $Z[x]$. We conclude with **Eisenstein's Irreducibility Criterion**. It remains the best such irreducibility criteria we have.

Theorem 2.4.6 *(Eisenstein) Let $z(x) = z_0 + z_1x + \cdots + z_nx^n$ be a polynomial in $Z[x]$ with $n \geq 2$. Suppose there exists a prime p such that $p \mid z_i$, $0 \leq i \leq n - 1$, $p^2 \nmid z_0$, and $p \nmid z_n$. Then $z(x)$ is non-factorable in $Z[x]$, hence $z(x)$ is irreducible in $Q[x]$.*

Proof. The proof has the same flavor as the proof that the product of primitive polynomials is primitive. Suppose $z(x)$ satisfies the hypothesis of the theorem but that $z(x) = w(x)y(x) = (w_0 + w_1x + \cdots + w_kx^k)(y_0 +$

$y_1 x + \cdots + y_l x^l) \in Z[x]$ with $k, l \geq 1$. We'll derive a contradiction. Since $p \nmid z_n = w_k y_l$, then $p \nmid w_k$ and $p \nmid y_l$. Since $p \mid z_0 = w_0 y_0$ but $p^2 \nmid z_0$, then p divides exactly one of w_0, y_0, say that $p \mid w_0$ and $p \nmid y_0$. Suppose that $p \mid w_i$, $0 \leq i \leq s$, but $p \nmid w_{s+1}$. Note that $s < k$ since $p \nmid w_k$. Hence $s+1 < n$. Now consider the equation $z_{s+1} = (w_0 y_{s+1} + \cdots + w_s y_1) + w_{s+1} y_0$. As in the proof of Lemma 2.4.1, we conclude that $p \mid w_{s+1} y_0$. But p divides neither of w_{s+1}, y_0, a contradiction. ∎

For example, by Eisenstein with $p = 2$, the polynomial $z(x) = 2 + 6x - 32x^2 + 10x^3 - 8x + 15x^5$ is irreducible in $Q[x]$.

Actually, there is one last tool in our toolbox. In Problem 2.4.2 below, we ask you to prove the **rational root theorem**. This theorem gives us necessary conditions for a rational number a/b to be the root of $z(x) \in Z[x]$. Hence, by substituting in the possibilities, we can determine when $z(x) \in Z[x]$ has a linear factor $x - a/b$ in $Q[x]$.

Fourteenth Problem Set

Problem 2.4.1 *Find $q(x), r(x)$ in the division algorithm for $f(x) = 3 - 2x + 5x^2 + x^3, g(x) = x - 2$. Here all polynomials are in $(Z/7Z)[x]$.*

Problem 2.4.2 *Prove the **rational root theorem**. Let $r/s \in Q$ be a root of the integral polynomial $z(x) = z_0 + z_1 x + \cdots + z_n x^n$. Assume that r/s is written in reduced form, that is $(r, s) = 1$. Prove that $r \mid z_0$ and $s \mid z_n$.*

Problem 2.4.3 *Factor $p(x) = x^3 + x^2 - x - 10$ into irreducibles in $Q[x]$.*

Problem 2.4.4 *Let q be a fixed prime and, for $n \in Z$, let $\bar{n}$ be $n \in (Z/qZ)$; for example, if $q = 7, \overline{19} = 5$. (a) Prove that the map $z_0 + z_1 x + \cdots + z_n x^n \rightarrow \bar{z}_0 + \bar{z}_1 x + \cdots + \bar{z}_n x^n$ is a ring homomorphism from $Z[x]$ onto $(Z/qZ)[x]$. (b) Prove that $x^4 - 7x^3 + x + 1$ is irreducible in $Q[x]$.*

The kind of trick illustrated above in 2.4.4 (b) often works to show polynomials in $Z[x]$ are irreducible in $Q[x]$.

2.5   I[x] is a ufd*

In this brief section we outline a proof that, if an domain I is a ufd, so is $I[x]$. This result will allow us to give an example of a ufd that is not a pid. Since none of the material here is needed subsequently, the instructor feeling a little short on time may want to skip this section.

Recall that a domain I is a ufd if every nonzero $a \in I$ can be factored in I as $a = up_1 \cdots p_k$, where u is a unit and the p_i are irreducibles. Further, if a has a different factorization of the same form $a = vq_1 \cdots q_l$, then $k = l$ and, after rearrangement, for all i, $q_i = e_i p_i$ with each e_i a unit. Without further ado, here is the theorem.

Theorem 2.5.1 *If a domain I is a ufd so is $I[x]$*

Proof. Let $f(x)$ be a nonzero element of $I[x]$. If $f(x) = f_0 \neq 0 \in I$, then we can factor f_0 uniquely as a unit in I times a product of irreducibles in I. It is easy to see that this factorization is the only factorization of f_0 into irreducibles in $I[x]$. (See Problem 2.5.1 below.)

Suppose deg $f(x) \geq 1$. We'll prove that $f(x)$ has a factorization in $I[x]$ of the form $F : f(x) = u(i_1 \cdots i_t)p_1(x) \cdots p_k(x)$, where u is a unit in I, each i_j is an irreducible in I, hence in $I[x]$, and each $p_i(x)$ is a primitive irreducible polynomial in $I[x]$. Let F be the quotient field of I and regard $I \leq F$. In this way we can naturally regard $I[x] \leq F[x]$. Since $F[x]$ is a ufd, there is a factorization $f(x) = af_1(x) \cdots f_k(x)$, where $0 \neq a \in F$ and the $f_i(x)$ are all irreducible polynomials in $F[x]$. Without loss, we can assume that $a \in I$.

Now, exactly as in the proof of Lemma 2.4.2, write each $f_i(x) = c_i z_i(x)$ with the c_i elements of F and the $z_i(x)$ primitive polynomials in $I[x]$. Then write $f(x) = cz(x)$, where $c \in I$ and $c(x) \in I[x]$ is primitive.

We have $f(x) = cz(x) = a(c_1 \cdots c_k)[z_1(x) \cdots z_k(x)]$. By the extension of Gauss' Lemma from $Z[x]$ to $I[x]$, the product $z_1(x) \cdots z_k(x)$ is primitive. Since $z(x)$ is also primitive, then, modulo a unit, the content of $f(x)$ can be obtained as either c or as $a(c_1 \cdots c_k)$. Thus, $c = va(c_1 \cdots c_k)$ with v a unit in I. Substitution, then cancellation of the factor $a(c_1 \cdots c_k)$, gives us the equation $vz(x) = [z_1(x) \cdots z_k(x)]$. Multiplication by $u = v^{-1}$ yields $z(x) = u[z_1(x) \cdots z_k(x)]$. Then multiplication by c yields $f(x) = uc[z_1(x) \cdots z_k(x)]$. Finally, if c is not a unit, factor it into irreducibles in I, say $c = i_1 \cdots i_t$. We have constructed our desired factorization.

Suppose that we have factorizations of the form F above of a given $f(x) \in I[x]$:

$$(*) : u(i_1 \cdots i_t)p_1(x) \cdots p_k(x) = v(j_1 \cdots j_s)q_1(x) \cdots q_l(x).$$

Looking at these as factorizations in $F[x]$, we can conclude that $l = k$ and that, after rearrangement, for all i, $q_i(x) = (a_i/b_i)p_i(x)$, $0 \neq a_i, b_i \in I$. Thus, $b_i q_i(x) = a_i p_i(x)$. Since, for each i, the polynomials $q_i(x)$, $p_i(x)$ are primitive in $I[x]$, it follows that $a_i = e_i b_i$ with e_i a unit in I. Hence, for each i, $(a_i/b_i) = e_i$, a unit in I. Let $e = e_1 \cdots e_k$. After substitution, equation $(*)$ transforms to the equation

$$u(i_1 \cdots i_t)p_1(x) \cdots p_k(x) = ve(j_1 \cdots j_s)p_1(x) \cdots p_k(x).$$

Since the product of primitive polynomials is primitive, we argue as before to claim that $u(i_1 \cdots i_t) = (wve)(j_1 \cdots j_s)$ with w a unit. Note that u and (wve) are units in I and invoke unique factorization in I to complete the proof. ∎

Fifteenth Problem Set

Problem 2.5.1 *Let I be a ufd and i be a nonzero non unit element of I. If $i = ui_1 \cdots i_j$ is a factorization of i into the product of a unit times a collection of irreducibles in I, prove that this factorization remains a unique factorization of the same type, considering i as an element of $I[x]$.*

Problem 2.5.2 *Prove that $Z[x, y]$ is a ufd but not a pid.*

Problem 2.5.3 *Prove that if $\{a_1, ..., a_n\} \subseteq U$, U a ufd, and not all $a_i = 0$, then there exists a gcd for the set $\{a_1, ..., a_n\}$.*

Problem 2.5.4 *Find a ufd U and elements $a, b \in U$ such that $d = (a, b)$ cannot be written in the form $d = ra + sb$ for $r, s \in U$.*

Problem 2.5.5 *Prove that in a ufd any irreducible is prime.*

2.6 Euclidean domains*

We close our chapter on ring theory by considering a generalization of the polynomial ring $F[x]$. The treatment here pretty much follows that of Fraleigh [F]. As in the previous section, an instructor short on time might want to cover this material lightly or relegate some of it to homework.

Definition 2.6.1 *A domain I is called a **Euclidean domain** if there is a function v, called a **valuation on** I, mapping the nonzero elements of I into the set of nonnegative integers such that:*

(1) For all $a, b \in I$ with $b \neq 0$, there exist $q, r \in I$ such that $a = qb + r$ where either $r = 0$ or $v(r) < v(b)$.
(2) For all nonzero elements $a, b \in I$, $v(a) \leq v(ab)$.
For short, we say that (I, v) **is Euclidean** to mean that I is a Euclidean domain with valuation v.
A few minutes thought will show that all of the following are Euclidean domains.

Example 2.6.1 *The domain Z is a Euclidean domain with valuation $v(a) = | a |$ for $0 \neq a \in Z$.*

Example 2.6.2 *For F a field, the polynomial ring $F[x]$ is a Euclidean domain with valuation $v[p(x)] = \deg p(x)$ for $0 \neq p(x) \in F[x]$.*

Example 2.6.3 *Any field F is a Euclidean domain with the **trivial valuation** $v(a) = 1$ for $0 \neq a \in F$.*

A large part of the point of the above definition is in the following theorem.

Theorem 2.6.1 *Any Euclidean domain is a principal ideal domain.*

Proof. The proof is an exact analogue of the proof of Theorem 2.2.2. We assign it as Problem 2.6.1. ∎

Corollary 2.6.1 *Any Euclidean domain is a unique factorization domain.*

We note that not every pid is a Euclidean domain. Unfortunately, it takes a good bit of work to construct an example. For further investigation, we refer the reader to [R2], Example 3.19.

Let (I, v) be Euclidean. We can use the conditions of Definition 2.6.1 to characterize the units of I.

Theorem 2.6.2 *For (I, v) Euclidean, $v(1)$ is the minimal element of the set $v(I) = \{v(a) : 0 \neq a \in I\}$. Furthermore, $u \in I$ is a unit if and only if $v(u) = v(1)$.*

Proof. By valuation condition (2), $v(a) = v(a1) \geq v(1)$ for all nonzero $a \in I$. Hence, $v(1)$ is minimal in the set $v(I)$. If u is a unit of I then $uv = 1$ so that $v(1) \geq v(u)$. Thus, $v(u) = v(1)$. Conversely, suppose that $a \in I$ is such that $v(a) = v(1)$. By condition (1), we can write $1 = qa + r$ with $r = 0$ or $v(r) < v(a) = v(1)$. By minimality of $v(1)$, we must have $r = 0$ and a is a unit of I. ∎

Since every Euclidean domain I is a pid, hence a ufd, then every pair of elements $a, b \in I$, at least one of which is nonzero, has a gcd $d \in I$. Such a d (defined up to unit multiples) can be constructed from the factorizations of a, b. (See Problem 2.6.2.) Moreover, since I is a pid, we've seen that d will be a generator of the ideal generated by a and b. Thus, d will be of the form $d = ra + sb$.

But better yet, in the Euclidean situation, a gcd $d = (a, b)$ can be explicitly constructed by an algorithm using the valuation. Moreover, we can produce elements r, s with $d = ra + sb$. We think it will suffice to illustrate the algorithm in the following example. Also, see Problem 2.6.3.

Example 2.6.4 *We use the valuation in Z to construct $(85536, 4680)$, the gcd of the integers 4680 and 85536. First, we have (1):* $85536 = 18 \cdot 4680 + 1296$. *Equation (1) reveals that the set of divisors of 85536 and 4680 coincides with the set of divisors of 4680 and 1296. Hence, $(85536, 4680) = (4680, 1296)$. Applying Euclidean division again, we obtain (2):* $4680 = 3 \cdot 1296 + 792$, *whence $(4680, 1296) = (1296, 792)$. Continuing the process: (3)* $1296 = 1 \cdot 792 + 504$, $(1296, 792) = (792, 504)$; *(4)* $792 = 1 \cdot 504 + 288$, $(792, 504) = (504, 288)$; *(5)* $504 = 1 \cdot 288 + 216$, $(504, 288) = (288, 216)$; *(6)* $288 = 1 \cdot 216 + 72$, $(288, 216) = (216, 72) = 72$. *Now we can use back substitution, starting with equation (6), to express 72 in the form $r \cdot 85{,}536 + s \cdot 4{,}680$. Specifically:*

$$72 = 288 - 1 \cdot 216 = 288 - 1 \cdot (504 - 1 \cdot 288) = -1 \cdot 504 + 2 \cdot 288 =$$
$$-1 \cdot 504 + 2 \cdot (792 - 1 \cdot 504) = 2 \cdot 792 - 3 \cdot 504 = 2 \cdot 792 - 3 \cdot (1296 - 1 \cdot 792) =$$
$$-3 \cdot 1296 + 6 \cdot 792 = -3 \cdot 1296 + 6 \cdot (4680 - 3 \cdot 1296) = 6 \cdot 4680 - 21 \cdot 1296 =$$
$$6 \cdot 4680 - 21 \cdot (85{,}536 - 18 \cdot 4680) = -21 \cdot 85{,}536 + 384 \cdot 4680.$$

For our final topic, we present an interesting Euclidean ring together with an application.

Definition 2.6.2 *Let J, the **ring of Gaussian integers**, be the subring of C consisting of all complex numbers of the form $a + bi$ with $a, b \in Z$.*

Theorem 2.6.3 *The ring J is a Euclidean domain with respect to the valuation $v(a + bi) = a^2 + b^2$.*

Proof. Plainly, J is an domain and v maps the nonzero elements of J into the set of positive integers. We need to verify conditions (1) and (2) of Definition 2.6.1. A simple calculation reveals that v (being the square of the ordinary complex absolute value) is multiplicative. Hence, (2) holds.

The proof of (1) is a technical computation. Let $\alpha = a + bi, \beta = c + di$ be elements of J with $\beta \neq 0$. We must find $\sigma, \rho \in J$ with $\alpha = \sigma\beta + \rho$ and either $\rho = 0$ or $v(\rho) < v(\beta)$. Write the complex number α/β in the form $r + si$ with $r, s \in Q$. Let q_1, q_2 be integers as close as possible to r, s. Put $\sigma = q_1 + q_2 i$ and $\rho = \alpha - \beta\sigma$. If $\rho = 0$, we are done. Otherwise, by construction of σ, we have $\mid r - q_1 \mid \leq 1/2$ and $\mid s - q_2 \mid \leq 1/2$. Extending the valuation v to $Q[i]$, we have:

$$v(\alpha/\beta - \sigma) = v[(r+si) - (q_1 + q_2 i)] = v[(r - q_1) + (s - q_2)i] = (r - q_1)^2 + (s - q_2)^2.$$

Since the final displayed term is at most $1/2$, we obtain:

$$v(\rho) = v(\alpha - \beta\sigma) = v[\beta(\alpha/\beta - \sigma)] \leq v(\beta)v(\alpha/\beta - \sigma) \leq v(\beta)1/2 < v(\beta)$$

as desired. ∎

Here is the promised application.

Theorem 2.6.4 *(Fermat's $p = a^2 + b^2$ Theorem) Let p be an odd prime in Z. Then $p = a^2 + b^2$ for $a, b \in Z$ if and only if p is congruent to 1 mod 4.*

Proof. First suppose $p = a^2 + b^2$. Since p is odd, exactly one of a, b must be odd. Say $a = 2r + 1$ and $b = 2s$. Then $p = (2r + 1)^2 + (2s)^2$ is congruent to 1 mod 4. (This direction was easy enough.)

We pause to prove an interesting fact which will be generalized in Chapter 5. We consider the multiplicative group $(Z/pZ)^*$ of nonzero elements of the field Z/pZ. First, $(Z/pZ)^*$ is abelian of order $p-1$. In view of Theorem 1.12.1 and Lemma 1.12.3, any element a of maximal order m in a finite abelian group generates a cyclic summand of that group. Furthermore, the order of any other element must be a divisor of m. (See Problem 2.6.6 below.) Say $a \in (Z/pZ)^*$ has maximal order $m \leq p - 1$. Then, for all $b \in Z_p^*$, $b^m = 1$ so that $b^{m+1} = b$. Put another way, every element of the field Z/pZ is a root of the polynomial $(x^{m+1} - x)$. Since Z/pZ has p distinct elements, $p \leq m+1$. But $m+1 \leq p$ so that $m+1 = p$. Since $m = p-1 = |(Z/pZ)^*|$, we have shown that $(Z/pZ)^* = <a>$. ∎

Theorem 2.6.5 Proof. *Now suppose $p \equiv 1(4)$. Since $(Z/pZ)^*$ is cyclic of order $p - 1$ and 4 is a divisor of $p - 1$, $(Z/pZ)^*$ contains an element c of order 4. Then c^2 has order 2 in $(Z/pZ)^*$ so that $c^2 = -1$ in the field Z/pZ. (Why?) Thus, in Z, $c^2 \equiv -1 \pmod{p}$, equivalently $p \mid c^2 + 1$ in Z.* ∎

Proof. Viewing p and $c^2 + 1$ as integers in J, we have $p \mid c^2 + 1 = (c-i)(c+i)$. If p is prime in J, it must be that $p \mid (c - i)$ or $p \mid (c+i)$. Say that $p(a + bi) = c - i$. Then $pb = -1$, plainly an impossibility. Similarly, $p \nmid (c + i)$. It follows that the integral prime p does not remain prime regarded as an element of the pid J. Since primes and irreducibles coincide in J, p is not irreducible in J.

Write $p = (a+bi)(c+di)$, a factorization in J with neither factor a unit. Taking the valuation in J, yields a factorization in $Z : p^2 = v(p) = [v(a + bi)][v(c + di)] = [a^2 + b^2][c^2 + d^2]$. Furthermore, in view of Theorem 2.6.2, neither $a^2 + b^2$ nor $c^2 + d^2$ can be equal to 1. Hence, $p = a^2 + b^2 = c^2 + d^2$ and the proof is complete. ∎

Sixteenth Problem Set

Problem 2.6.1 *Look at the proof that $F[x]$ is a pid and then prove that any Euclidean domain is a pid. Note that condition (2) of the definition of Euclidean domain is not needed for the proof of this theorem.*

Problem 2.6.2 *Find the gcd $d(x)$ of $a(x) = x^5 - 3x^4 - 2x^3 + 6x^2 + x - 3$ and $b(x) = x^3 - x^2 - 4x + 4$ in $Q[x]$. Find $r(x), s(x) \in Q[x]$ such that $d(x) = r(x)a(x) + s(x)b(x)$.*

Problem 2.6.3 *Which primes $p \in Z$ remain prime (irreducible) in $Z[i]$? (Prove your answer.)*

Problem 2.6.4 *Show that $Z[\sqrt{-2}] = \{a + b\sqrt{-2} : a, b \in Z\}$ is a Euclidean domain with respect to the valuation $v(a + b\sqrt{-2}) = a^2 + 2b^2$.*

Problem 2.6.5 *Find the units in $Z[\sqrt{-2}]$.*

Problem 2.6.6 *Use Theorem 1.12.1 and Lemma 1.12.3 to prove that any element a of maximal order m in a finite abelian group G generates a cyclic summand of G. Furthermore, the order of any $b \in G$ must be a divisor of m.*

Further reading: Rowen's student version of his ring theory text [Ro] is an up-to-date, encylopedic introduction to modern ring theory. It's well worth the price.

Chapter 3

Modules

3.1 Elementary concepts

Definition 3.1.1 *Let R be a ring and $(M, +)$ an additive abelian group. Then M is called a (left) R-module if elements of M can be scalar multiplied by elements of R in a natural way. More precisely:*

1. For each $r \in R$ and $m \in M, rm \in M$.

2. Scalar multiplication by $1 \in R$ coincides with the identity map on M.

3. For each $r \in R$ and all $m, m' \in M$, $r(m + m') = rm + rm'$.

4. For each $m \in M$ and all $r, r' \in R$, $(r + r')m = rm + r'm$.

5. For each $m \in M$ and all $r, r' \in R$, $(rr')m = r(r'm)$.

For those of you who have seen the definition of a vector space (that should include everyone), the definition above should look familiar. The set M corresponds to the set of vectors with $+$ the vector addition, and R corresponds to the set of scalars. Our definition would correspond precisely to the definition of a vector space but for one point: the scalars are allowed to come from an arbitrary ring rather than from a field. This-perhaps at first glance minor-difference turns out to be profound. The notion of an R-module has proved to be extremely useful both in several branches of algebra and in algebraic topology.

An abelian group $(M, +)$ may perhaps be a module over several different rings. For example $(\mathcal{R}, +)$, the additive group of real numbers may be naturally regarded as a module over the ring $\mathcal{R}$, the ring Q or the ring Z. In all cases the scalar multiplication is just ordinary ring multiplication.

If we want to specify the ring R for which we are considering M to be an R-module, we write $_RM$.

Definition 3.1.2 *Let M be an R-module. A subgroup $(N, +)$ of $(M, +)$ such that $RN \subseteq N$ is called an R-submodule of M.*

Continuing with the notation of the previous chapters, we write $N \leq M$ or $_RN \leq_R M$ to indicate that N is an R-submodule of M.

Example 3.1.1 *Let $(M, +)$ be an abelian group. For $m \in M$ and $n \in Z$, we already have a natural definition for nm. It is trivial to check that, under this definition, M becomes a Z-module. Thus, the class of abelian groups coincides with the class of Z-modules. The submodules are just the subgroups.*

Example 3.1.2 *Let R be a ring. Put $M = (R, +)$, the underlying additive abelian group of the ring R. For $r \in R$ and $m \in M = R$, define rm to be the ring product. The ring axioms reveal that R can be regarded as a module over itself. Here the submodules are the **left ideals** of R, that is subgroups $(L, +)$ of $(R, +)$ such that $r \in R, l \in L \Longrightarrow rl \in L$.*

Exercise 3.1.1 *Let R be the subring of the ring of 2×2 real matrices consisting of all lower triangular matrices. Show that the submodules of $_RR$ are the trivial submodules, $R = \left\{ \begin{bmatrix} a & 0 \\ b & c \end{bmatrix} : a, b, c \in \mathcal{R} \right\}$ and $< 0 >= \left\{ \begin{bmatrix} 0 & 0 \\ 0 & 0 \end{bmatrix} \right\}$, together with the proper submodules*

$$L_1 = \left\{ \begin{bmatrix} a & 0 \\ b & 0 \end{bmatrix} : a, b \in \mathcal{R} \right\}, \quad L_2 = \left\{ \begin{bmatrix} 0 & 0 \\ 0 & c \end{bmatrix} : c \in \mathcal{R} \right\} \text{ and } N = \left\{ \begin{bmatrix} 0 & 0 \\ b & 0 \end{bmatrix} : b \in \mathcal{R} \right\}.$$

Definition 3.1.3 *Let M, N be R-modules. An R-**homomorphism** $\theta : M \to N$ is an additive group homomorphism such that $\theta(rm) = r\theta(m)$ for all $r \in R, m \in M$.*

If we think of R-modules as generalizations of vector spaces, then the R-homomorphisms are the corresponding generalizations of linear transformations.

Definition 3.1.4 *Two R-modules M and N are R-**isomorphic** if there if a bijective R-homomorphism from M onto N. We write $M \cong N$, or, if we wish to clarify the ground ring, $_RM \cong {}_RN$.*

We prove two theorems for R-modules analogous to those we've already proved for groups and then for rings. (Actually we prove one and ask you, in Problem 3.1.1, to prove the other.)

Theorem 3.1.1 *(factor modules)* Let $_RN \leq \, _RM$. Then the additive abelian group $M/N = \{N + m : m \in M\}$ becomes an R-module via the scalar multiplication $r(N + m) = N + rm$.

Proof. If $N + m = N + m'$, then $m - m' \in N$. Thus, for all $r \in R$, $rm - rm' = r(m - m') \in N$. We have just shown that scalar multiplication is well defined. Now to check that M/N is an R-module is trivial. ■

Theorem 3.1.2 *(first isomorphism theorem for modules)* Let $\theta : M \to N$ be an R-homomorphism. Then: (a) $Ker\ \theta = \{m \in M : \theta(m) = 0\}$ is an R-submodule of M. (b) $Im\ \theta$ is an R-submodule of N. (c) The factor module $M/Ker\ \theta$ is R-isomorphic to $Im\ \theta$.

Proof. See Problem 3.1.1. ■

We close this section with three definitions, all of which should look familiar.

Definition 3.1.5 Let $\{m_1, ..., m_k\} \subseteq \, _RM$. Then we write $< m_1, ..., m_k >$ for the smallest R-submodule of M containing the set $\{m_1, ..., m_k\}$. We call $< m_1, ..., m_k >$ the **R-submodule of M generated by** $\{m_1, ..., m_k\}$. An R-module M is said to be **finitely generated** if $M = < m_1, ..., m_k >$ for some finite set $\{m_1, ..., m_k\} \subseteq M$.

Exercise 3.1.2 Prove that $< m_1, ..., m_k > = \{\sum_{i=1}^{k} r_i m_i : \{r_1, ..., r_k\} \subseteq R\}$.

Definition 3.1.6 A module $_RM$ is **cyclic** if $M = < m >$ for some $m \in M$.

For $N_1, ..., N_k$ submodules of an R-module M, let $\sum_{i=1}^{k} N_i = \{n_1 + \cdots + n_k : n_i \in N_i, 1 \leq i \leq k\}$. It is easy to check that $\sum_{i=1}^{k} N_i$ is the smallest submodule of M that contains the submodules $N_1, ..., N_k$.

Definition 3.1.7 *(internal direct sum)* Let $N_1, ..., N_k$ be submodules of an R-module M. The module M is called the (internal) direct sum of the submodules N_i (written $M = \bigoplus_{i=1}^{k} N_i$) if $\sum_{i=1}^{k} N_i = M$ and if, for each fixed $j, 1 \leq j \leq k$, $N_j \cap \sum_{i=1, i \neq j}^{k} N_i = < 0 >$.

Author's Comment: There is no particular reason for favoring left over right modules. Everything in this chapter could be done, with appropriate changes, for right modules.

Seventeenth Problem Set

Problem 3.1.1 *Prove the first isomorphism theorem for modules.*

Problem 3.1.2 *Is the additive group $Z(4)$ a module over the ring $Z(6)$? If so, prove it. If not, explain why not.*

Problem 3.1.3 *Prove that $M = \bigoplus_{i=1}^{k} N_i$ if and only if each $m \in M$ has a unique representation as a sum $m = \sum_{i=1}^{k} n_i$ with $n_i \in N_i, 1 \leq i \leq k$.*

Problem 3.1.4 *Recall that, for a fixed prime p, the ring Z_p is the subring of Q consisting of those rationals which, when written in reduced form, have denominators prime to p. Let G be an additive abelian (not necessarily finite) group all of whose elements have order a power of p; that is, $g \in G \implies p^{k(g)}g = 0$. Prove that G can be made into a Z_p-module.*

3.2   Free and projective modules

In this section we investigate two important classes of R-modules. Some of the results here and in the next section will be fundamental for our proof of the structure theorem for finitely generated modules over a pid. At this point, we introduce the notion of the direct product and sum of a set of algebraic structures indexed by an infinite index set.

Definition 3.2.1 *Let $\{X_\alpha : \alpha \in A\}$ be a set of algebraic structures (groups, rings, modules, vector spaces) indexed by an infinite set A. Then the **direct product** of the X_α, denoted $\prod_{\alpha \in A} X_\alpha$, is the set of all functions $f : A \to \cup_{\alpha \in A} X_\alpha$ such that $f(\alpha) \in X_\alpha$ for each $\alpha \in A$. The **direct sum** of the X_α, denoted $\bigoplus_{\alpha \in A} X_\alpha$, is the set of all functions $f : A \to \cup_{\alpha \in A} X_\alpha$ such that $f(\alpha) \in X_\alpha$ for each $\alpha \in A$ and such that $f(\alpha) = 0$ for all but finitely many α.*

To construct a direct product or sum over an infinite index set, we need to use a logical axiom called the **Axiom of Choice**. The axiom of choice says precisely: Given $\{X_\alpha : \alpha \in A\}$, a set of nonempty sets indexed by a set A, there exist functions $f : A \to \cup_{\alpha \in A} X_\alpha$ such that $f(\alpha) \in X_\alpha$ for each $\alpha \in A$. (In other words, the direct product of nonempty sets is nonempty.) We will discuss this axiom in more detail in Chapter 4. We can think of the functions f as determining an "extended sequences" $(x_\alpha)_{\alpha\in A}$. A sequence in the direct sum has the additional property that $x_\alpha = 0$ for almost all α. I find it more intuitive to regard the elements of both $\prod_{\alpha\in A} X_\alpha$ and $\bigoplus_{\alpha\in A} X_\alpha$ as the collection of such extended sequences, and will do so hereafter. The direct sum and product become R-modules via coordinatewise addition and scalar multiplication: $(x_\alpha) + (y_\alpha) = (x_\alpha + y_\alpha)$ and $r(x_\alpha) = (rx_\alpha)$.

Exercise 3.2.1 *Prove that* $_R[\bigoplus_{\alpha\in A} X_\alpha] \leq_R [\prod_{\alpha\in A} X_\alpha]$.

If the index set A is finite, then the direct sum and product are the same. In this case, we use whichever notation is more appropriate; for example we would take the direct sum of the additive groups $Z(2)$ and $Z(4)$, written $Z(2) \oplus Z(4)$, and the direct product of the multiplicative groups S_3 and D_4, written $S_3 \times D_4$. This is the point of view we took in Chapter 1.

Definition 3.2.2 *An R-module F is the **free module over a set** A if F is isomorphic to a direct sum $\bigoplus_{\alpha\in A} R_\alpha$, where each R_α is a copy of $_RR$ (R regarded naturally as a module over itself).*

Theorem 3.2.1 *Every module $_RM$ is an epimorphic image of a free R-module.*

Proof. This proof is so easy it feels like cheating. For the index set A take the set M. Form the free R-module over M, $F = \bigoplus_{m\in M} R_m$. Let $\pi : F \to M$ be defined by $\pi(x) = (r_{m_1} m_1 + \cdots + r_{m_k} m_k) \in M$, where $x = (r_m)_{m\in M}$ and $\{m_1, ..., m_k\}$ is the finite subset of the index set M on which the coefficient $r_m \neq 0$. It is routine to check that π is an R-module epimorphism. ∎

We consider a property for R-modules that generalizes the property of being free.

Definition 3.2.3 *An R-module P is **projective** if, whenever $\theta : B \to P$ is an R-module epimorphism, then B decomposes into an R-module direct sum: $B = (Ker\ \theta) \oplus X$.*

The complementary summand X need not be uniquely determined. But note that, by the first isomorphism theorem, any such X will be isomorphic to P.

The projective property may look unnatural at first glance. The next theorem puts it into a more concrete setting.

Theorem 3.2.2 *A module $_RM$ is projective if and only if it is isomorphic to a direct summand of a free R-module.*

Proof. We do the "only if" part first, since it is easier. Suppose that $_RM$ is projective. Let $\theta : F \to M$ be an R-module epimorphism, with F a free module. By the projective property $F = (Ker\ \theta) \oplus X$. We have $X \cong F/(Ker\ \theta) \cong M$.

For the "if" part of the proof, we'll first show that a free module F is projective. Suppose that $\theta : B \to F = \bigoplus_{\alpha \in A} R_\alpha$ is an R-module epimorphism. For each $\alpha \in A$, Let $1_\alpha \in F$ be the extended sequence that is 1 in the α-th position and zero everywhere else. Since θ is onto, for each α, there is some $x_\alpha \in B$ such that $\theta(\ x_\alpha) = 1_\alpha$. Let X be the submodule of B generated by the elements $\{x_\alpha : \alpha \in A\}$. The module X will just be the set of all finite sums of the form $\sum_{i=1}^{k} r_i x_{\alpha_i}$.

We claim that $B = (Ker\ \theta) \oplus X$. To see this, let $b \in (Ker\ \theta) \cap X$. Then $b = \sum_{i=1}^{k} r_i x_{\alpha_i}$ and $\theta(b) = 0$. Thus, $0 = \theta(\sum_{i=1}^{k} r_i x_{\alpha_i}) = \sum_{i=1}^{k} r_i \theta(x_{\alpha_i}) = \sum_{i=1}^{k} r_i 1_{\alpha_i}$. But the only way for $\sum_{i=1}^{k} r_i 1_{\alpha_i}$ to be zero is for each r_i to be zero. Hence, $(Ker\ \theta) \cap X =< 0 >$. Now we show $B = (Ker\ \theta) + X$. Equivalently, we show that for each $b \in B$ there exists $x \in X$ such that $(b - x) \in (Ker\ \theta)$. Let $b \in B$. Then $\theta(b)$ can be written as a finite sum, $\theta(b) = \sum_{i=1}^{k} r_i 1_{\alpha_i}$. If $x = \sum_{i=1}^{k} r_i x_{\alpha_i}$, we have $\theta(b - x) = 0$, as desired. Thus, every free module is projective.

Next, we show that a direct summand of a projective is projective. Let $P = U \bigoplus V$ be projective. Suppose $\theta : B \to U$ is an epimorphism. Then, adopting the obvious notation, $(\theta \oplus 1_V) : B \bigoplus V \to U \bigoplus V = P$ is also an epimorphism. Since P is projective, $B \bigoplus V = Ker\ (\theta \oplus 1_V) \bigoplus X$. It is easy to see that, as a submodule of $B \bigoplus V$, $Ker\ (\theta \oplus 1_V) = Ker\ \theta \leq B$. To show that U is projective, we prove that $B = Ker\ \theta \bigoplus (B \cap X)$. Certainly, $(Ker\ \theta) \cap (B \cap X) = Ker\ (\theta \oplus 1_V) \cap X =< 0 >$. Let $b \in B \leq B \bigoplus V$. Write $b = k + x$, where $k \in Ker(\theta \oplus 1_V) = Ker\ \theta \leq B$ and $x \in X$. Then $x = (b - k) \in B \cap X$. We have proven that the class of projective modules, for a given ring R, is closed under taking direct summands.

The final step is to show that any module isomorphic to a projective is projective. This is Problem 3.2.1 below. ∎

Definition 3.2.4 *A **finite rank free** R-module is the direct sum of finitely many copies of $_RR$.*

Lemma 3.2.1 *Let R be a pid and I a nonzero ideal of R. Then $_RR \cong {}_RI$.*

Proof. See Problem 3.2.2, below. ∎

Lemma 3.2.1 above extends as follows.

Theorem 3.2.3 *Let R be a pid and $F = \bigoplus_{i=1}^{n} R_i$ a finite rank free R-module. Then any nonzero submodule of F is free.*

Proof. The proof is by induction on the natural number n (called the **rank** of F). The case $n = 1$ is Problem 3.2.2. Suppose $n > 1$ and that the theorem holds for all $1 \leq m < n$. Let $< 0 > \neq N \leq F = \bigoplus_{i=1}^{n} R_i$. We show N is free. Consider $\pi_n : F \to R_n$, the projection of F onto its n-th coordinate, defined by $\pi_n[(r_1, ..., r_n)] = r_n$. It is easy to check that π_n is an R-module epimorphism. Let π'_n be π_n restricted to the submodule N. If $\pi'_n(N) = < 0 >$, then $N \leq \bigoplus_{i=1}^{n-1} R_i$ and we're done by induction.

Otherwise, $\pi'_n(N) = I_n, I_n$ a nonzero ideal (submodule) of R_n. By Problem 3.2.2, $_R R_n \cong {}_R I_n$ so that $_R I_n$ is free, hence projective. Thus, $N = Ker\ \pi'_n \oplus I_n$. But $Ker\ \pi'_n = N \cap \bigoplus_{i=1}^{n-1} R_i$. By induction, $Ker\ \pi'_n$ is free. Hence, both $Ker\ \pi'_n$ and I_n are free. Therefore, N is free and the proof is complete. $\blacksquare$

Author's Comment: The conclusion of the above theorem holds for submodules of an infinite rank free module over a pid. The more general proof needs transfinite methods.

Eighteenth Problem Set

Problem 3.2.1 *Let P be a projective R-module which is isomorphic to an R-module P'. Prove that P' is projective.*

Problem 3.2.2 *Let R be a pid and I a nonzero ideal of R. Then $_R R \cong {}_R I$.*

Problem 3.2.3 *Show that the additive group $Z(3)$ is a projective but not free $Z(6)$-module.*

Problem 3.2.4 *Suppose $A \leq B$ and that B/A is projective. Prove that A is a direct summand of B*

Problem 3.2.5 *Let $F = \bigoplus_{\alpha \in A} R_\alpha$ be the free R-module on a set A. Let i be the injection mapping A into F defined by $i(\alpha) = 1_\alpha$. Prove that for any function $f : A \to {}_R M$, $_R M$ an arbitrary R-module, there exists a unique R-module homomorphism $\theta : F \to M$ such that $\theta i = f$.*

Problem 3.2.6 *Let F' be an R-module with an injection $j : A \to F'$ for some set A. Suppose that (F', j) satisfies the **universal mapping property** of the previous problem: for any function $f' : A \to {}_R M$, $_R M$ an arbitrary R-module, there exists a unique R-module homomorphism $\theta' : F' \to M$ such that $\theta' j = f'$. Prove that $F' \cong F$, where F is the free R-module on the set A. Hence, a free R-module on A could have been defined as a pair $(_R F, i)$, i an embedding of A into $_R F$, satisfying this particular universal mapping property.*
Hint: use symmetry and the unicity of the extension homomorphisms.

3.3   Tensor products

In this section R will be a commutative ring. We will construct a new R-module from R-modules M and N, called the **tensor product** over R of M, N, denoted $M \bigotimes_R N$. Our construction will be in three steps. First, we define an R-module T as one that satisfies a universal mapping property with respect to M and N. Then, we show that, if such a module T exists, it must be unique up to R-module isomorphism. Finally, we construct a concrete example of such a T, which we denote by $M \bigotimes_R N$. Having constructed $M \bigotimes_R N$ we close the section by computing some examples and verifying some of the tensor product properties.

We start by considering maps from the Cartesian product $M \times N$ into an R-module B.

Definition 3.3.1 *A map $f : M \times N \to {}_R B$ is called R-**bilinear** if $f(rm + r'm', n) = rf(m, n) + r'f(m', n)$ and $f(m, rn + r'n') = rf(m, n) + r'f(m, n')$ for all $m, m' \in M$, $n, n' \in N$ and $r, r' \in R$.*

In other words, f is an R-linear map in each variable with the other variable held fixed. Note that addition and scalar multiplication only take place in the R-modules M, N, B. The Cartesian product is just a set.

Definition 3.3.2 *Given R-modules M, N, a pair $({}_R T, i)$, where $i : M \times N \to T$ is a R-bilinear map, is **universal with respect to R-bilinear maps** if, whenever $j : M \times N \to {}_R L$ is another R-bilinear map from $M \times N$ into any R-module L, there exists a unique R-module homomorphism $\theta : T \to L$ such that $\theta i = j$.*

The following diagram illustrates this property:
$$\begin{array}{ccc} M \times N & \xrightarrow{\ i\ } & T \\ \downarrow j & \swarrow \theta & \\ L & & \end{array}$$

Theorem 3.3.1 *Let M, N be fixed R-modules. Suppose that $({}_R T, i)$ and $({}_R T', i')$ are universal with respect to R-bilinear maps, as defined above. Then there is an R-module isomorphism $\alpha : T \to T'$ such that $\alpha i = i'$.*

Proof. We will construct R-module homomorphisms $\alpha : T \to T'$ and $\beta : T' \to T$ such that $\beta \alpha = 1_T$, $\alpha \beta = 1_{T'}$. Since $i' : M \times N \to T'$ is R-bilinear and (T, i) is universal with respect to R-bilinear maps, there exists a unique R-module homomorphism $\alpha : T \to T'$ such that $\alpha i = i'$. Reversing the roles of (T, i) and (T', i'), we obtain a unique R-module homomorphism $\beta : T' \to T$ such that $\beta i' = i$. Then $\beta \alpha : T \to T$ is an R-module homomorphism such that $(\beta \alpha)i = \beta(\alpha i) = \beta i' = i$. But, if we put $L = T, j = i$ in Definition 3.3.2, we see that there must be a unique R-module homomorphism $\phi : T \to T$ such that $\phi i = i$. But, plainly, $\phi = 1_T : T \to T$ will do the job. Hence, $\beta\alpha = 1_T$. By symmetry, $\alpha\beta = 1_{T'}$. Thus, α and β are inverse isomorphisms. By construction, $\alpha i = i'$. ∎

Here is a pictorial version of our construction of the maps α, β :

$$
\begin{array}{ccc}
M \times N & \stackrel{i}{\to} & T \\
\downarrow i' & \nearrow \alpha & \\
\end{array}
\qquad
\begin{array}{ccc}
M \times N & \stackrel{i'}{\to} & T' \\
\downarrow i & \nearrow \beta & \\
T & &
\end{array}
\qquad \text{yields} \qquad
\begin{array}{ccc}
M \times N & \stackrel{i}{\to} & T \\
\downarrow i & \nearrow \beta\alpha & \\
T & &
\end{array}
$$

We say that the pair (T, i) is **unique up to isomorphism over** i.

Theorem 3.3.2 *For a commutative ring R and any R-modules M and N, there exists a pair $(M \bigotimes_R N, t)$ with $M \bigotimes_R N$ an R-module and $t : M \times N \to M \bigotimes_R N$ an R-bilinear map that is universal with respect to R-bilinear maps. (And thus, by the previous theorem, the pair $(M \bigotimes_R N, t)$ is unique up to isomorphism over t.)*

Proof. We first construct the R-module $M \bigotimes_R N$. We start with $F = \bigoplus_{(m,n) \in M \times N} R_{(m,n)}$, the free R-module defined on the index set $M \times N$. Then factor by the R-submodule $T \leq F$ that is generated by all the "tensor relations": $\{1_{(rm+r'm',n)} - r1_{(m,n)} - r'1_{(m',n)}, 1_{(m,rn+r'n')} - r1_{(m,n)} - r'1_{(m,n')} : m, m' \in M, n, n' \in N, r, r' \in R\}$. Here we continue our established notation, $1_{(m,n)}$ is the extended sequence in F that is 1 at the (m, n) coordinate and 0 otherwise.

Let $M \bigotimes_R N$ be the factor module F/T and let $m \otimes n$ denote the coset $T + 1_{(m,n)}$. We call $m \otimes n$ a **simple tensor**. Note that every element of F can be represented by a finite sum of the form $\sum r_i 1_{(m_i, n_i)}$. Thus, every element of $F/T = M \bigotimes_R N$ can be represented by a finite sum of the form $\sum r_i(m_i \otimes n_i)$, that is a finite R-combination of simple tensors. Also note that, by the choice of generators for T, we have the following R-bilinear relations: $(rm + r'm') \otimes n = r(m \otimes n) + r'(m' \otimes n)$ and $m \otimes (rn + r'n') = r(m \otimes n) + r'(m \otimes n')$. As a consequence of these, $r(m \otimes n) = rm \otimes n = m \otimes rn$. This final relation implies that every element of $x \in M \bigotimes_R N$ can be written as a finite sum of simple tensors: $x = \sum (r_i m_i) \otimes n_i$.

Next, we need a R-bilinear function $t : M \times N \to M \bigotimes_R N$. Let $t(m, n) = T + 1_{(m,n)} = m \otimes n$. By definition of T, it follows immediately that t is R-bilinear.

Finally, we need to show that the pair $(M \bigotimes_R N, t)$ is universal with respect to R-bilinear maps. Let $g : M \times N \to {}_R L$ be such a map. By Problem 3.2.5, there exists a unique R-module homomorphism $\theta' : F \to L$ such that $\theta'[1_{(m,n)}] = g(m, n)$ for all $(m, n) \in M \times N$. But, since g is R-bilinear, the map θ' sends every generator of T to $0 \in L$. (Check this.) Thus, $\theta'(T) = < 0 >$. It follows that θ' induces an R-module homomorphism $\theta : M \bigotimes_R N \to L$. It is easy to check (do so) that $\theta t = g$. The only remaining detail to check is that θ is unique. Suppose $\phi : M \bigotimes_R N \to L$ is an R-module homomorphism with $\phi t = g$. Then $\phi t = \theta t$, or, put another way, $\theta(m \otimes n) = \phi(m \otimes n)$ for all $(m, n) \in M \times N$. But the simple tensors generate $M \bigotimes_R N$ as an R-module (or abelian group). Therefore $\theta = \phi$, and the proof is complete. ∎

To summarize what we've just done: (1) For any R-modules M, N, there is a pair $(M \bigotimes_R N, t)$ universal with respect to R-bilinear maps. (2) The R-module $M \bigotimes_R N$ is unique up to isomorphism over t. (3) Any element $x \in M \bigotimes_R N$ can be written as a finite sum of simple tensors: $x = \sum m_i \otimes n_i$. (4) The R-module structure on $M \bigotimes_R N$ is given by $rx = \sum rm_i \otimes n_i$. The only thing to remember here is that, **while the simple tensors $m \otimes n$ look like elements, they're not.** (This is the most common mistake beginners make when working with the tensor product.) **Simple tensors designate cosets.** So we can't define a R-homomorphism on $M \bigotimes_R N$ by just specifying its action on a collection of simple tensors. We'd then have to go on prove that our definition was invariant under change of coset representative, the sort of thing we've had to do several times in the past. Fortunately, there's a much better way to define maps on $M \bigotimes_R N$. We illustrate this technique in the following examples.

Example 3.3.1 *For any ${}_R N$, $R \bigotimes N =_R R \bigotimes_R N \cong_R N$. As a general rule, the way to establish any isomorphism involving a tensor product is to construct maps α, β going in opposite directions whose composite is the appropriate identity map. We follow this plan.*

If you think for awhile, the only one reasonable choice for the map $\beta : N \to R \bigotimes N$ is to put $\beta(n) = 1 \otimes n$. If that map doesn't work, it's unlikely anything else will. By the tensor relations, β is an R-module homomorphism. Next, we'd like to define $\alpha : R \bigotimes_R N \to N$ by the formula $(\bigstar)$ $\alpha \sum_{i=1}^{k}(r_i \otimes n_i) = \sum_{i=1}^{k} r_i n_i \in N$. As long as this formula for α makes sense, it's easy to check that α is a R-homomorphism and that $\beta\alpha = 1_{R \otimes N}$, $\alpha\beta = 1_N$. However, heeding our warning remarks above, we won't make the mistake of trying to define α directly via the formula $(\bigstar)$.

Exercise 3.3.1 *Check that β is an R-module homomorphism. Check that, if the formula for α makes sense, then $\beta\alpha = 1_{R \otimes N}$ and $\alpha\beta = 1_N$.*

Example 3.3.2 *(Example 3.3.1 continued) What we do next is a model for constructing maps out of a tensor product. Let $f : R \times N \to N$ be the function defined by $f(r, n) = rn \in N$. The R-module axioms for $_R N$ suffice to show that f is a R-bilinear function. By the universality of $(R \bigotimes_R N, t)$, there is a unique R-homomorphism $\alpha : R \bigotimes_R N \to N$ such that $\alpha t = f$. Let $(r, n) \in R \times N$. Then $\alpha t(r, n) = \alpha(r \otimes n)$. But $f(r, n) = rn$. Thus, $\alpha(r \otimes n) = rn$ so that $\alpha \sum_{i=1}^{k}(r_i \otimes n_i) = \sum_{i=1}^{k} r_i n_i$, as desired.*

Exercise 3.3.2 *Prove that in $M \bigotimes_R N$, $m \otimes 0 = 0 \otimes n = 0$.*

Example 3.3.3 *Regard the abelian groups Q and $Z(n)$ as Z-modules. Let's compute a simple tensor $a/b \otimes x$, where $a/b \in Q$ and $x \in Z(n)$. We have $a/b \otimes x = n(a/bn) \otimes x = (a/bn) \otimes nx = (a/bn) \otimes 0$. By the previous exercise, $(a/bn) \otimes 0 = 0$. We have shown that $Q \otimes_Z Z(n) = <0>$.*

The preceding example shows that the tensor product of nonzero modules may be zero. After all, the tensor relations are complicated enough so that everything may collapse to zero. As a young assistant professor, I once asked an eminent algebraist "How can you show that the tensor product of two modules in nonzero?". He answered, only half facetiously, "Map it onto something that isn't zero.". The next example will be useful in the following section.

Example 3.3.4 *Let S be a subring of the ring R and let N be an S-module. Regard $R = {}_S R$ by letting the scalar multiple sr just be the product in the ring R. Form the S-module $R \bigotimes_S N$. We'd like to make this into an R-module in the natural way, by setting $r \sum_{i=1}^{k} r_i \otimes n_i = \sum_{i=1}^{k} rr_i \otimes n_i$-but we can't do this directly. As above, we have to start with an S-bilinear map and use the universality of the tensor product.*
   Fix an $r \in R$. Let $f_r : R \times N \to_S (R \bigotimes_S N)$ be given by $f_r(r', n) = rr' \otimes n$.

Exercise 3.3.3 *Prove that f_r is an S-bilinear map.*

Example 3.3.5 *(Continuation of Example 3.3.4) By universality of the above tensor product with respect to S-bilinear maps, there is a unique S-module homomorphism $m_r : R \bigotimes_S N \to R \bigotimes_S N$ such that $m_r t = f_r$. For $(r', n) \in R \times N$, $m_r(r' \otimes n) = m_r t(r', n) = f_r(r', n) = rr' \otimes n$. Thus, $m_r(\sum_{i=1}^{k} r_i \otimes n_i) = \sum_{i=1}^{k} rr_i \otimes n_i$. We have shown that left multiplication by a fixed $r \in R$ naturally induces an S-module endomorphism on $R \bigotimes_S N$. Now for $x \in R \bigotimes_S N$ and $r \in R$, we can define $rx = m_r(x)$. Hence, we can make $R \bigotimes_S N$ into an R-module in the natural way.*

At this point, the alert reader may be wondering why we wanted R to be a commutative ring. Tensor products can be constructed for a non-commutative rings R. In that situation, we have to start with a right R-module M_R and a left R-module $_RN$. A generalized R-bilinear function $f : M \times N \to L$, L an abelian group, will be one that satisfies the additive relations in each variable, $f(m + m', n) = f(m, n) + f(m', n)$, $f(m, n + n') = f(m, n) + f(m, n')$, plus the relation $f(mr, n) = f(m, rn)$ for all $r \in R$, $m, m' \in M$ and $n, n' \in N$. We then go through our construction pretty much as above with one crucial difference: F will be the free abelian group generated by the set $M \times N$ and T will be the additive subgroup of F generated by the additive relations plus all relations of the form $1_{(mr,n)} - 1_{(m,rn)}$. Thus, $M \otimes N = F/T$ will only be an abelian group, one that is universal with respect to generalized R-bilinear functions mapping $M \times N$ into abelian groups. There is no reasonable way to define $r(m \otimes n)$ here. Moreover, there is no reasonable way we could modify F and T in order to be able to put an R-module structure on F/T. Play around with possible definitions to see which R-module axiom gives the trouble.

Nineteenth Problem Set

Problem 3.3.1 *Compute the following tensor products of Z-modules: (a) $Z(15) \bigotimes Z(14)$ (b) $Z(6) \bigotimes Z(9)$.*

For the next three problems, return to the situation where the ring R is commutative. (The next problem would make no sense otherwise.)

Problem 3.3.2 *Prove for R-modules M, N that there is an R-isomorphism: $M \bigotimes_R N \cong N \bigotimes_R M$.*

Problem 3.3.3 *Prove for R-modules M, N, L that there is an R-isomorphism: $(M \bigotimes_R N) \oplus (M \bigotimes_R L) \cong M \bigotimes_R (N \oplus L)$.*

Problem 3.3.4 *Find a ring R and R-modules $A \leq A', B$ such that $A \otimes_R B \neq < 0 >$ but $A' \otimes_R B = < 0 >$.*

3.4   Finitely generated modules over a pid

Before we begin, to get yourself back into the pid frame of mind, try these exercises.

Exercise 3.4.1 *Let $a, b \in I$, I a domain. Then $(a) = (b)$ if and only if $b = ua$, u a unit of I.*

Exercise 3.4.2 *For $a, b \in I$, I a domain, define $b \sim a$ if $b = ua$, u a unit of I. Prove that $\sim$ is an equivalence relation on I.*

If $a \sim b$ as above, a and b are called **associates**.

Now, before reading any further, do the first problem in the Twentieth Problem Set below.

Since ± 1 are the only units in Z, if $a, b \in Z$ then $a \sim b \iff b = \pm a$. With this in mind we could restate our fundamental theorem of arithmetic in Z : Let $n \in Z$ with $\mid n \mid > 1$. Then $n = (\pm 1)p_1^{e_1} \cdots p_k^{e_k}$ where the $\{p_1, ..., p_k\}$ are distinct positive primes.

Exercise 3.4.3 *Describe the equivalence relation of being associate in the ring $F[x]$. Write the corresponding restatement of the fundamental theorem of arithmetic in $F[x]$.*

Definition 3.4.1 *Let $m \in M$. Then $(0 : m) = \{r \in R : rm = 0\}$ is called the **annihilator of** m.*

Note that $(0 : m)$ will be a left ideal in any ring R, thus a two-sided ideal in a domain.

We want to extend the notion of order from elements in an abelian group to elements in a module over a pid. Let M be an R-module with R a pid. For $m \in M$, we have two sensible choices for the definition of the order of m. We could just say that $\mid m \mid$ is the annihilator ideal $(0 : m)$-that way there would be no ambiguity. But, to be more in parallel with the theory of abelian groups, I prefer to say that $\mid m \mid = a \in R$, where $(0 : m) = (a)$. This looks to me more like an order should look. The only price we pay in adopting my chosen viewpoint is that, by Exercise 3.4.1, orders will only be defined up to associates. We'll live with this slight ambiguity.

There is one special exception, where the terminology is inconsistent, but sensible.

Definition 3.4.2 *If $(0 : m) = (0)$ we say that the element $m \in M$ is of **infinite order** . In this case, the natural left module $_R R$ is isomorphic to Rm, the cyclic submodule of M generated by m, via the isomorphism $r \to rm$.*

Note the parallel here with the theory of abelian groups. An element a of an abelian group has infinite order if and only if $Z \cong Za$ via the natural map $n \to na$. The following definition is also a natural extension from the abelian group case.

Definition 3.4.3 *An element $m \in M$ is of **prime power order** if $(0 : m) = (p^e)$, where p is a prime (equivalently irreducible) element of the pid R and $e \in N$. In this case $R/(p^e) \cong Rm$.*

We follow the same game plan that we used in proving the structure theorem for finite abelian groups. We first prove a direct sum decomposition theorem for finitely generated modules over a pid. Obviously, the proof of a decomposition theorem for finitely generated modules over a pid will be quite a bit more involved than the proof of the decomposition theorem for finite abelian groups. To begin preparation for the proof, we ask you to do a simple exercise.

Exercise 3.4.4 *Prove that a direct summand of a finitely generated module over any ring R is finitely generated.*

Definition 3.4.4 *The **annihilator of** M, denoted $(0 : M)$, is the intersection of the annihilators of all elements of M, $(0 : M) = \cap_{m \in M}(0 : m)$.*

It is easy to check that, even if R is a noncommutative ring, $(0 : M)$ will be a two-sided ideal of R. (This is Problem 3.4.2 below.) Our first lemma deals with the analogue of finitely generated torsion abelian groups.

Lemma 3.4.1 *Let $T \neq\, < 0 >$ be a finitely generated module over a pid R. Suppose $(0 : T) = (r), r \neq 0$. Write $r = up_1^{e_1} \cdots p_k^{e_k}$, where u is a unit and $\{p_1, ..., p_k\}$ are non-associate primes in R. For $1 \leq i \leq k$, let $T_{p_i} = \{t \in T : p_i^{e_i}t = 0\}$. Then $T = \bigoplus_{i=1}^{k} T_{p_i}$.*

Proof. Unlike the proof of Theorem 1.12.1, we can't employ any Sylow Theorems for modules over a pid. So this time we'll have to give an honest proof. First, note that there must be some primes present in the factorization of r. Otherwise, $(0 : M) = (u) = R$ and $T = 1T =< 0 >$, contrary to assumption. Also note that, since R is commutative, the T_{p_i} are R-submodules of T.

Suppose $t \in T_{p_j} \cap \bigoplus_{i=1, i \neq j}^{k} T_{p_i}$. Let $s = p_1^{e_1} \cdots p_{j-1}^{e_{j-1}} p_{j+1}^{e_{j+1}} \cdots p_k^{e_k}$. Then $p_j^{e_j}t = 0$ and $st = 0$. But, since r, s are relatively prime elements of the pid R, there exist $a, b \in R$ with $a\, p_j^{e_j} + bs = 1$. So $t = (a p_j^{e_j} + bs)t = 0$.

It remains to show that $\sum_{i=1}^{k} T_{p_i} = T$. For each fixed $1 \leq j \leq k$, let $s_j = r/p_j^{e_j}$. Then $1 \in R$ is a greatest common divisor of the set $\{s_1, ..., s_k\}$ (check this). Thus $\sum_{i=1}^{k} a_i s_i = 1$ for suitable elements $a_i \in R$. If $t \in T$, then $t = \sum_{i=1}^{k} (a_i s_i t)$. Note that each term $a_j s_j t \in T_{p_j}$. The proof that $T = \bigoplus_{i=1}^{k} T_{p_i}$ is complete. ∎

Lemma 3.4.2 *Let T be a finitely generated module over a pid R such that $p^e T =< 0 > \neq p^{e-1}T$ for p a prime in R and $e \in N$. Then T is a finite direct sum of cyclic submodules of p-power order.*

Proof. It's enough to show that:

$(*)$ If $t \in T$ is an element of maximal order p^e, then $T = < x > \bigoplus T'$.

Then, we simply apply $(*)$ to T', and continue the process. Since T is finitely generated, we can extract only finitely many cyclic summands (Exercise 3.4.3 below). This will show that T is a finite direct sum of cyclic summands of p-power order.

The proof of $(*)$ is just the proof of Lemma 1.12.3, verbatim except for replacing Z by R. ∎

The previous two lemmas, taken together, imply that, if T is a finitely generated torsion module (that is $T \neq 0$ and $(0 : T) \neq 0$) over a pid, then T is a finite direct sum of cyclic submodules, each of prime power order.

Twentieth Problem Set

Problem 3.4.1 *Let R be a pid and $p \in R$ a prime. Suppose p and x are associates. Prove that x is prime. Thus, an equivalence class induced by the equivalence relation of being associate, either contains all primes or no primes.*

Problem 3.4.2 *Let M be a module over an arbitrary ring R. Prove that $(0 : M)$ is an ideal of R. Give an example of a ring R, an R-module M and an element $m \in M$ such that $(0 : m)$ is a left, but not a two-sided ideal of R.*

Problem 3.4.3 *Prove that if $_R M$ is finitely generated, we can extract only a finite direct sum of cyclic summands as a summand of M.*

3.5 A structure theorem

We are almost ready to prove a decomposition and a uniqueness of decomposition theorem for finitely generated modules over a pid. We need one more definition and a final preparatory lemma.

Definition 3.5.1 *An R-module M is **torsion-free** if $(0 : m) = (0)$ for all $0 \neq m \in M$.*

Let I be a domain and M a torsion-free I-module. Regard I as a subring of its quotient field $F = q(I)$. We've already shown that $_I F \bigotimes_I {}_I M$ is an F-module in the natural way, that is via the action $f \sum f_j \otimes m_j = \sum f f_j \otimes m_j$.

But we've seen (Problem 3.3.4) that, for an arbitrary R-module M, if $_RA \leq_R A'$ we need not have that $A \otimes (_RM) \leq A' \otimes (_RM)$. Put another way, the natural map $\iota : A \otimes (_RM) \to A' \otimes (_RM)$ such that $\iota(y) = y$, y regarded as an element in $A' \otimes (_RM)$, need not be monic. But, in the case of a domain I with quotient field $F = q(I)$ and a torsion-free module $_IM$, the natural map $\iota : M \cong I \otimes_I M \to F \otimes {}_IM$ will be an I-module embedding. This is the content of the following lemma, whose proof is more or less standard.

Lemma 3.5.1 *Let I, F, M be as above. If $0 \neq y \in I \otimes M$ then $0 \neq y \in F \otimes {}_IM$.*

Proof. All tensors are over I, so we omit the subscript. We prove the contrapositive. Suppose $y = 0 \in F \otimes M$. We show that y is already zero regarded as an element of $I \otimes M$. Write y as a sum of simple tensors, $y = \sum_{j=1}^{s} i_j \otimes m_j$. The fact that $y = 0 \in F \otimes M$ means that in S, the free I-module on the set $F \times M$, $\sum_{j=1}^{s} 1_{(i_j, m_j)} = \sum_{finite} a_k t_k$, where the a_k are in I and the t_k are to be defined momentarily. As before, $1_{(a,m)}$ denotes the "sequence" in S with 1 in position (a, m) and with 0 in all other positions. We use t_k as a shorthand for one of the generators of $_IT$, either
$$t_k = 1_{(if+i'f',m)} - 1_{(if,m)} - 1_{(i'f',m)} \text{ or } t_k = 1_{(f,im+i'm')} - 1_{(f,im)} - 1_{(f,i'm')}.$$
Let B be the I-submodule of F generated by I and the finitely many elements $\{h_l\} \subseteq F \backslash I$, that occur as part of a subscript in the expression for a t_k. Then, since we've put in all the tensor relations we need, $y = 0 \in {}_IB \otimes {}_IM$. (This is the key point of the proof. Make sure that you understand it.)

Write each $h_l = c_l/d_l$ with $c_l, d_l \in I$ and let d be the least common multiple of the set $\{d_l\}$ (or just take the product of the elements d_l). Then $\{dh_l\} \subset I$. Since $_IB$ is torsion-free, multiplication by d is an I-module isomorphism mapping B onto dB. Therefore, $B \otimes M \cong dB \otimes M$ under the map induced by sending a simple tensor $b \otimes m$ to $db \otimes m$. It follows that $dy = 0 \in dB \otimes {}_IM$. But $dB \subseteq I$ as an I-submodule. So any element that is zero in $dB \otimes M$ is even more solidly zero in $I \otimes M$. Thus, $dy = 0 \in M \cong_I I \otimes {}_IM$. Since M is torsion-free, $y = 0$ and the proof is complete. ∎

Theorem 3.5.1 *(decomposition theorem) Let $_RM$ be finitely generated, R a pid. Then M is a finite direct sum of cyclic submodules of either infinite or prime power order.*

Proof. Let T be the **torsion submodule** of M; that is, $T = \{m \in M : (0, m) \neq (0)\}$. Since M is finitely generated, then $\bar{M} = M/T$ is a finitely generated torsion-free R-module. (Check this.) Say that $\bar{M} =< \bar{m}_1, ..., \bar{m}_t >$, where $\{m_1, ..., m_t\}$ is part of a R-generating set for M.

As above, regard $R \leq F = q(R)$ and let $F\bar{M}$ denote the F-module $F \otimes_R \bar{M}$ with the natural F-action. By Lemma 3.5.1, $\bar{M} \cong (R \otimes_R \bar{M}) \leq F\bar{M}$. It is easy to check that $F\bar{M}$ is generated as an F-module by the elements $\{1 \otimes \bar{m}_1, ..., 1 \otimes \bar{m}_t\}$. (Again, check this.)

Call a subset of $F\bar{M}$ of the form $\{1 \otimes \bar{x}_1, ..., 1 \otimes \bar{x}_k\}$, $\{x_1, .., x_k\} \subseteq M$, **independent** if $\sum_{i=1}^{k} f_i(1 \otimes \bar{x}_i) = 0 \implies$ all $f_i = 0$. We claim that, if $\{1 \otimes \bar{x}_1, .., 1 \otimes \bar{x}_k\}$ is independent, then $k \leq t$, the number of generators of $\bar{M}$. The claim follows from the well-known fact that an independent set in the finite dimensional F-vector space $F\bar{M}$ has no more vectors than a spanning (F-generating) set of that space. (We'll review the details of the proof in Chapter 4.)

Thus, we can choose $\{1 \otimes \bar{x}_1, ..., 1 \otimes \bar{x}_k\}$, a maximal independent set of $\bar{M}$, with $k \leq t$. Let $\bar{m} \in \bar{M}$ be arbitrary. We must have a dependence relation $f(1 \otimes \bar{m}) + \sum_{j=1}^{k} f_j(1 \otimes \bar{x}_j) = 0$ with $\{f, f_1, ..., f_k\} \subseteq F, f \neq 0$. This yields the F-module equation $1 \otimes \bar{m} = \sum_{j=1}^{k}(f_j/f)(1 \otimes \bar{x}_j)$. Since each $1 \otimes \bar{m}$ is in the F-linear span of the elements $\{1 \otimes \bar{x}_1, ..., 1 \otimes \bar{x}_k\}$, the set $\{1 \otimes \bar{x}_1, ..., 1 \otimes \bar{x}_k\}$ is an independent set of F-generators for the F-module $F\bar{M}$. Thus, we have a direct sum decomposition $F\bar{M} = F \otimes_R \bar{M} = \bigoplus_{j=1}^{k} F(1 \otimes \bar{x}_j)$. (In vector space terms, we've simply shown that the set $\{1 \otimes \bar{x}_1, ..., 1 \otimes \bar{x}_k\}$ is a basis for $F\bar{M}$.)

Consider the map $\alpha : \bar{M} \to \bigoplus_{j=1}^{k} F(1 \otimes \bar{x}_j)$, where α is the composite of the isomorphism $\bar{M} \cong R \otimes_R \bar{M}$ and the embedding $(R \otimes_R \bar{M}) \to (F \otimes_R \bar{M})$. As in the proof of Lemma 3.4.3, choose $0 \neq d \in R$ such that $d\{\alpha(\bar{m}_i) : 1 \leq i \leq t\} \subset \bigoplus_{j=1}^{k} R(1 \otimes \bar{x}_j)$. Since each $\bar{m} \in \bar{M}$ is of the form $\bar{m} = \sum_{i=1}^{t} r_i \bar{m}_i$ with the $r_i \in R$, it follows that $d\alpha$ is an embedding of $\bar{M}$ into the finite rank free R-module $\bigoplus_{j=1}^{k} R(1 \otimes \bar{x}_j)$. By Theorem 3.2.3, $\bar{M} = M/T$ is free. Hence, by Theorem 3.2.2, M/T is projective. We, thus, can obtain a direct sum decomposition $M = T \bigoplus X$, where $X \cong M/T \cong R^k$. Here R^k denotes the direct sum of $k \geq 0$ copies of R.

Now we focus on the summand T. By Exercise 3.4.4, T is finitely generated. If $T \neq< 0 >$, say that $(0 : T)$ is the principal ideal $(r) \subsetneq R$. We can apply Lemma 3.4.1 to conclude that T is a finite direct sum of submodules T_p, where the primes p are the non-associate primes occurring in a factorization $r = \prod p^{e_p}$ of r. Each T_p will be finitely generated with $p^{e_p} T_p =< 0 >$.

By Lemma 3.4.2, each T_p will be a finite direct sum of cyclic submodules of p-power order. At last! We've proved our decomposition theorem! ∎

We finally describe how to attach a finite set of nonnegative integers to a finitely generated module M over a pid R. As we construct them, we'll show these numbers are module isomorphism invariants of $_RM$. At that point it will be clear that they are a **complete set of invariants**; that is, that they determine M up to isomorphism. We ask the reader's indulgence in allowing us to introduce into evidence one additional-as yet unproved-fact: a finite dimensional vector space has a well-defined dimension. We'll prove a stronger result at the beginning of Chapter 4.

Let $_RM$ be finitely generated with R a pid. Apply the construction of the decomposition theorem to M to obtain $M = T \bigoplus R^j$. Then $\bar{M} = M/T$ is certainly an isomorphism invariant of M. Since $\bar{M} \cong R^j$, then the free module R^j is also an invariant of M. Let p be a prime (equivalently irreducible) element of R. Since p is irreducible, $(p) = pR$ is a maximal ideal of R. By Theorem 2.2.7, R/pR is a field. Now $R^j/(pR^j)$ is an R-module invariant of M and $R^j/(pR^j) \cong (R/pR)^j$. (You should check out this last isomorphism, but it's not hard.) Note that the module $(R/pR)^j$ is a vector space of dimension j over the field R/pR. Hence, given that the dimension is an invariant for finite dimensional vector spaces, the number j is an invariant of $_RM$. Set $f_0(M) = j$. We have shown that the number of free summands occurring in any direct sum decomposition of M is an invariant of M.

Next consider T, the torsion submodule of M. Certainly T is an isomorphism invariant of M. If $T \neq < 0 >$, let $0 \neq r \in R$ be a generator for $(0 : T)$. As in the proof of the decomposition theorem, write $r = up_1^{e_1} \cdots p_k^{e_k}$, where u is a unit and $\{p_1, ..., p_k\}$ are non-associate primes in R. We then obtain a corresponding direct sum decomposition $T = \bigoplus_{i=1}^k T_{p_i}$, with $p_i^{e_i} T_{p_i} = < 0 >$ for all i. Suppose that we had chosen a different generator r' for the ideal $(0 : T)$. Say we had a prime factorization $r' = vq_1^{f_1} \cdots q_l^{f_l}$, where v is a unit and the $q_i's$ are non-associate primes. This would lead to a direct sum decomposition $T = \bigoplus_{i=1}^l T_{q_i}$. But, since $(0 : T) = (r) = (r')$, it follows that $r' = wr$ with w a unit in R. By Problem 3.5.1 below (a trivial strengthening of the statement of unique factorization in a pid), $k = l$ and, after rearrangement, $e_i = f_i$ and q_i is an associate of p_i for all i. Hence, after rearrangement, $T_{p_i} = T_{q_i}$, $1 \leq i \leq k$. (Think about this.) Hence, the decomposition of T into primary components does not dependent on the original choice of a generator for the ideal $(0 : T)$.

Further, each T_{p_i}, the set of all elements of M of p_i-power order, is certainly an invariant of M. Finally, we use the facts that T_{p_i} is finitely generated and that $p_i^{e_i} T_{p_i} = <0>$ to obtain a decomposition of T_{p_i} into a finite direct sum of cyclic submodules of p_i-power order. For each p_i as above and $0 \le k \le e_i - 1$, let $f_k^{p_i}(M)$ be the number of cyclic summands of order p_i^{k+1} that occur in the decomposition of T_{p_i}. To prove that these numbers are invariants of T_{p_i}, and hence of M, just take the proof of Lemma 1.12.4 and replace the ring Z with the ring R and the prime $p \in Z$ with a prime $p \in R$. Having done that, the proof goes through verbatim. Putting everything together, we have shown that the numbers $f_0(M)$ and $f_k^{p_i}(M)$ are a finite set of invariants of M. But these numbers tell us exactly the number of the various kinds of cyclic summands in any direct sum decomposition of M into a direct sum of cyclic submodules of infinite or prime power order. Hence, they are a complete set of numerical invariants. We have proved the following unicity theorem.

Theorem 3.5.2 *(unicity of decomposition for a finitely generated module over a pid)* *Let M be a finitely generated module over a pid R. Let $f_0(M)$ be the number of infinite cyclic summands and let $f_k^p(M)$ the number of p^{k+1}-cyclic summands, ranging over a complete set P of non-associate primes p in R and nonnegative integers k, in any direct sum decomposition of M into infinite and prime power cyclic submodules. Then the set $\{f_0(M), f_k^p(M) : p \in P, k \ge 0\}$ is a complete set of cardinal invariants of M.*

Twenty-first Problem Set

Problem 3.5.1 *Suppose $r = u p_1^{e_1} \cdots p_k^{e_k}$ and $wr = v q_1^{f_1} \cdots q_l^{f_l}$, where u, v, w are units and $\{p_1, ..., p_k\}$ and $\{q_1, ..., q_l\}$ are sets of non-associate primes. Prove $k = l$ and, after rearrangement, $e_i = f_i$ and q_i is an associate of p_i for all i.*

Problem 3.5.2 *M, N be finitely generated modules over a pid R. Suppose $M^n \cong N^n$. (For $n \in N$, M^n means the direct sum of n copies of the module M.) Prove $M \cong N$.*

Problem 3.5.3 *Let M, N, X, Y be finitely generated modules over a pid R. Suppose that, as R-modules, $M \bigoplus X \cong N$ and $N \bigoplus Y \cong M$. What, if anything, can you conclude about M and N?*

Further reading: Anderson and Fuller [AF] provide an excellent introduction to modules and the accompanying categorical notions. I often refer to it.

Chapter 4

Vector spaces

4.1 Definitions and glossary

A **vector space** V is just a module over a field F. Since the study of vector spaces far predated that of modules, vector space theory has its own standard terminology. We'll use this established terminology, so here's a glossary. A **vector** is an element of a vector space. We use boldface type to distinguish vectors. A **subspace** is just an F-submodule. We retain the notation $W \leq V$ to mean that W is a subspace of V. The **subspace spanned by a set of vectors** $\{v_a : \alpha \in A\}$ is the F-submodule generated by this collection of vectors. We use $< \mathbf{v}_a : \alpha \in A >$ to denote this subspace. Recall that $< \mathbf{v}_a : \alpha \in A >$ is the set of all finite sums $\{\sum_{\alpha \in A_0} f_\alpha \mathbf{v}_\alpha : f_\alpha \in F, A_0 \text{ is a finite subset of } A\}$. An expression of the form $\sum_{\alpha \in A_0} f_\alpha \mathbf{v}_\alpha$ with A_0 a finite subset of A is called a **linear combination** of the vectors $\{\mathbf{v}_\alpha : \alpha \in A\}$. We say the set of vectors $\{\mathbf{v}_a : \alpha \in A\} \subset V$ is a **spanning set for** V if $< \mathbf{v}_a : \alpha \in A >= V$. If V, W are vector spaces over F, a **linear transformation** from V to W is an F-module homomorphism from V to W. The words isomorphism, monomorphism and epimorphism are imported from module theory to vector space theory without change, as are the definitions of kernel and image. We also retain the notation $_F V$ to indicate that V is a vector space (module) over the field F. If $W \leq V$ the **factor space** is just the factor F-module V/W. Note that a vector space $_F V$ is torsion-free as an F-module. To see this, let $\mathbf{v} \in V$ and $0 \neq f \in F$ with $f\mathbf{v} = \mathbf{0}$. Then $\mathbf{v} = f^{-1}f\mathbf{v} = f^{-1}\mathbf{0} = \mathbf{0}$. Thus, the annihilator of every $\mathbf{0} \neq \mathbf{v} \in V$ is zero; equivalently there are no nonzero torsion elements of V.

4.2   Time for a little set theory

If we want to deal with infinite dimensional vector spaces (to be defined momentarily), we need to use the axiom of choice, introduced in Section 3.2. Recall that the axiom of choice intuitively says that, if we have a collection of nonempty sets indexed by some fixed index set, then we can choose an element from each of the sets.

Mathematicians do not agree on whether or not this is a reasonable axiom to assume. To me it seems reasonable enough, at least on first glance. One disturbing feature though is that this axiom has been proved to be independent of the ordinary axioms of what's called ZF set theory (axioms powerful enough to construct the real numbers). Thus, there would be no contradiction in supplementing ordinary ZF set theory with either the axiom of choice or its negation.

Another disturbing feature is that the axiom of choice is logically equivalent to another set theoretic axiom mysteriously called "Zorn's Lemma". The assertion made in Zorn's Lemma is certainly not so clear. To state Zorn's Lemma, we need a few definitions.

Definition 4.2.1 *A **partially ordered set** (abbreviated poset) is a set S, together with a relation, usually written as $\leq$, satisfying the following properties for all $s, t, l \in S$: (1) $s \leq s$ (reflexive property), (2) ($s \leq t$ and $t \leq s$) $\iff$ $s = t$ (antisymmetric property), and (3) ($s \leq t$ and $t \leq l$) $\implies s \leq l$ (transitive property).*

A partial ordering mimics the ordinary relation of $\leq$ on, say, the reals except for one crucial point. If $s, t \in S$, it need not be true that either $s \leq t$ or $t \leq s$. In such case, we call s, t **incomparable** elements of S. One can easily draw a graph representing a finite poset with incomparable elements, for instance

has incomparable elements s, t, w. Here the arrows represent the order relation.

Exercise 4.2.1 *Let S be the set $\{1, 2, 3\}$ and $P(S)$ be the set of all subsets of S. Check that $P(S)$ is a poset via the partial order $A \leq B$ if and only if $A \subseteq B$. Draw a graph representing the order relations in $P(S)$.*

In the following three definitions, S is always a poset.

Definition 4.2.2 *A subset T of S is called a **chain** if all the elements of T are comparable; that is, $t, t' \in T \Longrightarrow t \leq t'$ or $t' \leq t$.*

Example 4.2.1 *The set $\{\phi, \{1,2\}, \{1,2,3\}\}$ is a chain in the poset $P(S)$ above.*

Definition 4.2.3 *An element $x \in S$ is called an **upper bound** for subset $B \subset S$ if $b \in B \Longrightarrow b \leq x$.*

Note that x is not necessarily required to be in B.

Definition 4.2.4 *An element $y \in S$ is called **maximal** if $(z \in S, z \geq y) \Longrightarrow z = y$.*

Note that we're not saying y is bigger than or equal to everything in S, just that nothing in S is strictly greater than y. In the graph after Definition 4.2.1 u and w are incomparable maximal elements.

Finally, we're ready to state Zorn's Lemma: Let S be a poset in which every chain has an upper bound. Then S contains a maximal element. The existence of maximal elements is certainly clear for finite sets. But, in general, Zorn's Lemma-though equivalent to the Axiom of Choice-doesn't seem at all so clear. For a proof of the logical equivalence of Zorn's Lemma and the Axiom of Choice, see the appendix of [K]. All I can say is, that if you want to study infinite algebraic structures, your life is going to be miserable without the axiom of choice. Hence, being pragmatic in such matters, I will assume it.

We often will want to work with a set of vectors indexed by some index set A. To avoid logical difficulties, fix some large set U. Let $\mathcal{P}(U)$ be the power set of U, that is the set of all subsets of U. All of our index sets will be elements of this fixed $\mathcal{P}(U)$. For $A, B \in \mathcal{P}(U)$, define $A \sim B$ if there is a bijection from A onto B. It is immediate that $\sim$ defines an equivalence relation on $\mathcal{P}(U)$. For $A \in \mathcal{P}(U)$, let $|A|$, called the **cardinality** of A, be the equivalence class of A under this equivalence relation. Let $\mathcal{K}$ be the set of all cardinals (equivalence classes in $\mathcal{P}(U)$) under discussion.

Let $\gamma, \lambda \in \mathcal{K}$. Say that $\gamma = |C|$ and $\lambda = |L|$; $C, L \in \mathcal{P}(U)$. We define $\gamma \leq \lambda$ if there is a monomorphism $f : C \to L' \subset L$. Problem 4.3.1 below asks you to prove that $\leq$ is a well-defined relation; that is, is independent of the choice of representative sets C, L. We have the following theorem.

Theorem 4.2.1 *The set $\mathcal{K}$ with the relation $\leq$ is a poset.*

Proof. Let $\gamma = |C|$. Since the identity map 1_C is a monomorphism from C into itself, $\gamma \leq \gamma$ for all $\gamma \in \mathcal{K}$. It is a routine exercise to prove that

$\alpha \leq \beta, \beta \leq \gamma \Longrightarrow \alpha \leq \gamma$. Given Problem 4.3.1, the only remaining thing to check is the antisymmetric property: $\alpha \leq \beta, \beta \leq \alpha \Longrightarrow \alpha = \beta$. If you write out the appropriate definitions, you will find that this assertion is far from clear if $\alpha =\mid A \mid, \beta =\mid B \mid$, with A, B infinite sets . The proof that $\leq$ is antisymmetric is the content of the Schroder-Bernstein Theorem. ∎

Theorem 4.2.2 *(Schroder-Bernstein) If there exist monomorphisms f : $A \rightarrow A' \subset B$ and $g : B \rightarrow A' \subset A$, then there is a bijection $\theta : A \rightarrow B$.*

Proof. For each $a \in A$, we create a path $P_a = \{a, f(a), gf(a), fgf(a), ...\}$ going back and forth from A to B via the maps f and g. There are two possibilities for P_a. (1) If all elements of the set P_a are distinct, let $A_1 = \{a, gf(a), gfgf(a), ...\}$, $B_1 = \{f(a), fgf(a), fgfgf(a), ...\}$ and let $\theta \mid_{A_1} = f$. Then $\theta \mid_{A_1}$ is a bijection between the countable subsets $A_1 \subset A, B_1 \subset B$. (2) It could be that P_a is finite. Then P_a must be of the form $\{a, f(a), gf(a), fgf(a), ..., x = f(gf)(gf) \cdots (gf)(a)\}$ where $g(x) = a$. (We're using the fact that f and g are monic; draw a picture and think about this for a minute.) Reading from left to right, number the elements in P_a. Here the set P_a has an even number of elements. Further, the subset $A_1 = P_a \cap A$ coincides with the elements with odd indices and $B_1 = P_a \cap B$ coincides with the elements with even indices In this case, let $\theta \mid_{A_1}$ be any bijection from the finite set $A_1 \subset A$ to the finite set $B_1 \subset B$.

It's not hard to check that the sets $\{P_a : a \in A\}$ are either disjoint or identical and that A can be partitioned as a disjoint union: $A = \cup_{a \in A' \subset A} (P_a \cap A)$. With the obvious notation, define $\theta : A \rightarrow B$ as $\theta = \cup_{a \in A'} \theta \mid_{(P_a \cap A)}$. Now let $b \in B$. Then $g(b)$ is a member of one of the partitioning sets $P_a \cap A$. It follows that $b \in P_a \cap B$. A little thought now shows that $\theta : A \rightarrow B$ is a bijection. ∎

4.3   A structure theorem for vector spaces

Definition 4.3.1 *A set of vectors $\beta = \{\mathbf{v}_\alpha : \alpha \in A\} \subset {}_F V$ is called independent if, for all finite subsets $A_0 \subset A$, $\sum_{\alpha \in A_0} f_\alpha \mathbf{v}_\alpha = 0 \Longrightarrow f_\alpha = 0$ for all $\alpha \in A_0$. Otherwise, β is called dependent.*

Definition 4.3.2 *A basis β for a vector space V is a maximal linearly independent subset of V; equivalently, a basis is a maximal element in the poset $\mathcal{I}$ with elements linearly independent subsets of V and order the inclusion relation*

Note that, if β' is any subset of V that properly contains a basis β, then β' is dependent. (Up to this point we have used lower case Greek letters for homomorphisms. In this chapter, following more or less established notation, we will use $\beta, \beta', \gamma, \gamma'$ for sets of vectors, usually bases.)

Theorem 4.3.1 *Every vector space V has a basis.*

Proof. Let $\mathcal{I}$ be as above. If $C = \{S_\alpha : \alpha \in A\}$ is a chain in $\mathcal{I}$, put $B = \cup_{\alpha \in A} S_\alpha$. If we show that B is linearly independent, it certainly will be an upper bound in $\mathcal{I}$ for C. Suppose $\sum_{i=1}^n f_i \mathbf{b}_i = \mathbf{0}$, where $\{f_1, ..., f_n\} \subset F$ and $\{\mathbf{b}_1, ..., \mathbf{b}_n\} \subset B$. Each $\mathbf{b}_i$, being an element of the union, is an element of some S_{α_i}. Since the sets S_α form a chain under set inclusion, any finite subset of the S_α will have a largest element, a set which contains all the others. Hence, the equation $\sum_{i=1}^n f_i \mathbf{b}_i = 0$ is taking place in some independent set S_α. It follows that all $f_i = 0$, so that B is independent.

We have shown that every chain in $\mathcal{I}$ has an upper bound. Now we can apply Zorn's Lemma (Axiom) to conclude that $\mathcal{I}$ has a maximal element. By definition, this maximal element is a basis for V. ∎

Somewhat surprisingly, the Axiom of Choice, Zorn's Lemma and the statement that every vector space has a basis are all logically equivalent. The proof that the axiom that every vector space has a basis logically implies the other two axioms is relatively recent (see [B]). (I'd like to thank Jim Schmerl for providing this reference.)

Definition 4.3.3 *A set $\beta = \{\mathbf{v}_\alpha : \alpha \in A\} \subset V$ is called a **minimal spanning set** for V if $< \mathbf{v}_a : \alpha \in A >= V$ and if, for all $\alpha_0 \in A$, $< \mathbf{v}_a : \alpha \in A, \alpha \neq \alpha_0 > \neq V$.*

Theorem 4.3.2 *The following are equivalent for a set $\beta = \{\mathbf{v}_\alpha : \alpha \in A\} \subset V$: (a) β is a basis for V (b) each $\mathbf{v} \in V$ can be uniquely expressed as a linear combination of elements from β (c) β is a minimal spanning set for V.*

Proof. We give a cyclical proof starting with $(a) \implies (b)$. Assuming (a), let $\mathbf{v} \in V$. We first show $\mathbf{v}$ is a linear combination of the vectors $\{\mathbf{v}_\alpha : \alpha \in A\}$. If $\mathbf{v} = \mathbf{v}_\alpha$ for some α, then, certainly $\mathbf{v} \in < \mathbf{v}_a : \alpha \in A >$. If $\mathbf{v} \notin \beta = \{\mathbf{v}_\alpha : \alpha \in A\}$, then, by the maximality of β in the set of all linearly independent subsets of V, $\beta \cup \{\mathbf{v}\}$ must be dependent. The equation exhibiting dependence, $\sum_{i=1}^n f_i \mathbf{v}_{\alpha_i} + f \mathbf{v} = 0$, must have $f \neq 0$ lest we contradict the independence of β. Hence, $\mathbf{v} = (-f)^{-1} \sum_{i=1}^n f_i \mathbf{v}_{\alpha_i} \in < \mathbf{v}_a : \alpha \in A >$. Now suppose we have two possibly different representations of $\mathbf{v}$, say $\mathbf{v} = \sum_{i=1}^n f_i \mathbf{v}_{\alpha_i} = \sum_{i=1}^n f_i' \mathbf{v}_{\alpha_i}$. (It does no harm to assume that $\mathbf{v}$ is two possibly different linear combinations of the same finite subset of β; we can set some of the f_i or f_i' equal to zero if necessary.) Then $\sum_{i=1}^n (f_i - f_i') \mathbf{v}_{\alpha_i} = \mathbf{0}$. By the independence of β, $f_i = f_i'$ for all i.

$(b) \implies (c)$ If we assume (b), the set β is certainly a spanning set for V. We need to prove that, if any vector $\mathbf{v}_\alpha$ is removed from β, the resulting set β' is no longer a spanning set. Assume the contrary. Then $\mathbf{v}_\alpha = 1\mathbf{v}_\alpha$ and $\mathbf{v}_\alpha$ is also a linear combination of vectors from β'. This contradicts the unicity assumption from (b).

$(c) \implies (a)$. Assume (c). We need to prove that β is a maximal independent set. Suppose we have a linear combination of vectors from β equal to the zero vector, say $\sum_{i=1}^{n} f_i \mathbf{v}_{\alpha_i} = \mathbf{0}$. If some $f_i \neq 0$, then, as in the proof of $(a) \implies (b)$, the corresponding vector $\mathbf{v}_{\alpha_i}$ can be expressed as a linear combination of the others. But then $\mathbf{v}_{\alpha_i}$ can be deleted from β and we'd still have a spanning set. This contradicts the minimality assumption from (c). Thus β is independent. Let $\mathbf{v} \in V, \mathbf{v} \notin \beta$. Since β is a spanning set $\beta \cup \{\mathbf{v}\}$ is dependent. Hence, β is a maximal independent set. ∎

The next theorem expresses the key fact about bases of a vector space.

Theorem 4.3.3 *Let* $\beta = \{\mathbf{v}_\alpha : \alpha \in A\}$ *and* $\beta' = \{\mathbf{w}_\gamma : \gamma \in C\}$ *be bases for the vector space* $_F V$. *Then* $\mid A \mid = \mid C \mid$.

Proof. Suppose one of the index sets is finite. We'll show that the other is finite and of the same cardinality. This first part of the proof is standard and can be found in any book on linear algebra.

Suppose $\mid A \mid = n$ and $\mid A \mid \leq \mid C \mid$. Without loss, let $A = \{1, ..., n\}$ so that one of our bases can be written $\beta = \{\mathbf{v}_1, ..., \mathbf{v}_n\}$. Choose a vector $\mathbf{w}_1 \in \beta'$. Since β spans V, $\mathbf{w}_1$ is a linear combination of vectors from β. Say that $\mathbf{w}_1 - \sum_{i=1}^{n} f_i \mathbf{v}_i = \mathbf{0}$ with not all coefficients zero. Some f_i must be nonzero since $\mathbf{w}_1 \neq \mathbf{0}$. As above, the corresponding $\mathbf{v}_i$ can be expressed as a linear combination of the set of vectors $\beta_1 = \{\mathbf{w}_1, \mathbf{v}_1, ..., \mathbf{v}_{i-1}, \mathbf{v}_{i+1}, ..., \mathbf{v}_n\}$. Since β is a spanning set, so is β_1.

Choose a vector $\mathbf{w}_2 \in \beta', \mathbf{w}_2 \neq \mathbf{w}_1$. The vector $\mathbf{w}_2$ is a linear combination of vectors from β_1 so that the set $\{\mathbf{w}_1, \mathbf{w}_2, \mathbf{v}_1, ..., \mathbf{v}_{i-1}, \mathbf{v}_{i+1}, ..., \mathbf{v}_n\}$ is dependent. Again, a coefficient of some $\mathbf{v}_j$ in the resulting dependence relation must be nonzero. Otherwise we would have an equation of the form $g_1 \mathbf{w}_1 + g_2 \mathbf{w}_2 = \mathbf{0}$, with at least one of g_1, g_2 nonzero, contradicting the independence of β'. Without loss assume $j > i$. We can remove the vector $\mathbf{v}_j$ and still have a spanning set $\beta_2 = \{\mathbf{w}_1, \mathbf{w}_2, \mathbf{v}_1, ..., \mathbf{v}_{i-1}, \mathbf{v}_{i+1}, .., \mathbf{v}_{j-1}, \mathbf{v}_{j+1}, .., \mathbf{v}_n\}$.

If we continue this process for n steps, we obtain a spanning set $\beta_n = \{\mathbf{w}_1, ..., \mathbf{w}_n\}$, a subset of n distinct vectors from β'. Since β' is independent, we must have $\beta' = \beta_n$. Therefore, $\mid A \mid = \mid C \mid = n$. So our theorem is proved if either of the index sets is finite. Call a vector space with a finite basis (equivalently with a finite spanning set) **finite dimensional**. We have shown that any finite dimensional vector space V has a well-defined dimension, namely the cardinality of any basis for V. Those only interested in the finite dimensional result can stop here.

The remainder of the proof is from [J-1], Chapt. 9, Sect. 2. Suppose that both of the sets A and C are infinite. Express each $\mathbf{v}_\alpha \in \beta$ as a linear combination of vectors from β'; say that, for each $\alpha \in A$, we have the equation: $(\natural_\alpha)$ $\mathbf{v}_\alpha = \sum_{i=1}^{k(\alpha)} f_{\alpha_i} \mathbf{w}_{\gamma_i}$, where all coefficients are nonzero elements of the field F. Now each $\mathbf{w}_\gamma \in \beta'$ must occur in some equation $(\natural_\alpha)$. This is because, if some $\mathbf{w}_0$ does not occur, since $\mathbf{w}_0$ is a linear combination of finitely many $\mathbf{v}_\alpha$ and each of these $\mathbf{v}_\alpha$ are linear combinations of finitely many $\mathbf{w}_\gamma \neq \mathbf{w}_0$, we would contradict the linear independence of the set β'. This is the key point in this part of the proof.

We now define a map $\theta : C \to A$. For each $\gamma \in C$, let $\theta(\gamma)$ be one of the $\alpha \in A$ whose equation $(\natural_\alpha)$ involves γ. (Note that we're using the axiom of choice here.) Let $A' = \theta(C)$. For each $\alpha \in A'$, $\theta^{-1}(\alpha)$ consists of some of those $\gamma \in C$ which occur in the equation $(\natural_\alpha)$. Thus, each $\theta^{-1}(\alpha)$ is a finite subset of C. We have a one to one correspondence between the set A' and the collection of inverse images $\theta^{-1}(\alpha), \alpha \in A'$. The collection $\Gamma = \{\theta^{-1}(\alpha) : \alpha \in A'\}$ is a partition of C into a collection of disjoint finite subsets. Moreover, since C is infinite, so is Γ. It is a standard theorem from the theory of cardinal numbers that C and Γ have the same cardinality. Since $\mid A' \mid = \mid \Gamma \mid$ and $\mid \Gamma \mid = \mid C \mid$, it follows that $\mid A' \mid = \mid C \mid$. But $A' \subset A$. We've shown that $\mid C \mid \leq \mid A \mid$. By symmetry, $\mid A \mid \leq \mid C \mid$. By the Schroder-Bernstein Theorem, $\mid A \mid = \mid C \mid$ and the proof is complete. ∎

Definition 4.3.4 *Let* $_F V$ *have basis* $\beta = \{v_\alpha : \alpha \in A\}$. *Define the dimension of* V *over* F *($dim_F V$) to be* $\mid A \mid$.

The preceding lemma shows that the dimension of a vector space V over a field F is a well defined cardinal number. The reason we call attention to the field F, is that an abelian group V could be a module (vector space) over different fields F. The dimension is the cardinality of the basis β, a basis for V over F. Thus, changing the field would change the basis and possibly also the dimension. For example, if $V = (C, +)$, the additive group of complex numbers, then V can be regarded in the natural way as a vector space over any of the fields $Q, \mathcal{R}$ or C. Plainly, $\dim_C V = 1$ while $\dim_{\mathcal{R}} V = 2$. Since Q is countable, $\dim_Q V = \mid C \mid$.

The following theorem is the high point of the section.

Theorem 4.3.4 *(structure theorem for vector spaces) Let* V, W *be vector spaces over* F. *Then* $V \cong W$ *if and only if* $dim_F V = dim_F W$.

Proof. Suppose $\theta : V \cong W$. It is not hard to prove (See Problem 4.3.3 below), that if $\{v_\alpha : \alpha \in A\}$ is a basis for V, then $\{\theta(v_\alpha) : \alpha \in A\}$ is a basis for W. Thus, $\dim_F V = \dim_F W = \mid A \mid$.

Conversely, suppose that $\dim_F V = \dim_F W$. Then, V and W have bases $\{v_\alpha : \alpha \in A\}$ and $\{\mathbf{w}_\gamma : \gamma \in C\}$ with $\mid A \mid = \mid C \mid$. Let $\phi : A \to C$ be a bijection. In view of Theorem 4.3.2 (b), the map $\theta : V \to W$ defined by $\theta(\sum_{\alpha \in A_0} f_\alpha v_\alpha) = \sum_{\phi(\alpha) \in \phi(A_0)} f_\alpha \mathbf{w}_{\phi(\alpha)}$, where A_0 is an arbitrary finite subset of A, is easily seen to be a vector space isomorphism. (Also, see Problem 4.3.4 below.) ∎

Thus, the structure of vector spaces over a given field F is completely determined. The set consisting of the single cardinal number $\dim_F V$ is a complete set of isomorphism invariants for $_F V$.

Notation 4.3.1 *For a natural number n and field F, let F^n denote the set of n-tuples $\{(f_1, ..., f_n) : f_i \in F, 1 \leq i \leq n\}$. The set F^n, equipped with the natural (componentwise) addition and scalar multiplication, is the simplest example of an n-dimensional vector space over F. The **standard basis** for F^n is the basis $\{\mathbf{e}_1, ..., \mathbf{e}_n\}$, where $\mathbf{e}_i$ is the n-tuple with 1 in the i-th position and 0 elsewhere. By the above theorem, every n-dimensional F-vector space is just a copy of F^n.*

Corollary 4.3.1 *Let $\dim_F V = n$. Then $V \cong F^n$.*

Corollary 4.3.2 *Every finite dimensional F-vector space is a free F-module. (See Problem 4.3.2 directly below for a generalization.)*

Twenty-second Problem Set

Problem 4.3.1 *Prove that $\leq$ is a well-defined relation on $\mathcal{K}$.*

Problem 4.3.2 *Prove that a vector space $_F V$ is a free F-module.*

Problem 4.3.3 *Let V, W be vector spaces over F. Prove that if $\theta : V \cong W$ and $\{v_\alpha : \alpha \in A\}$ is a basis for V, then $\{\theta(v_\alpha) : \alpha \in A\}$ is a basis for W.*

Problem 4.3.4 *Let V, W be vector spaces over F. Suppose that $\{v_\alpha : \alpha \in A\}$ is a basis for V and $\{\mathbf{w}_\alpha : \alpha \in A\}$ is an arbitrary subset of vectors from W. Prove there is a unique linear transformation $T : V \to W$ such that $T(v_\alpha) = \mathbf{w}_\alpha$ for all $\alpha \in A$.*

4.4   Finite remarks on finite dimensional spaces

For the remainder of the chapter we'll deal only with finite dimensional vector spaces, those spaces with a basis $\beta = \{\mathbf{v}_1, ..., \mathbf{v}_n\}$. Throughout, V, W will be finite dimensional vector spaces over some arbitrary fixed field F. (Although many of our results, including the next theorem, could be suitably generalized to the infinite dimensional case.) Our next simple but useful theorem tells us that every subspace is a direct summand.

Theorem 4.4.1 *Let $V' \leq V$. Then $V = V' \oplus V''$, where $V'' \cong V/V'$.*

Proof. Let π be the natural factor map from V onto the factor space V/V'. We know that every vector space is free, hence projective. Since V/V' is projective, we have $V = Ker\pi \oplus V''$. But $Ker\pi = V'$ so that $V = V' \oplus V''$. Thus, $V/V' \cong V''$. ∎

Corollary 4.4.1 *Let $V' \leq V$. Then $dimV = dimV' + dimV/V'$.*

Proof. See Problem 4.4.1. ∎

Let $T : V \to W$ be a linear transformation. Recall that $KerT = \{v \in V : T(v) = 0\}$ and $Im\,T = \{w \in W : w = T(v) \text{ for some } v \in V\}$. By Theorem 3.1.2, $KerT$ is a subspace of V and $Im\,T$ is a subspace of W.

Definition 4.4.1 *Let $T : V \to W$ be a linear transformation. Then* **rank** *$T = dim(ImT)$ and* **nullity** *$T = dim(KerT)$.*

The three isomorphism theorems for vector spaces are just specializations of the corresponding theorems for modules. I'll formally restate them, since in the vector space context two of them have interesting corollaries.

Theorem 4.4.2 *Let $T : V \to W$ be linear. Then $V/KerT \cong ImT$.*

Corollary 4.4.2 With notation as above, rank $T+$ nullity $T = \dim V$.

Proof. By the definition and Theorem 4.4.2, rank $T = dim(ImT) = dim(V/KerT)$. Then, by Corollary 4.4.1, $dim(V/KerT) + dim(KerT) = dimV$. Thus, rank $T +$ nullity $T = \dim V$. ∎

Corollary 4.4.3 *Let $T : V \to V$. Then the following are equivalent: (a) T is an isomorphism (b) T is monic (c) T is epic.*

Proof. See Problem 4.4.3. ∎

Theorem 4.4.3 *Let $U \leq W \leq V$. Then $(V/U)/(W/U) \cong V/W$.*

Theorem 4.4.4 *Let W, W' be subspaces of V. Then $(W + W')/W' \cong W/(W \cap W')$.*

Corollary 4.4.4 *As above,* $dim(W+W') = dimW + dimW' - dim(W \cap W')$.

Proof. By Theorem 4.4.3, $\dim(W+W')/W' = \dim W/(W \cap W')$. Since all dimensions are finite, we have $\dim(W+W') - \dim W' = \dim W - \dim(W \cap W')$. ∎

Definition 4.4.2 *Let V have basis $\beta = \{v_1, ..., v_n\}$ and let $\mathbf{v} \in V$. The* **coordinates** *of $\mathbf{v}$ with respect to β (denoted $[\mathbf{v}]_\beta$) are the (unique) scalars $(\alpha_1, ..., \alpha_n) \in F^n$ such that $\mathbf{v} = \sum_{i=1}^{n} \alpha_i v_i$.*

If we use the standard basis, the coordinates of a vector $\mathbf{v} \in F^n$ can be identified with the vector $\mathbf{v}$ itself. Of course, when working with an arbitrary basis β for F^n, we have to distinguish between a vector and its coordinate sequence.

Remark 4.4.1 *Strictly speaking, we should say "the coordinates of $\mathbf{v}$ with respect to the ordered basis β". Clearly, not only the set β but also its ordering determine the coordinates of a vector.*

As mentioned at the end of the previous section, the vector space F^n can be thought of as the canonical representative of F-vector spaces of dimension n. The proof of the theorem below is assigned as Problem 4.4.2.

Theorem 4.4.5 *Let V have basis $\beta = \{v_1, ..., v_n\}$. The map $\mathbf{v} \rightarrow [\mathbf{v}]_\beta$ is a vector space isomorphism from V to F^n.*

Twenty-third Problem Set

Problem 4.4.1 *Give two different proofs that $dimV = dimV' + dimV/V'$. Here V is a finite dimensional vector space, but, if you know a little about addition of cardinal numbers, the result makes sense in general.*

Problem 4.4.2 *Prove Theorem 4.4.5.*

Problem 4.4.3 *Prove the following theorem. Let $T : V \rightarrow V$. Then the following are equivalent: (a) T is an isomorphism (b) T is monic (c) T is epic.*

Problem 4.4.4 *Give an example of an infinite dimensional vector space V and linear transformations $T, S : V \rightarrow V$ such that T is monic but not epic and S is epic but not monic.*

Problem 4.4.5 *Let V be a vector space over F with basis $\beta = \{v_1, ..., v_n\}$ and let F' be a field containing F. (a) Show that the set $1 \otimes \beta = \{1 \otimes v_1, ..., 1 \otimes v_n\}$ is a spanning set of the F'-vector space $V' = F' \otimes_F V$. (b) Show that $1 \otimes \beta$ is a basis for V' over F'. Hint: Construct an appropriate map from $F' \otimes_F V$ to $(F')^n$. (c) Show that $V' \cong (F')^n$.*

Problem 4.4.6 *Prove that the map $v \rightarrow 1 \otimes v$ is a F-vector space embedding from V into $F' \otimes_F V$.*

4.5 Matrices and systems of equations

In this brief section we review some standard results on matrices and systems of linear equations. We assume that everyone is familiar with some version of this material. So we'll be a little terse. The first chapter of Hoffman and Kunze [HK] has a nice discussion of these ideas.

Definition 4.5.1 *For natural numbers m, n an $m \times n$ **matrix** A over F is a rectangular array of m rows and n columns of entries from F. For $1 \leq i \leq m, 1 \leq j \leq n$, we let a_{ij} be the entry of A in row i, column j. We often denote a matrix by displaying its entries; we write $A = [a_{ij}]$.*

At times it will be convenient to identify both $1 \times n$ row matrices $r = \begin{bmatrix} \alpha_1 & \cdots & \alpha_n \end{bmatrix}$ and $n \times 1$ column matrices $c = \begin{bmatrix} \alpha_1 \\ \cdot \\ \cdot \\ \cdot \\ \alpha_n \end{bmatrix}$ with elements of F^n. It will sometimes be notationally convenient to indicate an column matrix c as $c = r^t$, the transpose of a $1 \times n$ row matrix r. In any event, we often will omit the unnecessary row or column subscript in the individual entries.

Definition 4.5.2 *Let $A = [a_{it}]$ be an $m \times n$ matrix and $C = [c_{tj}]$ be an $n \times l$ matrix. Then the **matrix product** $[\alpha_{it}][\gamma_{tj}]$ is the $m \times l$ matrix $[d_{ij}]$ defined by the equations $d_{ij} = \sum_{t=1}^{n} a_{it}b_{tj}, 1 \leq i \leq m, 1 \leq j \leq l$*

The number of columns of A must equal the number of rows of C in order for the matrix AC to be defined. What we are doing here is regarding the i-th row of A and the j-th column of C as entries of F^n. Then we take the "dot product" of the i-th row of A with the j-th column of C to obtain the i-j entry of AC. For example:

$$\begin{bmatrix} 2 & -1 & 3 \\ 1 & 0 & -4 \end{bmatrix} \begin{bmatrix} 1 & -1 & 0 & 4 \\ -2 & 0 & 1 & 1 \\ 3 & 2 & -1 & 0 \end{bmatrix} = \begin{bmatrix} 13 & 4 & -4 & 7 \\ -11 & -9 & 4 & 4 \end{bmatrix}.$$

Definition 4.5.3 *An $m \times n$ matrix A is in* **reduced form** *if: (1) all rows consisting of only zeros are at the bottom of the matrix; (2) the leading nonzero entry in any nonzero row is 1 (called the leading 1 of the row); and (3) the leading ones move strictly to the right as you go down the matrix.*

Conditions (1)-(3) imply that, below any leading one, the remainder of its column consists of all zeros.

Example 4.5.1 *The following matrix is in reduced form:*

$$
\begin{bmatrix}
1 & -2 & 4 & -3 & 1 \\
0 & 0 & 1 & -1 & 2 \\
0 & 0 & 0 & 1 & 5 \\
0 & 0 & 0 & 0 & 0
\end{bmatrix}.
$$

Definition 4.5.4 *The* **row operations** *that we will perform on a matrix are: R_1 interchange any two rows, R_2 multiply any row by a nonzero constant and R_3 add a constant multiple of any row to a different row.*

Theorem 4.5.1 *Any matrix with entries in a field F can be transformed to reduced form by a finite sequence of row operations.*

Proof. We'll just outline the reduction algorithm. Once you've used the procedure on one or two examples, that should be sufficient. Let A be an $m \times n$ matrix and, without loss, assume that A has at least one nonzero entry. Reading from left to right, find the first column -say column j-with a nonzero entry. Take any nonzero entry a in this column and interchange its row with the first row so that the resulting matrix A_1 has a in position $(1, j)$. Multiply the first row of A_1 by $a^{-1} \in F$ so that we now have a matrix A_2 with 1 in the $(1, j)$ position. Then repeatedly use R_3 to put a zero in all j-th column entries of A_2 below row one. We now have a matrix B whose first $j - 1$ columns are all zero and whose j-th column can be identified with the first standard basis vector $\mathbf{e}_1 \in F^m$. If B has zeros in all other positions, B is reduced and we are done.

Otherwise, read to the right and find the next column, say column k with $k > j$, of B with a nonzero entry b in some row below row one. Interchange this row with the second row to obtain a matrix B_1 with b in position $(2, k)$. Then multiply by b^{-1} to put the next leading 1 in position $(2, k)$. As above, use this leading 1 to put zeros below it in column k. The resulting matrix C will have the zero vector in the first $j - 1$ columns, $\mathbf{e}_1$ in the j-th column, then possibly some more columns of zeros if $k > j + 1$. Column k will have a leading one as its second entry, followed by all zeros.

(Its first entry will be more or less arbitrary.) At this point, as far as the first two rows are concerned, the matrix C is in reduced form. If all rows below the second are zero, we're done.

Otherwise find the next column, reading left to right, with a nonzero entry in some row. Move this row to row three and repeat the procedure outlined above. ∎

We can use the matrix reduction procedure outlined above to solve $m \times n$ **linear homogeneous systems** of equations in F. By an $m \times n$ linear homogeneous system in F, we mean a system of m equations in n unknowns with coefficients $a_{ij} \in F$ of the form

$$a_{11}x_1 + a_{12}x_2 + \cdots a_{1n}x_n = 0$$
$$a_{21}x_1 + a_{22}x_2 + \cdots a_{2n}x_n = 0$$
$$\cdots$$
$$a_{m1}x_1 + a_{m2}x_2 + \cdots a_{mn}x_n = 0.$$

First, we rewrite the system in matrix form: $AX = \mathbf{0}$. Here $X = \begin{bmatrix} x_1 & \cdots & x_n \end{bmatrix}^t$ is the $n \times 1$ column matrix of unknowns, $A = [a_{ij}]$ is the $m \times n$ **coefficient matrix** and $\mathbf{0} = \begin{bmatrix} 0 & \cdots & 0 \end{bmatrix}^t \in F^m$.

We make the elementary observation that, if A_1 is a matrix obtained from A by performing a row operation, the matrix equation $A_1 X = \mathbf{0}$ has exactly the same set of solutions in F^n as the original equation $AX = \mathbf{0}$. This is because each of the row operations is reversible, taking us from one system of equations to a logically equivalent system.

Thus, if E is the reduced matrix obtained from A by row operations, the solutions of $AX = \mathbf{0}$ and $EX = \mathbf{0}$ coincide as subsets of F^n. The matrix E has been arranged so that the solutions to $EX = \mathbf{0}$ can be simply obtained by substitution. Two examples should suffice to illustrate the point.

Example 4.5.2 (HK) *We find all solutions to the 3×4 system of homogeneous equations with rational coefficients*

$$2x - y + 3z + 2w = 0$$
$$x + 4y - w = 0$$
$$2x + 6y - z + 5w = 0.$$

Our matrix equation $AX = \mathbf{0}$ is

$$\begin{bmatrix} 2 & -1 & 3 & 2 \\ 1 & 4 & 0 & -1 \\ 2 & 6 & -1 & 5 \end{bmatrix} \begin{bmatrix} x \\ y \\ z \\ w \end{bmatrix} = \begin{bmatrix} 0 \\ 0 \\ 0 \end{bmatrix}.$$

Using row operations, we reduce the coefficient matrix A to the matrix

Example 4.5.3 $E = \begin{bmatrix} 1 & 4 & 0 & -1 \\ 0 & 1 & -1/3 & -4/9 \\ 0 & 0 & 1 & -11/3 \end{bmatrix}$.

The equivalent system $EX = 0$ is

$$\begin{aligned} x + 4y - w &= 0 \\ y - (1/3)z - (4/9)w &= 0 \\ z - (11/3)w &= 0. \end{aligned}$$

By back substitution, we obtain $x = -(17/3)w, y = (5/3)w, z = (11/3)w$.

In the previous example w is an **independent variable**. That is, we can choose it freely. This choice will determine the **dependent variables** x, y, z. The set of solutions to our system above is the subspace

$$\{(-(17/3)w, (5/3)w, (11/3)w, w) : w \in Q\} \leq Q^4.$$

Example 4.5.4 *Note that the position of the dependent variables of the system will correspond to positions of the leading ones in the reduced matrix E. For example, if E is the matrix*

$$\begin{bmatrix} 1 & -2 & 3 & 2 & 7 & 0 \\ 0 & 0 & 1 & -1 & 0 & 4 \\ 0 & 0 & 0 & 0 & 0 & 0 \end{bmatrix}$$

then variables x_1, x_3 are dependent. Variables x_2, x_4, x_5, x_6 can take on arbitrary values. The solution set S here can be represented as the set of 6-tuples $(x_1, ..., x_6) \in F^6$ such that (1) $x_1 = 2x_2 - 3x_3 - 2x_4 - 7x_5$ and (2) $x_3 = x_4 - 4x_6$. Substitution of equation (2) into equation (1) yields the equivalent system (1') $x_1 = 2x_2 - 5x_4 - 7x_5 + 12x_6$ and (2) $x_3 = x_4 - 4x_6$. Thus, our solution set S can be written

$$S = \{(2x_2 - 5x_4 - 7x_5 + 12x_6, x_2, x_4 - 4x_6, x_4, x_5, x_6) : x_2, x_4, x_5, x_6 \in F\}.$$

Exercise 4.5.1 *Let A be an $m \times n$ matrix over a field F. Prove that the solution space $\{X \in F^n : AX = 0\}$ is a subspace of F^n.*

Twenty-fourth Problem Set

Problem 4.5.1 *Find all solutions to $AX = 0$, where A is the matrix*

$$\begin{bmatrix} 2 & -3 & -7 & 5 & 2 \\ 1 & -2 & -4 & 3 & 1 \\ 2 & 0 & -4 & 2 & 1 \\ 1 & -5 & -7 & 6 & 2 \end{bmatrix}.$$

Problem 4.5.2 *Let A be an $m \times n$ matrix over a field F with $m < n$. Prove that the system of equations: $AX = 0$ has a nonzero solution.*

Problem 4.5.3 *Let E be a reduced $m \times n$ matrix. Suppose that E has $r \leq m$ nonzero rows. Prove that the dimension of the solution space of $EX = 0$ has dimension $n - r$.*

4.6 Linear transformations and matrices

The structure of vector spaces having been completely determined in Section 4.3, our attention-both for mathematical reasons and future applications-turns to the study of linear transformations. For the remainder of the chapter, V, W will be finite dimensional vector spaces over a field F and $T : V \to W$ be a linear transformation.

Definition 4.6.1 *Let $T : V \to W$ and let $\beta = \{v_1, ..., v_n\}$, $\gamma = \{w_1, ..., w_m\}$ be bases for V and W, respectively. The **matrix of T with respect to the bases** β, γ (denoted $[T]_{\beta\gamma}$) is the $m \times n$ matrix whose j-th column is the vector of coordinates of $T(v_j)$ with respect to the basis γ.*

The following theorem gives the relationship between linear transformations and their matrices. Here we write $[v]_\beta$, the coordinates of a vector v with respect to the ordered basis β, as a column vector.

Theorem 4.6.1 *Let T, V, W, β, γ be as in Definition 4.6.1. For $v \in V$ we have $[T(v)]_\gamma = [T]_{\beta\gamma}[v]_\beta$.*

Proof. This is a typical proof in linear algebra. The statement to be proved may look complicated, but it's just a matter of writing things out. Moreover, at every step in the proof there's only one thing you can sensibly do. First, let's check that the dimensions are right: $[T(\mathbf{v})]_\gamma$ is an $m \times 1$ column vector, $[\mathbf{v}]_\beta$ is $n \times 1$ and $[T]_{\beta\gamma}$ is an $m \times n$ matrix. The product of an $m \times n$ matrix times an $n \times 1$ matrix is an $m \times 1$ matrix. So, at least the formula has a chance of being correct. The k-th entry of the column vector $[T(\mathbf{v})]_\gamma$ is the k-th coordinate of the vector $T(\mathbf{v})$ with respect to the basis γ. We need to compute the k-th entry of $[T]_{\beta\gamma}[\mathbf{v}]_\beta$ and show that it's the right thing. Say $[\mathbf{v}]_\beta$ is the $n \times 1$ column matrix $\begin{pmatrix} a_1 & \cdots & a_n \end{pmatrix}^t$ and $[T]_{\beta\gamma} = [\alpha_{ij}]$. By definition, the k-th entry of $[T]_{\beta\gamma}[\mathbf{v}]_\beta$ is $\sum_{j=1}^n \alpha_{kj} a_j$. Is this the k-th coordinate of $T(\mathbf{v})$ with respect to γ? Well, $\mathbf{v} = \sum_{j=1}^n a_j \mathbf{v}_j$ so that $T(\mathbf{v}) = T(\sum_{j=1}^n a_j \mathbf{v}_j) = \sum_{j=1}^n a_j T(\mathbf{v}_j)$. By definition of $[T]_{\beta\gamma}$, $T(\mathbf{v}_j) = \sum_{i=1}^m \alpha_{ij} \mathbf{w}_i$. Hence,

$$T(\mathbf{v}) = \sum_{j=1}^n a_j \sum_{i=1}^m \alpha_{ij} \mathbf{w}_i = \sum_{j=1}^n \sum_{i=1}^m a_j \alpha_{ij} \mathbf{w}_i = \sum_{i=1}^m \sum_{j=1}^n a_j \alpha_{ij} \mathbf{w}_i.$$

Here we've used the common tricks of sliding an indexed variable inside a sum not involving that index (the distributive law) and of interchanging the order of finite sums. Set $i = k$ in the last term of the displayed string of equalities to get the k-th coordinate of $T(\mathbf{v})$ with respect to γ. It is $\sum_{j=1}^n a_j \alpha_{ik}$. Thus, the k-th coordinate of $T(\mathbf{v})$ is as desired, and the proof is complete. ■

For the rest of the chapter, we'll study **linear operators** on V, that is linear transformations $T : V \to V$. We'll use the same basis β for V as the domain and as the codomain space of T. Then we can write the matrix of T simply as $[T]_\beta$ rather than as $[T]_{\beta\beta}$. In this format, the preceding theorem becomes: $[T(\mathbf{v})]_\beta = [T]_\beta[\mathbf{v}]_\beta$.

The basic idea of most of the rest of the chapter is to obtain special bases β such that $[T]_\beta$ reveals the structure of the operator T. The details of this procedure will be spelled out in Sections 4.8 and 4.9, after a preliminary section on determinants. We close this section with a theoretical result and then a discussion on the invertibility of matrices.

Definition 4.6.2 *An F-algebra A is a ring $(A, +, \cdot)$ such that the additive group $(A, +)$ also has the structure of an F-vector space, via an appropriate scalar multiplication. In addition to the ring and the vector space axioms, we also require that the algebra satisfy the following axiom, relating the scalar multiplication to the ring product: $\alpha(ab) = (\alpha a)b = a(\alpha b)$ for all $\alpha \in F; a, b \in A$.*

In this way we can regard $F \cong F 1_A$ as being embedded in the center of the ring A.

Example 4.6.1 *Let $M_n(F)$ be the set of all $n \times n$ matrices over a field F. We add and scalar multiply matrices componentwise: $[\alpha_{ij}] + [\beta_{ij}]$ is the matrix with $\alpha_{ij} + \beta_{ij}$ in the i-j position; for $c \in F$, $c[\alpha_{ij}]$ is the matrix with $c\alpha_{ij}$ in the i-j position. It is immediate that, with this addition and scalar multiplication, $M_n(F)$ is an F-vector space of dimension n^2. The zero vector is the matrix all of whose entries are zero. A convenient basis for $M_n(F)$ is $\{E_{ij} : 1 \le i, j \le n\}$, where E_{ij} is the **elementary matrix** with 1 in the i-j position and 0's elsewhere.*

Note that, if $A, B \in M_n(F)$, then their matrix product $AB \in M_n(F)$. We claim that, with our given sum, scalar multiplication and ring multiplication, $M_n(F)$ becomes an F-algebra. The multiplicative identity is the matrix I_n, the $n \times n$ matrix with 1's down the principal top-left bottom-right diagonal and 0's elsewhere. (The only tedious thing to check here is the associative law for matrix multiplication. I don't have the heart to ask you to write it out. Just realize that, with a gun to your head, you could.)

Example 4.6.2 *Let $_FV$ be a vector space of dimension n. Define $\mathcal{L}_F(V)$ to be the set of all linear operators on V. Addition and scalar multiplication are defined pointwise: $(T + S)(\mathbf{v}) = T(\mathbf{v}) + S(\mathbf{v})$ for all $\mathbf{v} \in V$; $(cT)(\mathbf{v}) = cT(\mathbf{v})$ for all $\mathbf{v} \in V$ and $c \in F$. Multiplication is composition of functions: $(TS)(\mathbf{v}) = T[S(\mathbf{v})]$ for all $\mathbf{v} \in V$. I will ask you to check that $\mathcal{L}_F(V)$ is an F-algebra (Problem 4.6.3 below.)*

Theorem 4.6.2 *Let $_FV$ be a vector space of dimension n with basis β. Then the mapping $m_\beta : T \to [T]_\beta$ is an F-algebra isomorphism from $\mathcal{L}_F(V)$ to $M_n(F)$.*

Proof. The proof is in two parts, Exercise 4.6.1 (easy) and Problem 4.6.4 (notationally a little technical, but not deep). ∎

Exercise 4.6.1 *Prove that $T \to [T]_\beta$ is an F-vector space isomorphism from $\mathcal{L}_F(V)$ to $M_n(F)$.*

Definition 4.6.3 *An operator $T \in \mathcal{L}_F(V)$ is called **invertible** if it has a 2-sided multiplicative inverse in the ring $\mathcal{L}_F(V)$.*

By the remarks immediately following the definition of a group, if an inverse of T exists, then it is unique. We denote the inverse of T (surprise) by T^{-1}. By Theorem 4.6.2, T is invertible if and only if, for any basis β, $[T]_\beta$ is an invertible matrix.

We now need a practical criterion for determining the invertibility of and, if possible, finding the inverse for an $n \times n$ matrix A. Recall that R_1, R_2, R_3 are the elementary row operations. Here is a simple but useful lemma.

Lemma 4.6.1 *Let A be an $n \times n$ matrix. Then, for $1 \leq i \leq 3, R_i(A) = R_i(I_n)A$.*

Proof. The proof is simply checking three matrix multiplications. See Problem 4.6.1. ∎

Definition 4.6.4 *Let $A \in M_n(F)$. The **row space** of A is the subspace of F^n spanned by the row vectors of A. The **row rank** of A is the dimension of the row space of A.*

Theorem 4.6.3 *Let $A \in M_n(F)$. Then A is invertible if and only if there is a sequence of row operations $\rho_1, ..., \rho_k$ such that $\rho_k \cdots \rho_1(A) = I_n$. The latter condition holds if and only if row rank $A = n$. Furthermore, in this case, $A^{-1} = \rho_k \cdots \rho_1(I_n)$.*

Proof. Let $A_l : F^n \to F^n$ be the map which left multiplies an element of F^n (written as a column vector) by the matrix A. It is elementary to check that A_l is a linear operator on F^n and that the matrix of A_l, computed with respect to the standard basis for F^n, is simply the matrix A. Hence, by Theorem 4.6.2, the matrix A is invertible if and only if the operator A_l is invertible. By Corollary 4.4.3, A_l is an invertible operator if and only if it is monic. This latter assertion is equivalent to the assertion that the matrix equation $AX = \mathbf{0}$ has only the solution $X = \mathbf{0}$.

Plainly, $AX = \mathbf{0}$ has only the trivial solution if and only if $EX = \mathbf{0}$ has only the trivial solution, where E is the reduced matrix obtained by row reducing A. But $EX = \mathbf{0}$ has only the trivial solution if and only if E has n leading $1's$. This is equivalent to the fact that E can be further row reduced, putting $0's$ above each leading 1, to I_n. (Think about this for a minute.) We have shown that A is invertible if and only if we can row reduce A to the identity matrix.

It is not hard to check that performing any row operation on a collection of row vectors does not change the subspace spanned by that set of vectors. Since row rank $I_n = n$, it follows that A is invertible if and only if A has row rank n.

If $\rho_k \cdots \rho_1(A) = I_n$, then, by Lemma 4.6.1, we have the matrix equation $\rho_k(I_n) \cdots \rho_1(I_n)(A) = I_n$. But then, since A is invertible, it follows that $A^{-1} = \rho_k(I_n) \cdots \rho_1(I_n) = \rho_k \cdots \rho_1(I_n)$. The proof is complete. ∎

Theorem 4.6.3 is reasonably easy to apply in small cases, using the following method. Write an $n \times 2n$ array consisting of the matrix A and the identity matrix placed to the right of it. Use row operations to try to row reduce A to the identity matrix. Do the same operations on the rows of I, regarded as extensions of the rows of A. If you get I on the left hand side, A is invertible and A^{-1} is sitting on the right hand side. If you don't get I on the left, A is not invertible and the matrix on the right is useless.

Exercise 4.6.2 *Try the above procedure for the following matrix*

$$\begin{bmatrix} 1 & -1 & 2 & 3 \\ 2 & 2 & 0 & 2 \\ 4 & 1 & -1 & -1 \\ 1 & 2 & 3 & 0 \end{bmatrix}.$$

Twenty-fifth Problem Set

Problem 4.6.1 *Prove Lemma 4.6.1.*

Problem 4.6.2 *Let $T : V \to W$. Prove that a basis β for V and γ for W can be found such that $[T]_{\beta\gamma}$ is as follows:*

$$\begin{bmatrix} 1 & & & & & \\ & \cdot & & & & \\ & & 1 & & & \\ & & & 0 & & \\ & & & & 0 & \\ & & & & & 0 \end{bmatrix}$$

here there are $r = \text{rank } T$ ones down the main diagonal and all other entries are zero.

Problem 4.6.3 *Prove that $\mathcal{L}_F(V)$ is an F-algebra.*

Problem 4.6.4 *Fill in the one non-obvious step in the proof of Theorem 4.6.2, the fact that $[ST]_\beta = [S]_\beta [T]_\beta$. Use the proof of Theorem 4.6.1 as your guide.*

Problem 4.6.5 *Let A be an $m \times n$ matrix with entries in a field F. Prove that the image of the map $A_l : F^n \to F^m$ induced by left multiplication of an element of F^m by A can be identified with the column space of A.*

Problem 4.6.6 *Define the **column rank** of an $m \times n$ matrix A with entries in F as the dimension of the subspace of F^m spanned by the columns of A. Prove that row rank $A = $ column rank A. I always found this result surprising. Hint: Use Problem 4.5.3 and a basic theorem.*

4.7   Determinants

4.7.1   The determinant function

Definition 4.7.1 *Let F be a field and n a natural number. A **rank n** **determinant function** is a function $D : M_n(F) \to F$ that satisfies the following three axioms, called the **determinant axioms**:*

1. $D(I_n) = 1$.

2. The function D has the **alternating property**. That is, if A is an $n \times n$ matrix with two equal rows, then $D(A) = 0$.

3. The function D is n-**linear**. By this we mean the following. Pick i with $1 \le i \le n$. Consider a matrix A with fixed entries in all rows except row i, the entries in the i-th row of A being regarded as variable. Since the set of possible i-th rows of A can be identified with F^n, the function D induces a function $D(A, i)$ from F^n into F. Our axiom says that, for each $A \in M_n(F)$ and $1 \le i \le n$, the induced map $D(A, i)$ will be a linear transformation.

The last axiom probably needs clarification; I think the easiest way to provide it is by writing out a specific case. Let $D : M_3(Q) \to Q$ be a determinant function. Axiom 3 says that, if we fix all rows but one of a 3×3 matrix and identify the variable row vectors in that position with Q^3, then, under that identification, the induced map from Q^3 to Q is a linear transformation. Say we focus our attention on the second row and consider all 3×3 rational matrices of the form:.

$$\left\{ \begin{bmatrix} -1 & 0 & 4 \\ a & b & c \\ 2 & -2 & 7 \end{bmatrix} : a, b, c \in Q \right\}$$

(The fixed numbers were chosen arbitrarily.) Then the determinant

$$D\left(\begin{bmatrix} -1 & 0 & 4 \\ a + a' & b + b' & c + c' \\ 2 & -2 & 7 \end{bmatrix} \right)$$

can be expressed as the sum of determinants

$$D\left(\begin{bmatrix} -1 & 0 & 4 \\ a & b & c \\ 2 & -2 & 7 \end{bmatrix} \right) + D\left(\begin{bmatrix} -1 & 0 & 4 \\ a' & b' & c' \\ 2 & -2 & 7 \end{bmatrix} \right).$$

Additionally, for all $q \in Q$,

$$D\left(\begin{bmatrix} -1 & 0 & 4 \\ qa & qb & qc \\ 2 & -2 & 7 \end{bmatrix}\right) = qD\left(\begin{bmatrix} -1 & 0 & 4 \\ a & b & c \\ 2 & -2 & 7 \end{bmatrix}\right)$$

Theorem 4.7.1 *Let $D : M_n(F) \to F$ be any function satisfying axioms 2 and 3 above. Let B be the matrix obtained from a matrix A by interchanging two rows of A. Then $D(B) = -D(A)$.*

Proof. Say that B is obtained from A by interchanging row i and row j. Displaying matrices in terms of their row vectors, we have

$$A = \begin{bmatrix} \cdot \\ \cdot \\ \rho_i \\ \cdot \\ \rho_j \\ \cdot \end{bmatrix}, \quad B = \begin{bmatrix} \cdot \\ \cdot \\ \rho_j \\ \cdot \\ \rho_i \\ \cdot \end{bmatrix}.$$

By axioms 2 and 3, $0 = D\begin{bmatrix} \cdot \\ \rho_i + \rho_j \\ \cdot \\ \rho_i + \rho_j \\ \cdot \end{bmatrix} = D\begin{bmatrix} \cdot \\ \rho_i \\ \cdot \\ \rho_i + \rho_j \\ \cdot \end{bmatrix} + D\begin{bmatrix} \cdot \\ \rho_j \\ \cdot \\ \rho_i + \rho_j \\ \cdot \end{bmatrix} =$

$$D\begin{bmatrix} \cdot \\ \rho_i \\ \cdot \\ \rho_i \\ \cdot \end{bmatrix} + D\begin{bmatrix} \cdot \\ \rho_i \\ \cdot \\ \rho_j \\ \cdot \end{bmatrix} + D\begin{bmatrix} \cdot \\ \rho_j \\ \cdot \\ \rho_i \\ \cdot \end{bmatrix} + D\begin{bmatrix} \cdot \\ \rho_j \\ \cdot \\ \rho_j \\ \cdot \end{bmatrix} = 0 + D(A) + D(B) + 0.$$

Hence, $D(B) = -D(A)$. ∎

Example 4.7.1 *Let's find all determinant functions $D : M_2(F) \to F$. Let $\{e_1, e_2\}$ be the standard basis for F^2. Then, by axiom 3,* .

$$D\begin{bmatrix} a_{11} & a_{12} \\ a_{21} & a_{22} \end{bmatrix} = D\begin{bmatrix} a_{11}e_1 + & a_{12}e_2 \\ a_{21} & a_{22} \end{bmatrix} = a_{11}D\begin{bmatrix} 1 & 0 \\ a_{21} & a_{22} \end{bmatrix} + a_{12}D\begin{bmatrix} 0 & 1 \\ a_{21} & a_{22} \end{bmatrix}.$$

Expanding further,

$$a_{11}D\begin{bmatrix} 1 & 0 \\ a_{21}e_1 + & a_{22}e_2 \end{bmatrix} = a_{11}a_{21}D\begin{bmatrix} 1 & 0 \\ 1 & 0 \end{bmatrix} + a_{11}a_{22}D\begin{bmatrix} 1 & 0 \\ 0 & 1 \end{bmatrix}$$

Example 4.7.2 *and*

$$a_{12}D\begin{bmatrix} 0 & 1 \\ a_{21}\mathbf{e}_1+ & a_{22}\mathbf{e}_2 \end{bmatrix} = a_{12}a_{21}D\begin{bmatrix} 0 & 1 \\ 1 & 0 \end{bmatrix} + a_{12}a_{22}D\begin{bmatrix} 0 & 1 \\ 0 & 1 \end{bmatrix}.$$

By axiom 2, the first and fourth terms in the sum are zero. By axiom 1 and Theorem 4.6.1,

$$D\begin{bmatrix} a_{11} & a_{12} \\ a_{21} & a_{22} \end{bmatrix} = a_{12}a_{22} - a_{12}a_{21}.$$

Thus, any determinant function $D : M_2(F) \to F$ must be given by the formula above. Moreover, it is not difficult to check that the function $D(A) = a_{12}a_{22} - a_{12}a_{21}$ actually does satisfy the rank two determinant axioms. Hence, for any field F, there is a unique determinant function on $M_2(F)$. We denote this function as d_2.

Exercise 4.7.1 *Show that there is at most one determinant function d_3 : $M_3(F) \to F$.*

We now will prove, for each field F and natural number n, there is exactly one determinant function $d_n : M_n(F) \to F$. The proof, while complex in notation, is simply an expanded version of Example 4.6.1. We use axiom 3 to "pull apart" the determinant on each row, ending with a sum of determinants of matrices whose rows are made up from vectors in the standard basis for F^n. We then use axiom 1 and Theorem 4.7.1 to complete the calculation.

Theorem 4.7.2 *For each natural number n, there is a unique function $d_n : M_n(F) \to F$ that satisfies the determinant axioms.*

Proof. Let $\{\mathbf{e}_1, ..., \mathbf{e}_n\}$ be the standard basis for F^n and let $A = [a_{ij}] \in M_n(F)$. Suppose that $D : M_n(F) \to F$ satisfies the determinant axioms. Then, using multilinearity on the first row, we have:

$$D(A) = D(\begin{bmatrix} a_{11}\mathbf{e}_1 + \cdots + a_{1n}\mathbf{e}_n \\ \cdots\cdots\cdots\cdots \\ \cdots\cdots\cdots\cdots \\ \cdots\cdots\cdots\cdots \end{bmatrix}) = \sum_{j_1=1}^{n} a_{1j_1} D(\begin{bmatrix} \mathbf{e}_{j_1} \\ \cdots\cdots\cdots\cdots \\ \cdots\cdots\cdots\cdots \\ \cdots\cdots\cdots\cdots \end{bmatrix}).$$

If we take the individual terms in the summation and, using multilinearity, "pull them apart along the second row", we have:

$$D(A) = \sum_{j_1=1}^{n} \sum_{j_2=1}^{n} a_{1j_1} a_{2j_2} D\left(\begin{bmatrix} \mathbf{e}_{j_1} \\ \mathbf{e}_{j_2} \\ \cdots \cdots \cdots \cdots \cdots \\ \cdots \cdots \cdots \cdots \cdots \end{bmatrix} \right).$$

Continuing the process for each row we have:

$$D(A) = \sum_{j_1=1}^{n} \sum_{j_2=1}^{n} \cdots \sum_{j_n=1}^{n} a_{1j_1} a_{2j_2} \cdots a_{nj_n} D\left(\begin{bmatrix} \mathbf{e}_{j_1} \\ \mathbf{e}_{j_2} \\ \cdots \cdots \cdots \cdots \\ \mathbf{e}_{j_n} \end{bmatrix} \right).$$

If we evaluate D on any matrix with repeated rows, we get zero. Thus, the only nonzero terms in the above summation are those where $\{\mathbf{e}_{j_1}, ..., \mathbf{e}_{j_n}\} = \{\mathbf{e}_1, ..., \mathbf{e}_n\}$. This means that the ordered set of integers $\{j_1, ..., j_n\}$ is obtained by a permutation $\sigma \in S_n$ of the integers $\{1, ..., n\}$. Moreover, if $\{j_1, ..., j_n\}$ is any permutation of $\{1, ..., n\}$, then the term

$$a_{1j_1} a_{2j_2} \cdots a_{nj_n} D\left(\begin{bmatrix} \mathbf{e}_{j_1} \\ \mathbf{e}_{j_2} \\ \cdots \cdots \cdots \cdots \\ \mathbf{e}_{j_n} \end{bmatrix} \right)$$

will appear as part of the summation. Thus, putting $j_i = \sigma(i)$, we can rewrite our summation as:

$$D(A) = \sum_{\sigma \in S_n} a_{1\sigma(1)} \cdots a_{n\sigma(n)} D\left(\begin{bmatrix} \mathbf{e}_{\sigma(1)} \\ \mathbf{e}_{\sigma(2)} \\ \cdots \cdots \cdots \cdots \\ \mathbf{e}_{\sigma(n)} \end{bmatrix} \right).$$

If σ is written as a product of transpositions $\sigma = \tau_1 \cdots \tau_s$, then it will take exactly s interchanges in the rows of the $n \times n$ identity matrix to obtain the matrix

$$\begin{bmatrix} \mathbf{e}_{\sigma(1)} \\ \mathbf{e}_{\sigma(2)} \\ \cdots \cdots \cdots \cdots \\ \mathbf{e}_{\sigma(n)} \end{bmatrix}$$

Then, by Theorem 4.6.1,

$$D\left(\begin{bmatrix} \mathbf{e}_{\sigma(1)} \\ \mathbf{e}_{\sigma(2)} \\ \cdots \cdots \cdots \cdots \\ \mathbf{e}_{\sigma(n)} \end{bmatrix} \right) = sgn(\sigma).$$

We finally obtain:

$$(\dagger)D(A) = \sum_{\sigma \in S_n} sgn(\sigma)a_{1\sigma(1)} \cdots a_{n\sigma(n)}.$$

At this point, we have shown that there is **at most** one determinant function $d_n : M_n(F) \to F$ and it is given by formula $(\dagger)$. The formula $(\dagger)$ for the rank n determinant tells us to sum all of the $n!$ products formed by taking exactly one element from each row and column of a given $n \times n$ matrix. Each term in the sum is multiplied by the sign of the permutation which, in forming each product, chooses the column we take for each row.

We now need to check that the function $d_n(A)$, defined by $(\dagger)$ above, is a determinant function. The fact that $d_n(I_n) = 1$ is a simple calculation, left as an exercise. Suppose that the matrix A has two equal rows, say rows r and s. List the $n!/2$ left cosets of the subgroup $< e, (r, s) >$ in S_n. Each coset contains two permutations σ and $\sigma' = \sigma(r, s)$ of opposite sign. Furthermore, since rows r and s are equal, a few minute's thought reveals $a_{1\sigma(1)} \cdots a_{n\sigma(n)} = a_{1\sigma'(1)} \cdots a_{n\sigma'(n)}$. Thus, the terms in formula $(\dagger)$ can be divided into pairs with opposite sign, and, hence, $d_n(A) = 0$. To show multilinearity, fix the i-th row and suppose that it is a linear combination of two row vectors, say $b\rho + c\lambda$, $\rho = (\rho_1, ..., \rho_n)$ and $\lambda = (\lambda_1, ..., \lambda_n)$. Then

$$d_n\left(\begin{bmatrix} \cdots \\ \cdots \\ b\rho + c\lambda \\ \cdots \end{bmatrix} \right) = \sum_{\sigma \in S_n} sgn\sigma[a_{1\sigma(1)} \cdots a_{i-1\sigma(i-1)}]a_{i\sigma(i)}[a_{i+1\sigma(i+1)} \cdots a_{n\sigma(n)}].$$

Each term a_{ij} with $j = \sigma(i)$ that occurs in each product is of the form $a_{ij} = b\rho_j + c\lambda_j$. For each $j = \sigma(i)$, if we substitute this expression for a_{ij} into formula $(\dagger)$ and simplify, we see that d_n is a linear function of the i-th row. Henceforth, we let $d_n(A) = \sum_{\sigma \in S_n} sgn(\sigma)a_{1\sigma(1)} \cdots a_{n\sigma(n)}$ and call d_n the rank n determinant. ∎

Exercise 4.7.2 *Show that $d_n(I_n) = 1$.*

4.7.2   The determinant and invertibility

Theorem 4.7.3 *Let $A \in M_n(F)$. Then A is invertible if and only if $d_n(A) \neq 0$.*

Proof. We have already shown in Theorem 4.6.3 that A is invertible if and only if A can be transformed to I_n by a finite sequence of row operations. It follows directly from the determinant axioms that no row operation will change the determinant of a matrix from zero to nonzero or from nonzero to zero.

Suppose that A is invertible. Since the reduced form of A is the matrix I_n and $d_n(I_n) = 1$, then, in view of the previous paragraph, $d_n(A) \neq 0$. Conversely, if A is not invertible, then A can be reduced by row operations to some reduced matrix $B \neq I_n$. But a moment's thought reveals that any $n \times n$ reduced matrix $B \neq I_n$ must contain at least one row of zeros at the bottom. Hence, by formula (†), $d_n(B) = 0$. It follows that $d_n(A) = 0$ as well. The proof is complete. ∎

4.7.3   Computation formulas

To compute the determinant, formula (†) is not convenient. The next theorem will give us a more tractable set of computation formulas. For an $n \times n$ matrix $A, n \geq 2$, and $1 \leq i, j \leq n$, let $D(i \mid j)$ be the determinant of the $n - 1 \times n - 1$ matrix obtained by deleting the i-th row and j-th column of A. For $1 \leq j \leq n$, define $D_j(A)$ by the formula $D_j(A) = \sum_{i=1}^{n}(-1)^{i+j}a_{ij}D(i \mid j)$. Computing the summation for $D_j(A)$ is called **expanding the determinant down the j-th column**.

Example 4.7.3 $D_2\left(\begin{bmatrix} -1 & 2 & 3 \\ 4 & -1 & 2 \\ 0 & -2 & -3 \end{bmatrix}\right) = -1 \cdot 2 \cdot d_2\left(\begin{bmatrix} 4 & 2 \\ 0 & -3 \end{bmatrix}\right) + 1 \cdot (-1) \cdot$

$d_2\left(\begin{bmatrix} -1 & 3 \\ 0 & -3 \end{bmatrix}\right) - 1 \cdot (-2) \cdot d_2\left(\begin{bmatrix} -1 & 3 \\ 4 & 2 \end{bmatrix}\right) = -2(-12) - 1(3) + 2(-14) = -7.$

We have the following result.

Theorem 4.7.4 *For $A \in M_n(F), n \geq 2$ and any fixed $j, 1 \leq j \leq n$, $D_j(A) = d_n(A)$.*

Proof. Fix j with $1 \leq j \leq n$. It is easy to check (Problem 4.7.1) that $D_j(I_n) = 1$. An induction on $n \geq 2$ shows that D_j is multilinear (Problem 4.7.2). We will verify that D_j has the alternating property. Suppose that A has two equal rows, say row r and row s. By performing a series of switches we can suppose that $r = 1, s = 2$; all that we have done is possibly change the sign of the determinant. Since d_{n-1} is alternating, $D(i \mid j) = 0$ for $i \geq 3$. Hence, $D_j(A) = \pm[(-1)^{1+j}a_{1j}D(1 \mid j) + (-1)^{2+j}a_{2j}D(2 \mid j)]$. But, since the first two rows are equal, $a_{1j} = a_{2j}$ and $D(1 \mid j) = D(2 \mid j)]$. Thus, $D_j(A) = 0$ and the proof is complete. ∎

As a consequence of Problem 4.7.3 below, a similar formula holds for expansion along a fixed row.

Twenty-sixth Problem Set

Problem 4.7.1 *Use induction on $n \geq 2$ to prove that $D_j(I_n) = 1$ for each fixed j.*

Problem 4.7.2 *Use induction to prove that D_j is a multilinear function on $n \times n$ matrices.*

Problem 4.7.3 *Let A^t, the **transpose** of the $n \times n$ matrix A, be the matrix obtained from A by interchanging its rows and columns. Use formula (†) to prove that $d_n(A^t) = d_n(A)$.*

Problem 4.7.4 *Let A be an $n \times n$ **lower triangular** matrix, that is $a_{ij} = 0$ if $i > j$. Prove that $d_n(A) = a_{11}a_{22} \cdots a_{nn}$.*

Problem 4.7.5 *Let $A, B \in M_n(F)$. Prove $d_n(AB) = d_n(A)d_n(B)$. Hint: first verify the formula if B is not invertible. Then suppose B is invertible and define an appropriate function $D : M_n(F) \to F$. Use the unicity of the determinant function.*

4.8   Characteristic values, vectors, basis change

Let $T : V \to V$ and suppose V has two bases, an "old" basis $\beta = \{v_1, ..., v_n\}$ and a "new" basis $\beta' = \{v'_1, ..., v'_n\}$. We want to relate the matrices $[T]_\beta$ and $[T]_{\beta'}$. Let P be the $n \times n$ matrix whose j-th column, for each j, is the column vector of coordinates of the new basis vector v'_j with respect to the old basis β. We start with the following simple but interesting result.

Lemma 4.8.1 *Let P_l be left multiplication by the matrix P, regarded as a linear operator on F^n. Then P_l is an isomorphism whose action on F^n can be identified as that of taking the β' coordinates of a vector $v \in V$ to the β coordinates of that vector.*

   Proof. Let e_j be the j-th standard basis vector for F^n. As a column vector, e_j can be regarded as the column of coordinates of the vector v'_j with respect to β'. Now Pe_j coincides with the j-th column of P, which by definition is the column of coordinates of v'_j with respect to β. Since P_l is a linear operator on F^n, it follows that the action of P_l on any column vector $c \in F^n$ is the following: Let $v \in V$ be the unique vector such that $[v]_{\beta'} = c$. Then $Pc = [v]_\beta$. Think about this for a minute.

Since the vectors in β' are independent, so are the columns of P, regarded as vectors in F^n. Hence, column rank $P = n$; equivalently, the linear transformation P_l maps F^n onto F^n. It follows that P_l is an isomorphism. ∎

Theorem 4.8.1 *With notation as above,* $[T]_{\beta'} = P^{-1}[T]_\beta P$.

Proof. Note that left multiplication by P^{-1} takes the β coordinates of a vector to the β' coordinates of that same vector. Thus, for each $\mathbf{v} \in V$, $(P^{-1}[T]_\beta P)[\mathbf{v}]_{\beta'} = P^{-1}[T]_\beta[\mathbf{v}]_\beta = P^{-1}[T(\mathbf{v})]_\beta = [T(\mathbf{v})]_{\beta'}$. But, by Theorem 4.6.1, $[T]_{\beta'}[\mathbf{v}]_{\beta'} = [T(\mathbf{v})]_{\beta'}$ for each $\mathbf{v} \in V$. Hence, $(P^{-1}[T]_\beta P)[\mathbf{v}]_{\beta'} = [T]_{\beta'}[\mathbf{v}]_{\beta'}$ for all coordinate columns $[\mathbf{v}]_{\beta'}$. In particular, if we put $\mathbf{v} = \mathbf{v}'_j$, so that $[\mathbf{v}]_{\beta'}$ is the column vector $\mathbf{e}_j$, we can conclude that the j-th column of $[T]_{\beta'}$ is equal to the j-th column of $P^{-1}[T]_\beta P$ for all j. The proof is complete. ∎

Example 4.8.1 *Let* $T : Q^3 \to Q^3$ *be the linear transformation given by the formula* $T(x, y, z) = (2x - z, y + 3z, x + y)$. *Find* $[T]_{\beta'}$, *where* β' *is the basis* $\{(1, 1/2, 1/3), (1/2, 1/3, 1/4), (1/3, 1/4, 1/5)\}$.

We could compute $[T]_{\beta'}$ *directly from its definition, but we'll use our change of basis theorem. If we let* $\beta = St$, *the standard basis, the matrix* P *above is simply*

$$\begin{bmatrix} 1 & 1/2 & 1/3 \\ 1/2 & 1/3 & 1/4 \\ 1/3 & 1/4 & 1/5 \end{bmatrix}.$$

To find the matrix of T *with respect to* St *we simply compute* $T(1, 0, 0) = (2, 0, 1), T(0, 1, 0) = (0, 1, 1), T(0, 0, 1) = (-1, 3, 0)$, *we obtain*

$$[T]_{St} = \begin{bmatrix} 2 & 0 & -1 \\ 0 & 1 & 3 \\ 1 & 1 & 0 \end{bmatrix}.$$

We have $[T]_{\beta'}$. *We next find* P^{-1} *by row reduction obtaining.*

$$P^{-1} = \begin{bmatrix} 9 & -36 & 30 \\ -36 & 192 & -180 \\ 30 & -180 & 180 \end{bmatrix}.$$

Hence, we can compute $[T]_{\beta'}$ *as follows:*

$$P^{-1}\begin{bmatrix} 2 & 0 & -1 \\ 0 & 1 & 3 \\ 1 & 1 & 0 \end{bmatrix} P = \begin{bmatrix} 9 & -36 & 30 \\ -36 & 192 & -180 \\ 30 & -180 & 180 \end{bmatrix}\begin{bmatrix} 2 & 0 & -1 \\ 0 & 1 & 3 \\ 1 & 1 & 0 \end{bmatrix}\begin{bmatrix} 1 & 1/2 & 1/3 \\ 1/2 & 1/3 & 1/4 \\ 1/3 & 1/4 & 1/5 \end{bmatrix}.$$

(The alert reader will realize that I don't feel like doing the matrix multiplication.)

Two square matrices A, B are called **similar** if $B = P^{-1}AP$ for some invertible matrix P. We've just shown that matrices of the same transformation with respect to different bases are similar.

Exercise 4.8.1 *Prove that similarity is an equivalence relation on $M_n(F)$.*

If we are trying to get a basis such that the matrix of T has a simple form, a good thing to do would be to include a vector **v** such that $T(\mathbf{v}) = c\mathbf{v}$ for some scalar c. Then the column corresponding to that vector would contain only a c in the diagonal position. This leads to the following definition.

Definition 4.8.1 *A nonzero vector* **v** *such that $T(\mathbf{v}) = c\mathbf{v}$ for some scalar c is called a **characteristic vector** (or **eigenvector**) for T. The scalar c is called a **characteristic value** (or **eigenvalue**) for T. We say that the characteristic vector* **v** *and the characteristic value c are **associated**.*

Note that, if **v** is a characteristic vector associated with the characteristic value c, then so is $\lambda\mathbf{v}$ for any nonzero scalar λ.

Definition 4.8.2 *Let $T \in \mathcal{L}_F(V)$ and let A be the matrix of T with respect to **any** basis β for* **v**. *Then the **characteristic polynomial** of T, $c_T(x)$, is the determinant of the matrix $(xI_n - A)$. Here $n = \dim_F(V)$ and xI_n is the $n \times n$ matrix with x down the main diagonal and zeros elsewhere.*

It's clear that the degree of $c_T(x)$ is equal to the dimension of V over F. For this definition to make sense $c_T(x)$ must be independent of the choice of the basis β, thus depend only on the operator T. I ask you to check this independence in Problem 4.8.1. (The proof, using Theorem 4.8.1, is not difficult.)

Here is an extremely useful fact.

Theorem 4.8.2 *A scalar $c \in F$ is a characteristic value for the operator $T \in \mathcal{L}_F(V)$ if and only if c is a root of $c_T(x)$.*

Proof. A scalar $c \in F$ is a characteristic value for the operator T if and only if there exists $0 \neq \mathbf{v} \in V$ such that $T(\mathbf{v}) = c\mathbf{v}$. The latter condition holds if and only if the operator $(c1_\mathbf{v} - T)$ is not invertible, where $1_\mathbf{v}$ is the identity operator on V. But $(c1_\mathbf{v} - T)$ is not invertible if and only if the determinant of the matrix $(cI_n - A)$ is zero, where $A = [T]_\beta$ computed with respect to any basis β for V. This final claim is saying precisely that c is a root of $c_T(x)$. ∎

We now are ready to consider the simplest kind of operator.

Definition 4.8.3 *An operator $T \in \mathcal{L}_F(V)$ is* **diagonalizable** *if V has a basis consisting of characteristic vectors for T.*

The reason for the terminology is clear. If β is a basis of V consisting of characteristic vectors for T then $[T]_\beta$ will be a matrix with the associated characteristic values down the diagonal. Most operators, of course, fail to be diagonalizable. There are two reasons an operator can fail to be diagonalizable, one easily rectified and one fundamental.

Example 4.8.2 *Consider the transformation $T : Q^4 \to Q^4$ whose matrix in the standard basis is*

$$[A]_{St} = \begin{bmatrix} 0 & -2 & 0 & 0 \\ -1 & 0 & 0 & 0 \\ 0 & 0 & 0 & -1 \\ 0 & 0 & 1 & 0 \end{bmatrix}.$$

Then

$$c_T(x) = \det \begin{bmatrix} x & 2 & 0 & 0 \\ 1 & x & 0 & 0 \\ 0 & 0 & x & 1 \\ 0 & 0 & -1 & x \end{bmatrix} = (x^2 - 2)(x^2 + 1)$$

has no roots in Q. Hence, T has no characteristic values, considered as an element of $\mathcal{L}_Q(Q^4)$. But this difficulty is easily remedied. We can enlarge $F = Q$ to the field $F' = \{q_0 + q_1\sqrt{2} + q_2 i + q_3\sqrt{2}i : q_j \in Q\} \subset C$. (Exercise 4.8.2 asks you to check that F' is a subfield of the field of complex numbers.) In $F'[x]$, $c_T(x)$ factors completely into linear factors, $c_T(x) = (x - \sqrt{2})(x + \sqrt{2})(x - i)(x + i)$. Now we replace the Q-vector space $V = Q^4$ with the F'-vector space $V' = (F' \otimes_Q V)$. (Refer back to Problem 4.4.5). Let $\{e_i : 1 \le i \le 4\}$ be the standard basis for Q^4 over Q. Then, by Problem 4.4.5, $\beta' = \{1 \otimes e_i : 1 \le i \le 4\}$ is a basis for V' over F'. Furthermore, the balanced bilinear function $(f', v) \to [f' \otimes T(v)] \in V'$ induces a linear operator T' on V', denoted naturally by $T' = (1 \otimes T)$. Using the fact that $(1 \otimes T)\sum_{finite} f'_j \otimes v_j = \sum_{finite} f'_j \otimes T(v_j)$, we see that $[T']_{\beta'} = [T]_{St}$. Working in V', we can now find four independent characteristic vectors for T' corresponding to the four distinct characteristic values in F'. The extended operator T' has become diagonalizable; the only price we had to pay was that we had to enlarge our field of scalars a little bit.

We find a characteristic vector for the characteristic value i. This means *finding* $\mathbf{v} \neq \mathbf{0}$ *such that* $T'(\mathbf{v}) = i\mathbf{v}$. *We solve*

$$
\begin{bmatrix}
0 & -2 & 0 & 0 \\
-1 & 0 & 0 & 0 \\
0 & 0 & 0 & -1 \\
0 & 0 & 1 & 0
\end{bmatrix}
\begin{pmatrix} a \\ b \\ c \\ d \end{pmatrix}
= i
\begin{pmatrix} a \\ b \\ c \\ d \end{pmatrix}.
$$

The matrix equation above is equivalent to the simple system (no need for row reduction here) $-2b = ia, -a = ib, -d = ic, c = id$. *By inspection, $a = d = 1, b = (-1/2)i, c = i$ is a solution. Hence,* $\mathbf{v} = (1, (-1/2)i, i, 1)$ *will serve as an eigenvector. So will any nonzero multiple of* $\mathbf{v}$.

Exercise 4.8.2 *Prove that the set F' above is a subfield of C.*

Exercise 4.8.3 *Solve the appropriate matrix equations to find three remaining independent characteristic vectors for T'. (For the independence part, see Problem 4.8.3.)*

In contrast to the previous example, our next example illustrates a fundamental difficulty in the attempt at diagonalization.

Example 4.8.3 *Let $T : C^2 \to C^2$ have matrix $A = \begin{bmatrix} 0 & 0 \\ 1 & 0 \end{bmatrix}$ with respect to the standard basis for C^2 over C. Then $c_T(x) = x^2$. Hence, 0 is the only characteristic value for T. Solving the equation $A \begin{pmatrix} c_1 \\ c_2 \end{pmatrix} = 0 \begin{pmatrix} c_1 \\ c_2 \end{pmatrix} = \begin{pmatrix} 0 \\ 0 \end{pmatrix}$, yields $c_1 = 0$. Hence, any characteristic vector for T will be a vector of the form $(0, c), c \neq 0$. It follows that we cannot find a basis for C^2 consisting of characteristic vectors. This is an inherent failing of T and we can't do a thing about it.*

In the next section we will deal with the problem of finding decent matrix representations for non-diagonalizable operators.

Notation 4.8.1 *If $p(x) = \sum_{i=0}^{n} c_i x^i \in F[x]$, put $p(T) = \sum_{i=0}^{n} c_i T^i \in \mathcal{L}_F(V)$.*

Suppose $T \in \mathcal{L}_F(V)$ with $\dim_F V = n$ and let $0_V, 1_V$ be the zero and identity operators on V. Since $\mathcal{L}_F(V) \cong M_n(F)$, an algebra of dimension n^2 over F, the set $\{1_V, T, T^2, ..., T^{n^2}\}$ will be a linearly dependent subset of $\mathcal{L}_F(V)$. Thus, the ideal $A_T = \{p(x) \in F[x] : p(T) = 0_V\}$ is nonzero. Since $F[x]$ is a pid, A_T is a principal ideal. In fact we know that A_T will be generated by the polynomial of least degree among the set of monic polynomials in A_T. (At this point, we know that this least degree is less than or equal to n^2, but we'll see soon that this degree is actually bounded by n.)

Definition 4.8.4 *The **minimal polynomial** $m_T(x)$ for $T \in \mathcal{L}_F(V)$ is the monic generator for the ideal $A_T = \{p(x) \in F[x] : p(T) = 0_V\}$.*

It will be convenient to put off examples of the computation of $m_T(x)$ until the next section. Here is one easy fact we can give now.

Lemma 4.8.2 *If $c \in F$ is a root of $c_T(x)$ then c is a root of $m_T(x)$.*

Proof. We first ask you to check the following fact as an exercise: If $\mathbf{v}$ is a characteristic vector with characteristic value c and $p(x) \in F[x]$, then $p(T)(v) = p(c)\mathbf{v}$. Now suppose that $c \in F$ is a root of $c_T(x)$, equivalently that c is a characteristic value of T. Then $T(\mathbf{v}) = c\mathbf{v}$ for some $\mathbf{v} \neq \mathbf{0}$. We have $\mathbf{0} = m_T(T)(\mathbf{v}) = m(c)\mathbf{v}$. Since $\mathbf{v} \neq \mathbf{0}$, it must be that $m(c) = 0$. ∎

The following is a nice criterion for diagonalizability.

Theorem 4.8.3 *An operator $T \in \mathcal{L}_F(V)$ is diagonalizable if and only if $m_T(x)$ factors into distinct linear factors in $F[x]$.*

Proof. Let $\{c_1, ..., c_k\}$ be the set of distinct characteristic values for T. Suppose that T is diagonalizable. Then V has a basis $\beta = \{\mathbf{v}_{11}, ..., \mathbf{v}_{1n_1}, \mathbf{v}_{21}, ..., \mathbf{v}_{2n_2}, ..., \mathbf{v}_{k1}, ..., \mathbf{v}_{kn_k}\}$, where each $\mathbf{v}_{ij}$ is a characteristic vector with associated characteristic value c_i. By permuting its linear factors, we see that the polynomial $m(x) = (x - c_1) \cdots (x - c_k)$ has the property that $m(T)$ annihilates all the vectors in β. Thus, $m(x) \in A_T$ and so $m_T(x) \mid m(x)$. But, by Lemma 4.8.2, each characteristic value c_i is a root of $m_T(x)$. Hence, $m_T(x) = m(x)$, a product of distinct linear factors.

Conversely, suppose that $m_T(x) = (x - c_1) \cdots (x - c_k)$ with the $c_1, ..., c_k$ distinct scalars in F. For $1 \leq i \leq k$, let $V_i = \{\mathbf{v} \in V : (T - c_i)\mathbf{v} = \mathbf{0}\}$. Then each V_i is a subspace of V and, exactly as in the proof of Lemma 3.4.1, $V = \bigoplus_{i=1}^{k} V_i$. Any basis β_i for V_i will consist of characteristic vectors with characteristic value c_i. Thus, $\beta = \cup \beta_i$ will be a basis for V consisting of characteristic vectors for T; that is T will be diagonalizable. ∎

Twenty-seventh Problem Set

Problem 4.8.1 *Show that $c_T(x)$ depends only on the operator T.*

Problem 4.8.2 *Show that Example 4.8.1 can be generalized as follows. Let $T \in \mathcal{L}_F(V)$ and $\beta = \{\mathbf{v}_1, ..., \mathbf{v}_n\}$ be a basis for V over F. Let F' be a field containing F. Then $[1 \otimes T]_{1 \otimes \beta} = [T]_\beta$.*

Problem 4.8.3 *Let $\{v_1, ..., v_k\}$ be characteristic vectors for the distinct characteristic values $c_1, ..., c_k$. Prove that the set $\{v_1, ..., v_k\}$ is independent.*

Problem 4.8.4 *Let Φ be a multiplicative homomorphism from a finite multiplicative group G into $\mathcal{L}_F(V)$. Show for each $g \in G$ that $\Phi(g)$ is diagonalizable.*

Problem 4.8.5 *Without using a computer or calculator, compute A^{100} where $A = \begin{bmatrix} 0 & 3 & -1 \\ -6 & -17 & 6 \\ -17 & -51 & 18 \end{bmatrix}$.*

4.9   Canonical forms

Let V be a vector space over F and let T be a fixed linear transformation in $\mathcal{L}_F(V)$. It is easy to check that we can make V into an $F[x]$-module by defining $(f_0 + f_1 x + \cdots + f_n x^n)v = (f_0 + f_1 T + \cdots + f_n T^n)v$. In this context we have $(0 : V) = A_T = (m_T(x))$.

Now suppose $m_T(x)$ factors completely in $F[x]$, say $m_T(x) = (x - c_1)^{e_1} \cdot \cdots (x - c_k)^{e_k}$, where $c_1, ..., c_k$ are distinct scalars in F. Since this factorization is a prime factorization of $m_T(x)$ in $F[x]$, we can employ Lemma 3.4.1 to obtain a direct sum decomposition, $V = \bigoplus_{i=1}^{k} V_i$, where each V_i is the subspace consisting of all vectors v such that $(x - c_i)^{e_i} v = (T - c_i)^{e_i} v = 0$.

Using Lemma 3.4.2, we can further decompose each V_i into $F[x]$ cyclic submodules $V_i = \bigoplus_{j=1}^{l_i} F[x]v_{ij}$ with order $v_{ij} = (x - c_i)^{e_{ij}}$, $e_i = e_{i1} \geq e_{i2} \geq \cdots \geq e_{il_i}$. The orders of the cyclic generators v_{ij}, namely the polynomials $(x - c_1)^{e_{11}}, ..., (x - c_1)^{e_{1l_1}}; (x - c_2)^{e_{21}}, ..., (x - c_2)^{e_{2l_2}}; ...; (x - c_k)^{e_{k1}}, ..., (x - c_k)^{e_{kl_k}}$ are called the **invariant factors** of T.

Here is an important point. Theorem 3.5.2 applied to the module $_{F[x]}V$ shows that the invariant factors are determined by $_{F[x]}V$ and, hence, are determined by the linear operator T, not by any basis choice.

For each subspace $F[x]v_{ij}$ let $\beta_{ij} = \{v_{ij}, (T - c_i)v_{ij}, ..., (T - c_i)^{e_{ij} - 1}v_{ij}\}$. Since order $v_{ij} = (x - c_i)^{e_{ij}}$, it follows easily that β_{ij} is an F-basis for the $F[x]$ cyclic submodule $F[x]v_{ij}$. Furthermore, each $F[x]v_{ij}$ is invariant under the linear transformation T. If T_{ij} is the restriction of T to $F[x]v_{ij}$, plainly $[T_{ij}]_{\beta_{ij}} = [c_i]_{\beta_{ij}} + [T_{ij} - c_i]_{\beta_{ij}}$. It follows that

$$
\begin{bmatrix}
c_i & 0 & 0 & 0 & 0 & 0 \\
0 & c_i & 0 & 0 & 0 & 0 \\
0 & 0 & \cdot & 0 & 0 & 0 \\
0 & 0 & 0 & \cdot & 0 & 0 \\
0 & 0 & 0 & 0 & \cdot & 0 \\
0 & 0 & 0 & 0 & 0 & c_i
\end{bmatrix}
+
\begin{bmatrix}
0 & 0 & 0 & 0 & 0 & 0 \\
1 & 0 & 0 & 0 & 0 & 0 \\
0 & 1 & 0 & 0 & 0 & 0 \\
\cdot & \cdot & \cdot & \cdot & \cdot & \cdot \\
0 & 0 & 0 & 1 & 0 & 0 \\
0 & 0 & 0 & 0 & 1 & 0
\end{bmatrix}
=
\begin{bmatrix}
c_i & 1 & 0 & 0 & 0 & 0 \\
1 & c_i & 1 & 0 & 0 & 0 \\
0 & 1 & \cdot & 0 & 0 & 0 \\
0 & 0 & 0 & \cdot & 0 & 0 \\
0 & 0 & 0 & 1 & c_i & 0 \\
0 & 0 & 0 & 0 & 1 & c_i
\end{bmatrix}.
$$

We see that on $F[x]\mathbf{v}_{ij}$ the operator T_{ij} decomposes into a sum of two parts, a scalar multiplication by c_i and a "shift map" N_{ij} sending each basis vector in β_{ij} to the next one and the last one to zero. This results in an $e_{ij} \times e_{ij}$ matrix with the characteristic value c_i down the diagonal and 1 down the sub-diagonal. We obtain the e_{ij} from the invariant factors.

The operator N_{ij} is called **nilpotent** meaning that N_{ij} raised to a positive power, in this case e_{ij}, is zero. The matrix $[T_{ij}]_{\beta_{ij}}$ is called the **Jordan Block** for the transformation T with respect to the basis β_{ij}. Finally, if we let $\beta = \cup_{i=1}^{k} \cup_{j=1}^{l_k} \beta_{ij}$, we have produced a basis for V such that $J = [T]_\beta$ consists of Jordan Blocks. This matrix is called the **Jordan Form** for T. If we arrange the individual blocks corresponding to each eigenvalue c_i such that $e_{i1} \geq \cdots \geq e_{il_i}$, then the matrix J is invariant, modulo our original ordering of the eigenvalues.

Before we give some much needed examples, we have to prove one additional very useful result. I think this result is, at first glance, surprising. First we need a lemma, showing that extending the field of scalars of a vector space does not change the minimal and characteristic polynomials of a transformation.

Lemma 4.9.1 *Let $T \in \mathcal{L}_F(V)$ have minimal polynomial $m_T(x)$ and characteristic polynomial $c_T(x)$ and let F' be a field containing F. As in Example 4.8.2, extend V to the F'-vector space $V' = F' \otimes_F V$ and extend T to $T' = 1 \otimes T \in \mathcal{L}_{F'}(V')$. Let $m_{T'}(x)$ and $c_{T'}(x)$ be the minimal and characteristic polynomials of T'. Then $m_{T'}(x) = m_T(x)$ and $c'_T(x) = c_T(x)$.*

Proof. Recall that the map $\mathbf{w} \to 1 \otimes \mathbf{w}$ is a monomorphism from V into V' (Problem 4.4.6). It follows that, for any $q(x) \in F[x]$, $q(T) = 0 \in \mathcal{L}_F(V)$ if and only if $1 \otimes q(T) = q(1 \otimes T) = 0 \in \mathcal{L}_{F'}(V')$. Thus, extending T to T' does not change the minimal polynomial.

Let β be an F-basis for V. Then we know that $\beta' = 1 \otimes \beta$ is an F' basis for V' and that $[T]_\beta = [T']_{\beta'}$. Denote this common matrix by A and let I_n be the $n \times n$ identity matrix, where $n = \dim_F V = \dim_{F'} V'$. Then $c_{T'}(x) = \det(xI_n - A) = c_T(x)$. ∎

Theorem 4.9.1 *(Cayley-Hamilton) Let $T \in \mathcal{L}_F(V)$ have characteristic polynomial $c_T(x) \in F[x]$. Then $c_T(T) = 0$ or, equivalently $m_T(x) \mid c_T(x)$.*

Proof. Let $F \subset F'$, where F' is a field such that $m_T(x) = (x - c_1)^{e_1} \cdots (x - c_k)^{e_k}$ in $F'[x]$. (We will prove in the next chapter that we can always find such an F'. For the moment, we just ask you to accept this fact, since using it here makes for a much cleaner proof.) Extend V to the F'-vector space $V' = F' \otimes_F V$ and T to $T' = 1 \otimes T$. By the previous discussion at the beginning of the section, there exists a basis β' for V' such that $J = [T']_{\beta'}$ is in Jordan form.

Using the basis β' to compute $c_{T'}(x)$, we have $c_{T'}(x) = \det(xI_n - J)$. It is easy to check (or see Problem 4.7.4) that, for any matrix J in Jordan Form with notation as in the previous discussion , $\det(xI_n - J) = (x - c_1)^{f_1} \cdots (x - c_k)^{f_k}$, where $f_i = \sum_{j=1}^{l_i} e_{ij} \geq e_{i1} = e_i$ for each i. (Check this.) Hence, $m_{T'}(x) = \prod_{i=1}^{k} (x - c_i)^{e_i}$ divides $c_{T'}(x) = \prod_{i=1}^{k} (x - c_i)^{f_i}$ in $F'[x]$. We now apply Lemma 4.9.1 to complete the proof. ∎

Note that, in the course of proving our theorem, we have given a proof that **every** root of $m_T(x)$ shows up as a root of $c_T(x)$. Refer back to Lemma 4.8.2.

At last, some examples!

Example 4.9.1 *Let $T : \mathcal{R}^4 \to \mathcal{R}^4$ be given by the formula $T(a, b, c, d) = (a + b, -a - b, -2a - 2b + 2c + d, a + b - c)$. With respect to the standard basis T has matrix*

$$A = \begin{bmatrix} 1 & 1 & 0 & 0 \\ -1 & -1 & 0 & 0 \\ -2 & -2 & 2 & 1 \\ 1 & 1 & -1 & 0 \end{bmatrix}.$$

Expansion down the fourth column gives the characteristic polynomial $c_T(x) = \det(xI_4 - A) = x^2(x - 1)^2$. At this point, we know $m_T(x)$ will have both $0, 1$ as roots. Also, by the Cayley Hamilton Theorem, $m_T(x) \mid c_T(x)$. Hence, $m_T(x)$ must be one of the following polynomials: $x(x - 1), x^2(x - 1), x(x-1)^2$ or $x^2(x-1)^2$. Plugging A into the first three candidates doesn't produce the zero matrix. Thus, $m_T(x) = c_T(x) = x^2(x - 1)^2$. The Jordan Form J for T will consist of two blocks, $\begin{bmatrix} 0 & 0 \\ 1 & 0 \end{bmatrix}$ and $\begin{bmatrix} 2 & 0 \\ 1 & 2 \end{bmatrix}$. We have

$$J = \begin{bmatrix} 0 & 0 & 0 & 0 \\ 1 & 0 & 0 & 0 \\ 0 & 0 & 2 & 0 \\ 0 & 0 & 1 & 2 \end{bmatrix}.$$

For a basis β such that $[T]_\beta = J$, we can use $\{\mathbf{v}, T(\mathbf{v}), \mathbf{w}, (T-1)\mathbf{w}\}$, where $\mathbf{v}, \mathbf{w}$ are any vectors such that $T^2(\mathbf{v}) = \mathbf{0} \neq T(\mathbf{v}), (T-1)^2(\mathbf{w}) = \mathbf{0} \neq (T-1)(\mathbf{w})$. One can use either the operator T or the matrix A and look at the appropriate linear systems to find suitable $\mathbf{v}, \mathbf{w}$.

Exercise 4.9.1 *Find suitable vectors $\mathbf{v}, \mathbf{w}$.*

Example 4.9.2 *Let V be the real vector space of real polynomials in y of degree ≤ 4. By freshman calculus, D, the differentiation operator, is an element of $\mathcal{L}_\mathcal{R}(V)$. The matrix of D with respect to the basis $\beta = \{1, y, y^2, y^3, y^4\}$ is*

$$A = \begin{bmatrix} 0 & 1 & 0 & 0 & 0 \\ 0 & 0 & 2 & 0 & 0 \\ 0 & 0 & 0 & 3 & 0 \\ 0 & 0 & 0 & 0 & 4 \\ 0 & 0 & 0 & 0 & 0 \end{bmatrix}.$$

Example 4.9.3 *We have $c_D(x) = x^5$ so that D is nilpotent and has the single characteristic value 0. Since $m_D(x) \mid c_D(x)$, to find $m_D(x)$ we only need compute powers of A or D to see the first time we get zero. Because $D^4(y^4) = 24 \neq 0$, we have $m_D(x) = x^5$.*

Thus, as an $F[x]$-module V will be a cyclic submodule generated by an element $\mathbf{v}$ of order x^5. The vector $\mathbf{v} = y^4$ will do nicely. The basis γ with respect to which D will be in Jordan form is $\gamma = \{y^4, 4y^3, 12y^2, 24y, 24\}$. To obtain the matrix P such that $[D]_\gamma = P^{-1}[D]_\beta P$ we take the matrix whose columns are the coordinates of the vectors in γ with respect to β. Thus,

$$J = \begin{bmatrix} 0 & 0 & 0 & 0 & 0 \\ 1 & 0 & 0 & 0 & 0 \\ 0 & 1 & 0 & 0 & 0 \\ 0 & 0 & 1 & 0 & 0 \\ 0 & 0 & 0 & 1 & 0 \end{bmatrix}, \quad P = \begin{bmatrix} 0 & 0 & 0 & 0 & 24 \\ 0 & 0 & 0 & 24 & 0 \\ 0 & 0 & 12 & 0 & 0 \\ 0 & 4 & 0 & 0 & 0 \\ 1 & 0 & 0 & 0 & 0 \end{bmatrix}.$$

There is one more standard canonical form, called the **rational canonical form**, used when we don't want to go to an extension field to completely factor $m_T(x)$.

Example 4.9.4 *Suppose $T : Q^6 \to Q^6$ has $c_T(x) = m_T(x) = (x^4 + x^2 + 1)(x^2 - 2)$. Since $x^4 + x^2 + 1$ and $x^2 - 2$ are irreducible in $Q[x]$, we have a $Q[x]$-module decomposition, $Q^6 = Q[x]\mathbf{v} \oplus Q[x]\mathbf{w}$, where $\mid \mathbf{v} \mid = x^4 + x^2 + 1$*

and $\mid \mathbf{w} \mid = x^2 - 2$. *In fact, since the orders of* $\mathbf{v}, \mathbf{w}$ *are relatively prime, then the element* $\mathbf{u} = (\mathbf{v} \oplus \mathbf{w}) \in Q[x]\mathbf{v} \bigoplus Q[x]\mathbf{w}$ *will have order* $(x^4 + x^2 + 1)(x^2 - 2) = x^6 - x^4 - x^2 - 2$. *It follows that* $_{F[x]}Q^6 = F[x]\mathbf{u}$. *Consider the basis* $\beta = \{\mathbf{u}, T\mathbf{u}, T^2\mathbf{u}, T^3\mathbf{u}, T^4\mathbf{u}, T^5\mathbf{u}\}$. *We have*

$$[T]_\beta = \begin{bmatrix} 0 & 0 & 0 & 0 & 0 & 2 \\ 1 & 0 & 0 & 0 & 0 & 0 \\ 0 & 1 & 0 & 0 & 0 & 1 \\ 0 & 0 & 1 & 0 & 0 & 0 \\ 0 & 0 & 0 & 1 & 0 & 1 \\ 0 & 0 & 0 & 0 & 1 & 0 \end{bmatrix}.$$

For the last column, note that $T(T^5\mathbf{u}) = T^6\mathbf{u} = 2\mathbf{u} + T^2\mathbf{u} + T^4\mathbf{u}$. *Here the rational form consists of a single "block" corresponding to one element* $\mathbf{u}$, *a generator of maximal order of a cyclic direct summand of the module* $_{F[x]}Q^6$. *We call this order the first (and in this case the only)* **elementary divisor** *of* T.

Here is an example to show what happens in computing a rational form with more than one block or elementary divisor. It's the sort of computation that everyone should do once, but certainly not more than twice.

Example 4.9.5 *Suppose we have an operator* $T \in \mathcal{L}_F(V)$ *whose minimal polynomial* $m_T(x)$ *has an irreducible factorization in* $F[x]$ *of the form* $m_T(x) = (x^2 + 2)^3(\,x^2 - 2)^2(x - 1)$. *Further, suppose that* $\dim_F V = 16$. *Given this information, we examine the possibilities for the rational form of* T.

First, looking back at Theorem 3.5.2, we see that the module $_{F[x]}V$ *must have cyclic summands* $F[x]\mathbf{u}, F[x]\mathbf{v}, F[x]\mathbf{w}$ *generated by elements* $\mathbf{u}, \mathbf{v}, \mathbf{w}$ *of prime power orders* $(x^2 + 2)^3, (\,x^2 - 2)^2, (x - 1)$. *Otherwise,* $m_T(x)$ *would not be of the given form. As in the previous example, since these orders are relatively prime, the sum* $\mathbf{s} = \mathbf{u} + \mathbf{v} + \mathbf{w}$ *will be an element of order*
$(x^2 + 2)^3(x^2 - 2)^2(x - 1) = x^{11} - x^{10} + 2x^9 - 2x^8 - 8x^7 + 8x^6 - 16x^5 + 16x^4 + 16x^3 - 16x^2 + 32x - 32$.

Note that $F[x]\mathbf{s}$ *will be a* T-*invariant,* $F[x]$-*cyclic summand of* $_{F[x]}V$ *of* F-*dimension 11. Also note, by the prime power decomposition of* $_{F[x]}V$ *and the form of* $m_T(x)$, *that* $F[x]\mathbf{s}$ *will be a summand with* $\mid \mathbf{s} \mid$ *maximal among the set of generators of cyclic summands of* $_{F[x]}V$.

As in the previous example, we use the basis $\beta_1 = \{\mathbf{s}, T\mathbf{s}, ..., T^{10}\mathbf{s}\}$ *to produces the first block of the rational form, namely the matrix of* T *re-*

stricted to the invariant summand $F[x]\mathbf{s}$. Our first rational block is

$$R_1 = \begin{bmatrix} 0 & 0 & 0 & 0 & 0 & 0 & 0 & 0 & 0 & 0 & 32 \\ 1 & 0 & 0 & 0 & 0 & 0 & 0 & 0 & 0 & 0 & -32 \\ 0 & 1 & 0 & 0 & 0 & 0 & 0 & 0 & 0 & 0 & 16 \\ 0 & 0 & 1 & 0 & 0 & 0 & 0 & 0 & 0 & 0 & -16 \\ 0 & 0 & 0 & 1 & 0 & 0 & 0 & 0 & 0 & 0 & -16 \\ 0 & 0 & 0 & 0 & 1 & 0 & 0 & 0 & 0 & 0 & 16 \\ 0 & 0 & 0 & 0 & 0 & 1 & 0 & 0 & 0 & 0 & -8 \\ 0 & 0 & 0 & 0 & 0 & 0 & 1 & 0 & 0 & 0 & 8 \\ 0 & 0 & 0 & 0 & 0 & 0 & 0 & 1 & 0 & 0 & 2 \\ 0 & 0 & 0 & 0 & 0 & 0 & 0 & 0 & 1 & 0 & -2 \\ 0 & 0 & 0 & 0 & 0 & 0 & 0 & 0 & 0 & 1 & 1 \end{bmatrix}.$$

We need some more invariant subspaces to fill up the remaining 5 dimensions. There are several possibilities. In computing the rational form we take the second invariant subspace to be a cyclic summand $F[x]\mathbf{t}$, disjoint from $F[x]\mathbf{s}$, such that $\mid \mathbf{t} \mid$, the second invariant factor, is maximal among such remaining summands. The vector $\mathbf{t}$ will, as above, be a sum of cyclic generators of prime power orders, obtained via Theorem 3.5.1. Hence, $\mid \mathbf{t} \mid = (x^2 + 2)^e (x^2 - 2)^f (x - 1)^g$ with $e \le 3, f \le 2, g \le 1$. But $\dim_F F[x]\mathbf{t} \le 5$. A little thought reveals that the possibilities for the second invariant factor are as follows. First, we cannot have $e = 3$, since that would give rise to a summand $F[x]\mathbf{t}$ of F-dimension 6.

If $e = 2$, the second invariant factor could be either (a) $(x^2+2)^2(x-1) = x^5 - x^4 + 4x^3 - 4x^2 + 4x - 4$ or possibly (b) $(x^2 + 2)^2 = x^4 + 4x^2 + 4$. In case (a), we would have an $F[x]$-module direct sum decomposition $_{F[x]}V = F[x]\mathbf{s} \oplus F[x]\mathbf{t}$ and a rational form $R = \begin{bmatrix} R_1 & 0 \\ 0 & R_2 \end{bmatrix}$, where

$$R_2 = \begin{bmatrix} 0 & 0 & 0 & 0 & 4 \\ 1 & 0 & 0 & 0 & -4 \\ 0 & 1 & 0 & 0 & 4 \\ 0 & 0 & 1 & 0 & -4 \\ 0 & 0 & 0 & 1 & 1 \end{bmatrix}.$$

As above, the last column comes from the second invariant factor. A little more thought reveals that possibility (b) cannot actually occur. For then $\dim_F F[x]\mathbf{t} = 4$ so that, to complete our decomposition of $_{F[x]}V$, we would need one additional $F[x]$-cyclic summand, say $F[x]\mathbf{l}$, of F-dimension one. However, using our procedure, this could only happen if $\mid \mathbf{l} \mid = x-1$. But then $\mid \mathbf{t} + \mathbf{l} \mid = (x^2 + 2)^2(x - 1)$, which puts us back in case (a). (Remember that, when computing the invariant factors, we take the maximal possibility at each step.)

Similarly, $f = 2$ gives rise to only one possible second invariant factor,
$(x^2-2)^2(x-1) = x^5 - x^4 - 4x^3 + 4x^2 + 4x - 4$. *In this case, the corresponding
second and final block of our rational form is*

$$R_2 = \begin{bmatrix} 0 & 0 & 0 & 0 & 4 \\ 1 & 0 & 0 & 0 & -4 \\ 0 & 1 & 0 & 0 & -4 \\ 0 & 0 & 1 & 0 & 4 \\ 0 & 0 & 0 & 1 & 1 \end{bmatrix}.$$

*If either $e = 1$ or $f = 1$, the second invariant factor could be (c) $(x^2 +
2)(x^2 - 2)(x-1) = x^5 - x^4 - 4x + 4$ or (d) any proper divisor of (c). In case
(c), the rational form R again is a matrix with two blocks R_1, R_2, where*

$$R_2 = \begin{bmatrix} 0 & 0 & 0 & 0 & -4 \\ 1 & 0 & 0 & 0 & 4 \\ 0 & 1 & 0 & 0 & 0 \\ 0 & 0 & 1 & 0 & 0 \\ 0 & 0 & 0 & 1 & 1 \end{bmatrix}.$$

*Case (d) consists of four subcases, all resulting in more than two in-
variant factors. The possibilities for the additional invariant factors are
(d1) $(x^2 + 2)(x - 1), (x - 1), (x - 1)$ or (d2) $(x^2 + 2)(x - 1), (x^2 + 2)$ or
(d3) $(x^2 - 2)(x - 1), (x - 1), (x - 1)$ or (d4) $(x^2 - 2)(x - 1), (x^2 - 2)$. Problem 4.9.1 below asks you to explain why only these can be the remaining
invariant factors and to give the corresponding rational forms.*

Twenty-eighth Problem Set

Problem 4.9.1 *Finish Example 4.9.4 by working through case (d).*

Problem 4.9.2 *Let $T : V \to V$ have minimal polynomial $(x - 1)^2$ and
characteristic polynomial $(x - 1)^5$. What are the possibilities for the Jordan
form for T?*

Problem 4.9.3 *Let $T : C^6 \to C^6$ be given by the formula $T(u, v, w, x, y, z) =
(2u, u + 2v, -u + 2w, v + 2x, u + v + w + x + 2y, -z)$. Find the Jordan and
rational forms for T. Find the similarity matrices P that transform $[T]_{St}$
into the Jordan and rational forms.*

4.10   Dual spaces

In this short section we simply give the basic definitions and assign some standard (and not hard to prove) results as problems.

Definition 4.10.1 *Let V be a vector space of dimension n over a field F. The **dual space** of V, denoted by V^*, is the set of all linear transformations from V to F, F being regarded as a vector space over itself via left multiplication. The elements $f \in V^*$ are called **linear functionals** on V. If we define addition in V^* by the formula $(f + g)(\mathbf{v}) = f(\mathbf{v}) + g(\mathbf{v})$ and scalar multiplication by the formula $(cf)(\mathbf{v}) = c[f(\mathbf{v})]$, then it is easy to check that V^* becomes a vector space over F.*

Definition 4.10.2 *Let $\beta = \{\mathbf{v}_1, ..., \mathbf{v}_n\}$ be a basis for V over F. For $1 \leq i \leq n$, define a linear transformation $\mathbf{v}_i^* : V \to F$ by the requirement that $\mathbf{v}_i^*(\mathbf{v}_i) = 1, \mathbf{v}_i^*(\mathbf{v}_j) = 0, j \neq i$. The set $\beta^* = \{\mathbf{v}_1^*, ..., \mathbf{v}_n^*\} \subset V^*$ is called the **dual basis** of the basis β.*

Definition 4.10.3 *Let V, W be vector spaces over F and $T : V \to W$ be a linear transformation. Define $T^* : W^* \to V^*$ by the equation $T^*(\alpha)(\mathbf{v}) = (\alpha \circ T)(\mathbf{v})$ for $\alpha \in W^*, \mathbf{v} \in V$. The map T^* is called the **dual transformation** of the transformation T.*

Here is a picture, illustrating what's happening:

$$
\begin{array}{ccc}
V & \xrightarrow{T} & W \\
T^*\alpha \searrow & & \swarrow \alpha \\
& F &
\end{array}
$$

We will prove a little something.

Theorem 4.10.1 *The map T^* defined above is a linear transformation from W^* to V^*.*

Proof. Certainly, for each $\alpha \in W^*$, $T^*(\alpha)$ is a function from V into F. We claim that $T^*(\alpha)$ is a linear transformation from V into F; that is $T^*(\alpha) \in V^*$. This follows since: $T^*(\alpha)[c\mathbf{v}_1 + \mathbf{v}_2] = \alpha[T(c\mathbf{v}_1 + \mathbf{v}_2)] = \alpha[cT(\mathbf{v}_1) + T(\mathbf{v}_2)] = c\alpha T(\mathbf{v}_1) + \alpha T(\mathbf{v}_2) = cT^*(\alpha)[\mathbf{v}_1] + T^*(\alpha)[\mathbf{v}_2]$. Furthermore, we claim that the map T^* is a linear transformation from W^* to V^*. To see this we evaluate $T^*(c\alpha + \beta)$ for arbitrary elements $\alpha, \beta \in W^*$ and $c \in F$. Let $\mathbf{v} \in V$. Then: $T^*(c\alpha + \beta)[\mathbf{v}] = (c\alpha + \beta)T[\mathbf{v}] = c\alpha T[\mathbf{v}] + \beta T[\mathbf{v}] = cT^*(\alpha)[\mathbf{v}] + T^*(\beta)[\mathbf{v}]$. Since $\mathbf{v} \in V$ was also arbitrary, we have $T^*(c\alpha + \beta) = cT^*(\alpha) + T^*(\beta)$. We have shown that the map $T^* : W^* \to V^*$ is linear. ∎

Twenty-ninth Problem Set

Problem 4.10.1 *Let β be a basis for V. Prove that β^* is a basis for V^*.*

Problem 4.10.2 *For $\beta = \{\mathbf{v}_1, ..., \mathbf{v}_n\}$ a basis for V with dual basis $\beta^* = \{\mathbf{v}_1^*, ..., \mathbf{v}_n^*\}$ prove that, for each $\mathbf{v} \in V$, $\mathbf{v} = \sum_{i=1}^{n} \mathbf{v}_i^*(\mathbf{v})\mathbf{v}_i$.*

Problem 4.10.3 *Let V, W be vector spaces over F with bases β, γ respectively. Suppose that $[T]_{\beta\gamma} = A$. Prove that $[T^*]_{\gamma^*\beta^*} = A^t$ (the transpose of A).*

Problem 4.10.4 *Prove that the evaluation map $e : V \to (V^*)^*$ given by $e(\mathbf{v})[\alpha] = \alpha(\mathbf{v})$ for $\mathbf{v} \in V$, $\alpha \in V^*$ is an isomorphism.*

4.11   Inner product spaces[*]

In the final two sections of this chapter, we will consider only finite dimensional real or complex vector spaces. (Although it makes perfectly good sense to consider inner products in more abstract settings, in particular for infinite dimensional spaces.) Accordingly, $_FV$ will always be finite dimensional and F will be either the field of real or complex numbers. We closely follow the treatment in [HK]. The material in these sections will not be needed in the rest of the text. Some students may have seen it already in an undergraduate course. The instructor short on time may wish to omit these sections.

Definition 4.11.1 *An **inner product** on $_FV$ is a function $<,>: V \times V \to F$ such that, for all vectors $\mathbf{u}, \mathbf{v}, \mathbf{w} \in V$ and scalars $c \in F$,*

   1. $< \mathbf{u} + \mathbf{v}, \mathbf{w} > = < \mathbf{u}, \mathbf{w} > + < \mathbf{v}, \mathbf{w} >$;

   2. $< c\mathbf{u}, \mathbf{v} > = c < \mathbf{u}, \mathbf{v} >$;

   3. $< \mathbf{u}, \mathbf{v} > = \overline{< \mathbf{v}, \mathbf{u} >}$, the bar denoting complex conjugation;

   4. $< \mathbf{v}, \mathbf{v} >> 0$ if $\mathbf{v} \neq \mathbf{0}$ (note, by (3), $< \mathbf{v}, \mathbf{v} > \in \mathcal{R}$ for each vector $\mathbf{v} \in V$).

Conditions 1,2,3 together imply that $< \mathbf{u}, c\mathbf{v} + \mathbf{w} > = \bar{c} < \mathbf{u}, \mathbf{v} > + < \mathbf{u}, \mathbf{w} >$ for all $c, \mathbf{u}, \mathbf{v}, \mathbf{w}$. (See Problem 4.11.1.)

An **inner product space** is just a vector space with an inner product (surprise).

Exercise 4.11.1 *Prove that for* $\mathbf{v}, \mathbf{w} \in V$, *an inner product space,* $< 0, \mathbf{w} >=< \mathbf{v}, 0 >= 0$.

Example 4.11.1 *On* F^n *we have the* **standard inner product** *defined by the formula* $< (a_1, ..., a_n), (b_1, ..., b_n) >= \sum_{i=1}^{n} a_i \bar{b}_i$. *In the real case, of course, conjugation is unnecessary and the standard inner product is often called the* **dot product**.

Exercise 4.11.2 *Verify that the standard inner product is an inner product.*

Example 4.11.2 *In* R^2 *define* $< (a_1, a_2), (b_1, b_2) >= a_1 b_1 - b_2 a_1 - a_1 b_2 + 4 a_2 b_2$. *We claim that this definition provides an inner product. The first three inner product axioms are easy to check. For property 4, note that* $< (a_1, a_2), (a_1, a_2) >= (a_1 - a_2)^2 + 3 a_2^2$.

Definition 4.11.2 *Let* $A = [\alpha_{ij}]$ *be a complex matrix. Then the matrix* A^* *whose i-j entry is* $\bar{a}_{ji}$ *is called (for obvious reasons) the* **conjugate transpose** *of* A. *The matrix* A *is* **Hermitian** *if* $A = A^*$. *Thus, a real Hermitian matrix will simply be a matrix* A *such that* $A = A^t$. *We say that such an* A *is* **symmetric**.

Definition 4.11.3 *Let* $<, >$ *be an inner product on* $_F V$ *and* $\beta = \{\mathbf{v}_1, .., \mathbf{v}_n\}$ *be an (ordered) basis for* V *over* F. *Define a matrix* $A_\beta = [\alpha_{ij}] \in M_n(F)$, *called the* **matrix of** $<, >$ **with respect to** β, *by the stipulations* $\alpha_{ij} =< \mathbf{v}_j, \mathbf{v}_i >$.

By property 3 of the inner product, A_β is a Hermitian matrix. The next theorem, analogous to Theorem 4.6.1, makes explicit the relation between an inner product and its matrix. For $(b_1, ..., b_n) \in F^n$, let $(b_1, ..., b_n)^*$ be the **conjugate transpose** of $(b_1, ..., b_n)$, that is the $n \times 1$ column vector with entries $\bar{b}_1, ..., \bar{b}_n$.

Theorem 4.11.1 *Let* $<, >$ *be an inner product on* $_F V$ *and let* A_β *be the matrix of* $<, >$ *with respect to a basis* β. *Suppose* $\mathbf{v}, \mathbf{w} \in V$ *have* β-*coordinate rows* $[\mathbf{v}]_\beta = (a_1, ..., a_n)$, $[\mathbf{w}]_\beta = (b_1, ..., b_n)$. *Then* $< \mathbf{v}, \mathbf{w} >= (a_1, ..., a_n) A_\beta (b_1, ..., b_n)^*$.

Proof. The proof is a fairly simple computation, manipulating sums and using the properties of the inner product. With notation as above:
$$(\Delta) < \mathbf{v}, \mathbf{w} >=< \sum_{i=1}^{n} a_i \mathbf{v}_i, \sum_{j=1}^{n} b_j \mathbf{v}_j >= \sum_{i=1}^{n} a_i < \mathbf{v}_i, \sum_{j=1}^{n} b_j \mathbf{v}_j >= \sum_{i,j=1}^{n} a_i \bar{b}_j < \mathbf{v}_i, \mathbf{v}_j >.$$

Since $< \mathbf{v}_i, \mathbf{v}_j >= \alpha_{ji}$, a simple computation reveals that the final sum in (Δ) is equal to $(a_1, ..., a_n) A_\beta (b_1, ..., b_n)^*$. ∎

There is one big difference in the inner product setting as opposed to that of linear transformations. It is easy to see that, for any $n \times n$ Hermitian matrix A and basis β, the function $<,> : V \times V \to F$ defined by $< \mathbf{v}, \mathbf{w} >= [\mathbf{v}]_\beta A [\mathbf{w}]_\beta^*$ will satisfy properties 1-3 of the inner product definition. (Recall that F is either the real or complex number field.) But, we also need property 4.

Definition 4.11.4 *A matrix $A \in M_n(\mathcal{C})$ is* ***positive definite*** *if $xAx^* > 0$ for all nonzero row vectors $x \in C^n$.*

Remark 4.11.1 *Any positive definite matrix is invertible. We assign the proof as Problem 4.11.2.*

Exercise 4.11.3 *Prove that every diagonal entry in a positive definite Hermitian matrix must be real and positive.*

We now state the converse to Theorem 4.11.1.

Corollary 4.11.1 *Any inner product arises via the formula Δ of Theorem 4.11.1 for some basis β and some positive definite Hermitian matrix A.*

Proof. See Problem 4.11.3. ∎

Definition 4.11.5 *A set of nonzero vectors $\{\mathbf{v}_1, ..., \mathbf{v}_k\}$ in an inner product space is called* ***orthogonal*** *if $< \mathbf{v}_i, \mathbf{v}_j >= 0$ for all $i \neq j$. If, in addition, $< \mathbf{v}_i, \mathbf{v}_i >= 1$ for $1 \leq i \leq k$, an orthogonal set is called* ***orthonormal***.

The following theorem is classical. It illustrates what is called the **Gram-Schmidt orthonormalization process**.

Theorem 4.11.2 *Any basis $\{\mathbf{v}_1, ..., \mathbf{v}_n\}$ of an inner product space can be transformed into an orthonormal basis $\{\mathbf{u}_1, ..., \mathbf{u}_n\}$ such that the linear span of $\{\mathbf{u}_1, ..., \mathbf{u}_j\}$ is equal to the linear span of $\{\mathbf{v}_1, ..., \mathbf{v}_j\}$ for all $1 \leq j \leq n$.*

Proof. Define $\mid \mathbf{v} \mid$ called the **norm** of the vector $\mathbf{v}$, by the formula $\mid \mathbf{v} \mid = \sqrt{< \mathbf{v}, \mathbf{v} >}$. (We note that $\mid \mathbf{v} \mid$ plays the role of magnitude or length of the vector $\mathbf{v}$ and has all the properties we want for the magnitude of a vector. In particular, for each $\mathbf{v} \in V$, $\mid \mathbf{v} \mid$ is a nonnegative real number and $\mid \mathbf{v} \mid = 0$ if and only if $\mathbf{v} = \mathbf{0}$. (See the exercise immediately following this theorem and also Problem 4.11.6.)

We inductively define an **orthonormal basis** $\gamma = \{w_1, ..., w_n\}$ as follows. Since v_1 is nonzero, $\mid v_1 \mid > 0$. Put $u_1 = (1/\mid v_1 \mid)v_1$. Then $\mid u_1 \mid = 1$ (Exercise 4.11.4 below), so that the set $\{u_1\}$ is orthonormal. Suppose $j < n$ and an orthonormal set $\{u_1, ..., u_j\}$ has been defined such that the linear span of $\{u_1, ..., u_j\}$ is equal to the linear span of $\{v_1, ..., v_j\}$. Let $w_{j+1} = v_{j+1} - \sum_{i=1}^{j} < v_{j+1}, u_i > u_i$. For each index $l \leq j$,

$$< w_{j+1}, u_l > = < v_{j+1}, u_l > - \sum_{i=1}^{j} < v_{j+1}, u_i > < u_i, u_l > = < v_{j+1}, u_l > - < v_{j+1}, u_l > = 0.$$

Hence, the set $\{u_1, ..., u_j, w_{j+1}\}$ is orthogonal. Note that $w_{j+1} \neq 0$ since $v_{j+1} \notin span\{v_1, ..., v_j\} = span\{u_1, ..., u_j\}$. Now put $u_{j+1} = (1/\mid w_{j+1} \mid)w_{j+1}$. The set $\{u_1, ..., u_{j+1}\}$ will be as desired. ∎

Exercise 4.11.4 *Check that $\mid cv \mid = \mid c \mid \mid v \mid$ for each $v \in V, c \in F$. Here $\mid c \mid$ is the real or complex absolute value of c. Then check that, for nonzero vectors v, the vector $(1/\mid v \mid)v$ will have norm one. For a nice proof of the usual triangle equality for norms in a real inner product space, see Problem 4.11.6.*

Thirtieth Problem Set

Problem 4.11.1 *Prove that Conditions 1,2,3 of Definition 4.11.1 imply that $< u, cv + w > = \bar{c} < u, v > + < u, w >$ for all c, u, v, w.*

Problem 4.11.2 *Prove that any positive definite real or complex matrix is invertible.*

Problem 4.11.3 *Prove Corollary 4.11.1.*

Problem 4.11.4 *Find necessary and sufficient conditions for a real 2×2 matrix to be positive definite and Hermitian.*

Problem 4.11.5 *Prove that any orthogonal set is linearly independent.*

Problem 4.11.6 *Let v, w be vectors in a real inner product space. Define $q(x) = < v + xw, v + xw >$, regarding x as an arbitrary scalar. Then $q(x)$ can be regarded as a quadratic polynomial in $\mathcal{R}[x]$. Note that, for all x, $q(x) \geq 0$. Now use the quadratic formula to prove the triangle inequality: $\mid v + w \mid \leq \mid v \mid + \mid w \mid$.*

Problem 4.11.7 *Let $\{u_1, ..., u_n\}$ be an orthonormal basis for the inner product space V. Prove that, for all $v \in V$, $v = \sum_{i=1}^{n} < v, u_i > u_i$.*

Problem 4.11.8 *Let V be the real vector space of all $p(x) \in \mathcal{R}[x]$ with $\deg p(x) \leq 3$. Define an inner product on V by the formula $< p(x), q(x) >= \int_0^1 p(x)q(x)dx$. a) Verify the inner product axioms for this formula. b) Carry out the Gramm-Schmidt process for the basis $\{1, x, x^2, x^3\}$.*

Problem 4.11.9 *Let $_F V$ be an inner product space and W a subspace. Define $W^\perp = \{\mathbf{v} \in V :< \mathbf{v}, \mathbf{w} >= 0 \text{ for all } \mathbf{w} \in W\}$. a) Prove that $W^\perp$ is a subspace of W. b) Prove that $V = W \oplus W^\perp$.*

4.12   Linear functionals and adjoints*

As in the previous section, V is a finite dimensional real or complex inner product space. In this case, there is a simple relation between elements of the dual space $V^* = L_F(V, F)$ and the inner product.

Theorem 4.12.1 *Let $f \in V^*$. Then there is a unique vector $\mathbf{w} \in V$ such that $f(\mathbf{v}) =< \mathbf{v}, \mathbf{w} >$ for all $\mathbf{v} \in V$.*

   Proof. Let $\{\mathbf{u}_1, ..., \mathbf{u}_n\}$ be an orthonormal basis for V. We take $\mathbf{w} = \sum_{i=1}^n f(\bar{\mathbf{u}}_i)\mathbf{u}_i$ and let $\mathbf{v} \in V$ be arbitrary. By Problem 4.11.7, $\mathbf{v} = \sum_{i=1}^n < \mathbf{v}, \mathbf{u}_i > \mathbf{u}_i$. Hence, $f(\mathbf{v}) = \sum_{i=1}^n < \mathbf{v}, \mathbf{u}_i > f(\mathbf{u}_i)$. On the other hand, $< \mathbf{v}, \mathbf{w} >=< \mathbf{v}, \sum_{i=1}^n f(\bar{\mathbf{u}}_i)\mathbf{u}_i >= \sum_{i=1}^n f(\mathbf{u}_i) < \mathbf{v}, \mathbf{u}_i >= \sum_{i=1}^n < \mathbf{v}, \mathbf{u}_i > f(\mathbf{u}_i)$. Hence, our vector $\mathbf{w}$ is as desired.

   If $\mathbf{w}'$ were another such vector satisfying the claim of the theorem, necessarily $< \mathbf{v}, \mathbf{w} >=< \mathbf{v}, \mathbf{w}' >$ for all $\mathbf{v} \in V$. But then $< \mathbf{v}, \mathbf{w} - \mathbf{w}' >= 0$ for all $\mathbf{v} \in V$. In particular, $\mid \mathbf{w} - \mathbf{w}' \mid^2 =< \mathbf{w} - \mathbf{w}', \mathbf{w} - \mathbf{w}' >= 0$ so that $\mathbf{w} - \mathbf{w}' = 0$ and $\mathbf{w} = \mathbf{w}'$. $\blacksquare$

Thus, all linear functionals on finite dimensional real or complex inner product spaces are induced by inner products. We turn to a related concept.

Definition 4.12.1 *Let $T \in \mathcal{L}_F(V)$. A linear operator $T^* \in \mathcal{L}_F(V)$ is called the **adjoint** of T if $< T\mathbf{v}, \mathbf{w} >=< \mathbf{v}, T^*\mathbf{w} >$ for all $\mathbf{v}, \mathbf{w} \in V$.*

It is not clear at the moment which operators have an adjoint. Arguing as in the last paragraph of Theorem 412.1, if an adjoint for T exists, then this adjoint is unique. This justifies the use of the word "the" and the notation T^*.

Exercise 4.12.1 *Write a few lines justifying the claim that, if T^* exists, then it is unique.*

We next prove that, under our standing assumptions on V, each operator has an adjoint. **However**, see Problem 4.12.1 below. First we need a simple lemma.

Lemma 4.12.1 *Let* $\beta = \{\mathbf{u}_1, ..., \mathbf{u}_n\}$ *be an orthonormal basis for* $V, T \in \mathcal{L}_F(V)$ *and* $A = [\alpha_{ij}] = [T]_\beta$. *Then* $\alpha_{ij} = <T(\mathbf{u}_j), \mathbf{u}_i >$ *for all* i, j.

Proof. The matrix A is defined by the equations $T(\mathbf{u}_j) = \sum_{i=1}^n \alpha_{ij}\mathbf{u}_i$. But also, since β is orthonormal, $T(\mathbf{u}_j) = \sum_{i=1}^n <T(\mathbf{u}_j), \mathbf{u}_i > \mathbf{u}_i$. Hence, $\alpha_{ij} = <T(\mathbf{u}_j), \mathbf{u}_i >$ for all i, j. ∎

Theorem 4.12.2 *Every linear operator* T *on a finite dimensional real or complex inner product space* V *has an adjoint.*

Proof. Let V be such a space and let $T \in \mathcal{L}_F(V)$. Take $\beta = \{\mathbf{u}_1, ..., \mathbf{u}_n\}$ to be an orthonormal basis for V and put $A = [T]_\beta$. The adjoint T^* will then be the linear transformation uniquely defined by the condition $[T^*]_\beta = A^*$. It suffices to verify the adjoint condition for T^* on the members of β. Let $\mathbf{u}_i, \mathbf{u}_j \in \beta$. By Lemma 4.12.1, $< T\mathbf{u}_j, \mathbf{u}_i >= \alpha_{ij}$. On the other hand, since $[T^*]_\beta = A^*$, $< \mathbf{u}_j, T^*\mathbf{u}_i >=< \mathbf{u}_j, \sum_{t=1}^n \bar{a}_{it}\mathbf{u}_t >= \sum_{t=1}^n a_{it} < \mathbf{u}_j, \mathbf{u}_t >= \alpha_{ij}$, and the proof is complete. ∎

Exercise 4.12.2 *Let* $T : C^3 \to C^3$ *be the operator whose matrix with respect to the standard basis is*

$$\begin{bmatrix} -1 & i & 0 \\ 1-i & 2 & 3 \\ 1+i & 0 & 1 \end{bmatrix}.$$

(a) Find a formula for T^*. *(b) Check that* $< T\mathbf{v}, \mathbf{w} >=< \mathbf{v}, T^*\mathbf{w} >$ *for all* $\mathbf{v}, \mathbf{w} \in C^3$.

The adjoint map satisfies some standard and expected properties. See Problem 4.12.2.

We call a linear operator T **Hermitian** if its matrix with respect to some orthonormal basis is a Hermitian matrix. Put another way, we are simply saying that $T = T^*$. Thus, T is Hermetian if and only if $[T]_\beta$ is Hermetian with respect to any orthonormal basis β. The big theorem of this section is that every Hermitian operator on V is diagonalizable. To prove this we need a few preliminary results.

Theorem 4.12.3 *Let* T *be a Hermitian operator on* V. *Then each characteristic value of* T *is real and characteristic vectors associated with distinct characteristic values are orthogonal.*

Proof. Suppose $c \in C$ is such that $Tv = cv$ for some nonzero $v \in V$. Then $c < v, v >=< cv, v >=< Tv, v >=< v, T^*v >=< v, Tv >=< v, cv >= \bar{c} < v, v >$. Since $v \neq 0$, $< v, v >> 0$ so that $c = \bar{c}$; that is $c \in \mathcal{R}$.

Now suppose $Tw = dw$ with $w \neq 0$ and $d \neq c$. Then $c < v, w >=< cv, w >=< Tv, w >=< v, Tw >=< v, dw >= d < v, w >$. Since $d \neq c$, we must have $< v, w >= 0$. ∎

Lemma 4.12.2 *Let* $T \in \mathcal{L}_F(V)$ *with* $0 < \dim_F V < \infty$. *Then* T *has a characteristic value in* C.

Proof. This lemma has nothing to do with V being an inner product space or T being Hermitian, simply with the fact that V is finite dimensional. Let $n = \dim_F V$. Then the characteristic polynomial $c_T(x)$ is a non-constant polynomial in $C[x]$. Now we need to use the fact that the field of complex numbers is **algebraically closed**, that is every non-constant polynomial in $C[x]$ has a root in C. (Equivalently, every polynomial in $C[x]$ of degree $n \geq 1$ factors in $C[x]$ as a product of n linear factors.) Hence, $c_T(x)$ has a root $c \in C$. By Theorem 4.8.2, c is a characteristic value for T. ∎

The fact that the field C is algebraically closed is called the **Fundamental Theorem of Algebra**. There are at least six different proofs. The easiest one uses complex variables. The delightful little book [FR] has a nice discussion of these proofs together with the background material needed for them.

We are now ready for our main theorem.

Theorem 4.12.4 *Let* T *be a Hermitian operator on a real or complex inner product space* V *with* $0 < \dim_F V < \infty$. *Then there is an orthonormal basis* β *such that* $[T]_\beta$ *is diagonal with real entries.*

Proof. The proof is by induction on $\dim_F V$. Let c be a characteristic value for T and v an associated characteristic vector. Note that, by Theorem 4.12.3, $c \in \mathcal{R}$ and, by definition, $v \neq 0$. Put $u_1 = v / \mid v \mid$. If $\dim_F V = 1$, $\{u_1\}$ is the desired basis. Now suppose $\dim_F V = n > 1$ and that the theorem is true for inner product spaces of dimension less than n. Let U_1 be the one dimensional subspace of V spanned by u_1. Since $Tu_1 = cu_1$, U_1 is invariant under T. If $w \in U_1^\perp = \{w \in V :< w, u_1 >= 0\}$, then $< Tw, u_1 >=< w, T^*u_1 >= < w, Tu_1 >=< w, cu_1 >= 0$. This means that $U_1^\perp$ is also invariant under T. By induction, there is an orthonormal basis $\beta_1 = \{u_2, ..., u_n\}$ for $U_1^\perp$ such that the matrix of $T_{U_1^\perp}$ with respect to β_1 is diagonal with real entries. Since $V = U_1 \oplus U_1^\perp$ (Problem 4.11.9), with each summand T-invariant, we can put $\beta = \{u_1, ..., u_n\}$. ∎

Corollary 4.12.1 *Any real symmetric matrix is diagonalizable.*

Proof. If A is a real symmetric $n \times n$ matrix, the transformation $T = A_l$ (left multiplication by A) is a Hermitian operator on $\mathcal{R}^n$. ∎

We close the section and the chapter with an application to the above theorem and corollary.

Definition 4.12.2 *A nonzero polynomial $q(x,y) \in \mathcal{R}[x,y]$ of the form $q(x,y) = ax^2 + bxy + cy^2$ is called a **quadratic form**.*

The graphs of equations $q(x,y) = 1$ for various quadratic forms $q(x,y)$ give us some interesting examples of curves in the plane. We can use Theorem 4.12.4, together with some of our earlier change of coordinate results, to give a decent description of these curves. The existence of the coordinate transformation algorithm illustrated in the example below is called the **principal axis theorem**. I think one example will suffice to illustrate the procedure.

Example 4.12.1 *We describe the curve which is the graph of $3x^2 + 4xy - 2y^2 = 1$. First, dividing the xy-coefficient of $q(x,y)$ by 2, we write our equation in matrix form*

$$\begin{bmatrix} x & y \end{bmatrix} A \begin{bmatrix} x \\ y \end{bmatrix} = 1$$

with A a real symmetric (Hermitian) 2×2 matrix. In this particular case we have

$$\begin{bmatrix} x & y \end{bmatrix} \begin{bmatrix} 3 & 2 \\ 2 & -2 \end{bmatrix} \begin{bmatrix} x \\ y \end{bmatrix} = 1.$$

By Theorem 4.12.4, the matrix $A = \begin{bmatrix} 3 & 2 \\ 2 & -2 \end{bmatrix}$ will be diagonalizable with an orthonormal set of eigenvectors $\{\mathbf{u}_1, \mathbf{u}_2\}$. We have

$$c_T(\lambda) = \det \begin{bmatrix} 3-\lambda & 2 \\ 2 & -2-\lambda \end{bmatrix} = \lambda^2 - \lambda - 10.$$

The characteristic values for A are $r_1 = (1+\sqrt{41})/2$ and $r_2 = (1-\sqrt{41})/2$. A corresponding pair of characteristic vectors, which by Theorem 4.12.3 must be orthogonal, is $\mathbf{v}_1 = ((5+\sqrt{41})/4, 1)$, $\mathbf{v}_2 = ((5-\sqrt{41})/4, 1)$. Put $\mathbf{u}_1 = \mathbf{v}_1/ \mid \mathbf{v}_1 \mid$, $\mathbf{u}_2 = \mathbf{v}_2/ \mid \mathbf{v}_2 \mid$. Let β' be the orthonormal basis $\{\mathbf{u}_1, \mathbf{u}_2\}$ for $\mathcal{R}^2$ and note that $[T]_{\beta'} = D = \begin{bmatrix} r_1 & 0 \\ 0 & r_2 \end{bmatrix}$.

Now let P be the 2×2 coordinate change matrix P whose columns are the orthonormal vectors $\mathbf{u}_1, \mathbf{u}_2$. By Theorem 4.8.1, $P^{-1}AP = D$. Thus, $A = PDP^{-1}$. Further note that, since P has orthonormal columns $\mathbf{u}_1, \mathbf{u}_2$, then $P^{-1} = P^t = \begin{bmatrix} \mathbf{u}_1 \\ \mathbf{u}_2 \end{bmatrix}$.

Returning to our matrix equation, we have:

$$\begin{bmatrix} x & y \end{bmatrix} A \begin{bmatrix} x \\ y \end{bmatrix} = \begin{bmatrix} x & y \end{bmatrix} PDP^{-1} \begin{bmatrix} x \\ y \end{bmatrix} = 1.$$

Example 4.12.2 *Put $\begin{bmatrix} x' \\ y' \end{bmatrix} = P^{-1} \begin{bmatrix} x \\ y \end{bmatrix}$. Recall that multiplication by P^{-1} changes standard coordinates of a point in $\mathcal{R}^2$ to β' coordinates of that point. Thus, in x'-y' coordinates, our curve is*

$$\begin{bmatrix} x' & y' \end{bmatrix} \begin{bmatrix} r_1 & 0 \\ 0 & r_2 \end{bmatrix} \begin{bmatrix} x' \\ y' \end{bmatrix} = 1.$$

It follows that our curve is the ellipse: $r_1(x')^2 + r_2(y')^2 = 1$.

Exercise 4.12.3 *Use a calculator and elementary vector calculus to estimate the angle of rotation from the x-y axes to the x'-y' axes. Draw a sketch of the situation.*

Thirty-first Problem Set

Problem 4.12.1 *Let $V = \mathcal{R}[x]$, the real vector space of all real polynomials. Note that V is not finite dimensional. a) Prove that the definition $< p(x), q(x) > = \int_0^1 p(x)q(x)dx$ gives an inner product on V. b) Let $D : V \to V$ be the differentiation operator. Show that D has no adjoint.*

Problem 4.12.2 *Verify that, for $T, S \in \mathcal{L}_F(V)$ and $c \in F$: a) $(T+S)^* = T^* + S^*$; b) $(cT)^* = cT^*$; c) $(TS)^* = S^*T^*$; d) $(T^*)^* = T$.*

Problem 4.12.3 *Let $A \in M_n(\mathcal{C})$ be such that $A = A^*$ and such that the rows of A are an orthonormal basis for C^n. Prove that $A = A^{-1}$.*

Problem 4.12.4 *Following the procedure of Example 4.12.1, describe the plane curve whose equation is $(\sqrt{2}/3)x^2 + (1/3)xy - (\sqrt{2}/3)y^2 = 1$.*

Further reading: See [HK] for an alternate development of the canonical forms. For more advanced material, see [J] or [J2].

Chapter 5

Fields and Galois theory

For the organization of the first part of this chapter, we are guided by the excellent account by Jacobson in [J2]. Our proofs are, for the most part, slight revisions of his.

5.1   Preliminary results

Throughout, E, F, K will be a fields.

Definition 5.1.1 *If $(E, +, \cdot)$ is a field we say a subset $F \subset E$ is a **subfield** of E if $(F, +, \cdot)$, that is F with the addition and multiplication of E, is a field. Following our established convention we write $F \leq E$ to indicate that F is a subfield of E.*

Exercise 5.1.1 *Let E be a field with additive identity 0 and multiplicative identity 1 and let $F \leq E$. Show that $0, 1 \in F$; that is, the additive and multiplicative identities of the field F coincide with those of the containing field E. (The proof takes a few lines.)*

Definition 5.1.2 *The **characteristic** of F is the additive order of the multiplicative identity, except that, if $\mid 1 \mid = \infty$, we (illogically) say that characteristic $F = 0$.*

Example 5.1.1 *The characteristic of any of the ordinary fields of numbers, $Q, \mathcal{R}$ or C, is zero. The characteristic of Z/pZ is p.*

Exercise 5.1.2 *Prove that characteristic of any field F is either zero or a prime.*

Definition 5.1.3 *The **prime subfield** of F is the subfield of F generated by $\{0, 1\}$.*

Exercise 5.1.3 *Prove that the prime subfield is the unique smallest subfield of F. (Two or three lines will suffice.)*

Let $F \leq E$ and $a \in E \backslash F$. We want to consider the subfield of E generated by F and the new element a. We denote this subfield by $F(a)$. There are two cases to consider.

Case I) Suppose there exists a nonzero polynomial $p(x) \in F[x]$ such that $p(a) = 0$. In this case we call a **algebraic** over F. The monic generator $m(x)$ of the ideal $\{p(x) \in F[x] : p(a) = 0\}$ is called the **minimal polynomial** for a over F. Let $F[a] \leq E$ be the subring of E consisting of all polynomials in a with coefficients in F, $F[a] = \{f_0 + f_1 a + \cdots + f_k a^k : f_i \in F, k \geq 0\}$. In the algebraic case, the mapping $f(x) \rightarrow f(a)$ is a ring epimorphism from $F[x]$ onto $F[a]$ with kernel $(m(x))$ so that $F[x]/(m(x)) \cong F[a]$. Since $F[a] \leq E$ is an integral domain, it follows that $m(x)$ is irreducible. Hence, $(m(x))$ is a maximal ideal of the commutative ring $F[x]$ so that, by Theorem 2.2.7, $F[a] \cong F[x]/(m(x))$ is a field. Thus, $F[a] = F(a)$.

More precisely, if $m(x)$ has degree n, the subfield $F(a)$ coincides with the set of elements $\{f_0 + f_1 a + \cdots + f_{n-1} a^{n-1} : f_i \in F\}$. In Case I, we say that $F(a)$ is a **simple algebraic extension** of F and that a is **algebraic of degree** n over F.

Case II) Suppose that a is a root of no nonzero polynomial in $F[x]$. In this case we call a **transcendental** over F. Here the mapping $f(x) \rightarrow f(a)$ will be a ring isomorphism between $F[x]$ and $F[a]$. The field $F(a)$ will be the quotient field of the integral domain $F[a]$, constructed as a subfield of E. In Case II, it is easy to see that $F(a)$ can be identified with the subfield of E consisting of all rational functions in a with coefficients in F.

Example 5.1.2 *The field $Q(\sqrt[3]{2}) = Q[\sqrt[3]{2}] \leq \mathcal{R}$ coincides with the set of all real numbers which can be written in the form $q_0 + q_1(\sqrt[3]{2}) + q_2(\sqrt[3]{2})^2$ for arbitrary rational numbers q_0, q_1, q_2. The field $Q(\pi)$ consists of all rational functions in π with rational coefficients.*

Theorem 5.1.1 *Let $F \leq E$ and $a \in E \backslash F$. Then a is algebraic over F if and only if $F(a)$ is finite dimensional as a vector space over F.*

Proof. If a is algebraic of degree n over F, our prior discussion shows that $\{1, a, ..., a^{n-1}\}$ is a basis for $F(a)$ over F. Conversely, if $F(a)$ is finite dimensional as a vector space over F, then, for $k = \dim_F F(a)$, the set $\{1, a, ..., a^{k+1}\}$ is linearly dependent over F. Hence, a is algebraic over F. ∎

Notation 5.1.1 *If $F \leq E$ denote the dimension of E, regarded as a vector space over F, as $[E : F]$.*

Theorem 5.1.2 *Let $F \leq E \leq K$ be fields. Then $[K : F]$ is finite if and only if $[K : E]$ and $[E : F]$ are finite, in which case $[K : F] = [K : E][E : F]$.*

Proof. If $[K : F]$ is finite so is $[K : E]$, since $_EK$ is spanned by a basis for $_FK$. Also $[E : F]$ will be finite since E is an F-subspace of K. Now suppose that K has basis $\{k_1, ..., k_n\}$ over E, E has basis $\{e_1, ..., e_m\}$ over F. It is a routine exercise in linear algebra (manipulating finite sums) to show that $\{e_i k_j : 1 \leq i \leq m, 1 \leq j \leq n\}$ is a basis for $_FK$. See Problem 5.1.2 directly below. ∎

Definition 5.1.4 *A field E is called an **algebraic extension of a subfield** F if every element $a \in E$ is algebraic over F.*

The property of being an algebraic extension is transitive, i.e. if E is an algebraic extension of F and K is an algebraic extension of E, then K is an algebraic extension of F. The proof takes a little thought. See Problem 5.1.7 below.

Thirty-second Problem Set

Problem 5.1.1 *Prove that the prime subfield of any field F is either isomorphic to Q or to Z/pZ for some prime p.*

Problem 5.1.2 *To complete the proof of Theorem 5.1.2, show that the set $\{k_i e_j : 1 \leq i \leq n, 1 \leq j \leq m\}$ is a basis for $_FK$.*

Problem 5.1.3 *Explicitly describe the field $Q(\sqrt{2}, \sqrt{3}) = Q(\sqrt{2})(\sqrt{3}) = Q(\sqrt{3})(\sqrt{2})$ as a subfield of the reals.*

Problem 5.1.4 *For a prime p, show that the polynomial obtained by replacing x by $x+1$ in $x^{p-1} + x^{p-2} + \cdots + x + 1$ is irreducible in $Q[x]$. Hence, conclude that $x^{p-1} + x^{p-2} + \cdots + x + 1$ is also irreducible in $Q[x]$.*

Problem 5.1.5 *Let $\xi = e^{2\pi i/p} \in C$. Show that $[Q(\xi) : Q] = p - 1$.*

Problem 5.1.6 *If $[E : F] < \infty$ show that E is algebraic over F.*

Problem 5.1.7 *Prove that the property of being an algebraic extension is transitive.*

5.2   Straight edge and compass construction

In this section we consider the problem of Euclidean construction, that is construction with straight edge and compass. This problem had its origin in Greek mathematics. For the Greeks the major unsolved questions were as follows. Using a straight edge and compass is it possible to: (1) trisect any given angle (2) construct a line with length $\sqrt[3]{2}$ (and therefore be able to construct a cube with twice the volume of a given cube) (3) construct a regular 7-sided polygon (4) construct a square with area that of a given circle. These problems can be reformulated into questions about extension fields. The first three can easily be shown to have negative answers; the impossibility of the fourth comes from the fact that π is transcendental over Q, a fact proved by Lindemann in 1882.

If S is a set of points in the plane call a **line in** S a line containing two points of S and a **circle in** S a circle whose center is in S and which contains a point in S. We formulate Euclidean construction as follows. Given a set of points $\{P_0, P_1, ..., P_n\}$ in the plane, define sets S_k inductively by $S_1 = \{P_0, P_1, ..., P_n\}$ and $S_{k+1} = S_k \cup T_k$, where T_k consists of: (1) the set of points which are intersections of lines in S_k, (2) the set of points which are intersections of any line in S_k with any circle in S_k and (3) the set of points which are intersections of any two circles in S_k. We say that a point P **can be constructed from** $\{P_0, P_1, ..., P_n\}$ if $P \in \cup_{k=1}^{\infty} S_k$. Denote $\cup_{k=1}^{\infty} S_k$ by $C = C(P_0, P_1, ...P_n)$. The **constructibility problem** is, given $\{P_0, P_1, ..., P_n\}$, when is $P \in C$? A little thought shows that the constructibility problem can be reformulated in this way: Given a set of points$\{P_0, P_1, ..., P_n\}$ in the plane, when can a point P be constructed starting with these points and using a finite set of operations with a straight edge and compass?

We now formulate the constructibility problem algebraically. Suppose we have established coordinate axes in the plane by drawing perpendicular horizontal and vertical lines and marking off a unit length. Thereby, associated with every point is a complex number. Since, in establishing the axes, the real numbers $0, 1$ are marked off, we always assume that $P_0 = 0, P_1 = 1$.

We will prove that C is the unique smallest subfield of $\mathcal{C}$ containing $\{P_0, P_1, ..., P_n\}$ and closed under taking square roots. To see this, consider $z, z' \in C$. Plainly, by using vector addition, we can construct the points in the plane corresponding to $z \pm z'$. Thus, C will be an additive subgroup of $\mathcal{C}$. Next, given two lines of length r, r', we can use similar triangles (see Exercise 5.2.1 below) to construct a line of length r/r'. Since we already have the unit interval, a line of length 1, we can construct a line of length $1/r$ and of length $rr' = r/(1/r')$. It is also easy to see that, given angles of measure θ, θ', we can construct an angle of measure $\theta \pm \theta'$. Thus, if

$z = re^{i\theta}, z' = re^{i\theta'}$ are in C, then so is $zz' = rr'e^{i(\theta+\theta')}$ and $z/z' = (r/r')e^{i(\theta-\theta')}$. We have shown that C is a subfield of $\mathcal{C}$. Finally, given a line of length r, we draw a semicircle touching the x-axis at the points $(0,0)$ and $(1+r,0)$. We then erect a perpendicular to the x-axis at $(1,0)$. The line segment from $(1,0)$ to the intersection point with our semicircle has length $\sqrt{r}$. Since, using a straight edge and compass, we can bisect any given angle, it follows that if $z = re^{i\theta} \in C$ then $\sqrt{z} = \sqrt{r}e^{i(\theta/2)} \in C$. We have shown that C is a subfield of $\mathcal{C}$ closed under taking square roots.

Exercise 5.2.1 *Explain how, given lines of length r, r', we can construct a line of length r/r' using compass and straightedge.*

Exercise 5.2.2 *Prove that C is closed under complex conjugation.*

Now let L be any subfield of $\mathcal{C}$ containing $S_0 = \{P_0, P_1, ..., P_n\}$ and closed under taking square roots. We want to show that each $S_k \subset L$ so that $C = \cup_{k=1}^{\infty} S_k \subset L$. Given the inductive definition of the S_k, it suffices to show that a point of intersection of any lines in L, or a line in L with a circle in L, or of two circles in L is in L. Since L is a field it is immediate that the intersection of two lines in L is in L. If we solve algebraically for an intersection of a line and a circle in L or for the an intersection of two circles in L, we get an expression involving a rational combination of elements in L together with the square root of an element from L. Thus, such an intersection point is in L. We summarize our work in the following theorem.

Theorem 5.2.1 *Let $\{P_0, P_1, ..., P_n\}$ be a set of points in the plane with $P_0 = 0, P_1 = 1$. Identify points in the plane with complex numbers. Then the set of points which can be constructed from $\{P_0, P_1, ..., P_n\}$, using a finite sequence of operations with straight edge and compass, coincides with the unique smallest subfield of $\mathcal{C}$ containing $\{P_0, P_1, ..., P_n\}$ and closed under taking square roots.*

Definition 5.2.1 *Call a complex number **constructible** if it can be constructed starting with $S_0 = \{0, 1\}$.*

Let $\mathcal{K} \leq \mathcal{C}$ be the subfield of constructible complex numbers. Another characterization of a constructible number is one that we can construct with a finite sequence of compass and straight edge constructions, starting with the x-y coordinate axes in a plane and a unit length. Theorem 5.2.1, while interesting, does not provide a good criteria for deciding whether or not a point is in $\mathcal{K}$. We translate the condition for constructibility into a more tractable form.

Definition 5.2.2 *We call a proper ascending chain of subfields of* $\mathcal{C}$: $Q = F_0 \leq F_1 \leq \cdots \leq F_k$ *with* $F_i = F_{i-1}(\sqrt{\alpha_{i-1}}), \alpha_{i-1} \in F_{i-1}$ *for* $1 \leq i \leq k$ *a* **square-root tower** .

For example $Q \subset Q(\sqrt{2/3}) \subset Q(\sqrt{2/3})(\sqrt{1 + \sqrt{2/3}})$ is a square root tower.

Exercise 5.2.3 *Prove that in a square-root tower* $[F_k : Q] = 2^k$.

Theorem 5.2.2 *A point* $z \in \mathcal{C}$ *is constructible if and only if* $z \in F_k$ *for some square-root tower* $Q = F_0 \leq F_1 \leq \cdots \leq F_k$.

Proof. Plainly, $Q \leq \mathcal{K}$. Since $\mathcal{K}$ is closed under taking square roots, each field F_k at the top of any square-root tower is contained in $\mathcal{K}$. Conversely, if z is constructible, then it can be obtained starting with $\{0, 1\} \subset Q$ and doing a finite set of operations involving intersections of lines and circles. Each operation involves at most a rational combination of numbers already constructed together with a square root of a number already constructed. But then z is in a square-root tower. ∎

Our final translation (at last!) gives us a usable criterion for constructibility.

Corollary 5.2.1 *If a number* $c \in \mathcal{C}$ *is constructible then it is algebraic over* Q *and the degree of its minimal polynomial over* Q *is a power of two.*

Proof. If c is constructible then there is a square-root tower $Q = F_0 \leq F_1 \leq \cdots \leq F_k$ with $c \in F_k$. Thus, $Q(c) \leq F_k$ and $2^k = [F_k : Q] = [F_k : Q(c)][Q(c) : Q]$. It follows that $[Q(c) : Q] = 2^l$ so that c is algebraic over Q of degree 2^l. ∎

We now consider the constructibility questions mentioned at the beginning of this section. In light of the corollary and the previously mentioned result of Lindemann, we have already dealt with problem (4)

(1) Let $\theta = 60°$. We can certainly construct a 60° angle by constructing an equilateral triangle. So, if we could trisect 60°, we could construct a 20° angle. But then we could construct a line segment of length $r = \cos(20°)$. If we substitute $\theta = 20°$ into the trig identity $\cos(3\theta) = 4\cos^3(\theta) - 3\cos\theta$, we obtain $1/2 = 4r^3 - 3r$ or $8r^3 - 6r - 1 = 0$. The polynomial $8x^3 - 6x - 1$ is irreducible in $Q[x]$ by the rational root theorem. Therefore, the minimal polynomial for r over Q is $x^3 - (3/4)x - 1/8$ and r is not constructible. Hence, we cannot trisect 60° using only a compass and straight edge.

(2) We cannot construct a line of length $\sqrt[3]{2}$ since $\sqrt[3]{2}$ has minimal polynomial $x^3 - 2$.

(3) If we could construct a regular 7-gon then we could construct line segments of length $cos(2\pi/7)$ and $sin(2\pi/7)$. We could therefore construct the complex number $z = e^{2\pi i/7}$. The number z is a root of $x^7 - 1$ and so is a root of the polynomial $p(x) = x^6 + x^5 + \cdots + x + 1 = (x^7 - 1)/(x - 1)$. We have already noted (see Problem 5.1.4) that $p(x)$ is irreducible in $Q[x]$. Hence, z is algebraic of degree six over Q and is not constructible.

Thirty-third Problem Set

Problem 5.2.1 *For p a prime show that, if a regular p-gon is constructible, then $p - 1$ must be a power of two.*

5.3   Splitting fields

Definition 5.3.1 *A field E is an **extension field** of a field F if E contains an isomorphic copy of F.*

If $\theta : F \to E$ is a field embedding, sometimes it will be convenient to abuse notation slightly by suppressing the embedding and just writing $F \leq E$.

Definition 5.3.2 *If $\theta : F \to E$ is an embedding and $f(x) = \sum_{i=0}^{n} f_i x^i \in F[x]$, let $\theta f(x) = \sum_{i=0}^{n} \theta(f_i)x^i \in E[x]$.*

In this section we show that, for any polynomial $f(x) \in F[x]$ of degree $n > 0$, there exists an extension field E of F with associated embedding $\theta : F \to E$ such that, in $E[x]$, $\theta f(x) = (x - r_1) \cdots (x - r_n)$. (Without loss we will always assume that our polynomial $f(x)$ is monic.) Furthermore, E will be minimal in the sense that $E = (\theta F)(r_1, ..., r_n)$. That is, E will be generated as a field by its subfield $\theta(F)$ and roots $r_1, ..., r_n$ of $\theta f(x)$. Such a field E is called a **splitting field** for $f(x)$ **over** F. If we want to mention the embedding θ, we say that the pair (E, θ) is a splitting field.

Here is a simple but useful fact that we will use repeatedly. To be precise, we include the associated field embeddings into our statement.

Lemma 5.3.1 *Let $F \cong iF \leq K \cong jK \leq E$ and let $f(x) \in F[x]$ be a polynomial of positive degree. If (E, j) is a splitting field for $if(x) \in K[x]$ over K, then (E, ji) is a splitting field for $f(x)$ over F.*

Proof. See Problem 5.3.1. ∎

Theorem 5.3.1 *Any (monic) polynomial of positive degree has a splitting field.*

Proof. Let $f(x) \in F[x]$ be a monic polynomial with positive degree. Factor $f(x)$ into a product of irreducible factors in $F[x]$, say $f(x) = f_1(x) \cdots f_k(x)$. Plainly, $1 \le k \le n = \deg f(x)$. Our proof is by induction on the nonnegative integer $n - k$. If $n - k = 0$ then all the irreducible factors $f_i(x)$ are linear so that F itself is a splitting field. Assume that $n - k > 0$. Thus at least one irreducible factor, say $f_1(x)$, is of degree greater than one. Let $K = F[y]/(f_1(y))$. (The element y is just another polynomial variable introduced to, hopefully, avoid confusion.) Since $f_1(y)$ is irreducible, K is a field. Let $i : F \to K$ be the natural embedding $i(a) = (f_1(y)) + a$ for $a \in F$. Note that K contains a root of $if_1(x)$, namely the coset $e = (f_1(y)) + y$. (Check this.) Thus, in $K[x]$ the polynomial $if(x)$ factors as $if(x) = (x - e)f_1'(x)if_2(x) \cdots if_k(x)$. We now factor $f_1'(x)$ and $if_2(x), ..., if_k(x)$ into irreducible factors in $K[x]$ to obtain a factorization of $if(x)$ into $l \ge k + 1$ irreducible factors. By the inductive hypothesis, there is a splitting field (E, j) for $if(x)$ over K.

By Lemma 5.3.1, (E, ji) is a splitting field for $f(x)$ over F and the proof is complete. ■

Example 5.3.1 *We have already looked at this example. Let $p(x) = (x^2 - 2)(x^2 - 3) \in Q[x]$. Then $Q(\sqrt{2}, \sqrt{3}) = \{q_0 + q_1\sqrt{2} + q_2\sqrt{3} + q_3\sqrt{6}\} \subset \mathcal{R}$ is a splitting field for $p(x)$ over Q.*

Example 5.3.2 *Let $p(x) = (x^3 + x + 1) \in Q[x]$. We construct the splitting field E of $p(x)$. We can say "the splitting field" here (and in the previous example) since, when starting with a subfield F of the algebraically closed (Sect. 5.4) field C of complex numbers, we can always take E to be the subfield of C generated by F and the complex roots of $p(x)$. Since $p'(x) = 3x^2 + 1 > 0$ for all x, the real function $p(x)$ is strictly increasing everywhere. Thus, $p(x)$ has one real root r and, hence, two complex roots c_1, c_2. By the rational root theorem $r \notin Q$. Thus, $p(x)$ is irreducible in $Q[x]$. It follows that $[Q(r) : Q] = \deg p(x) = 3$. In $Q(r)$ the polynomial $p(x)$ factors as $p(x) = (x - r)q(x)$, where $q(x)$ is an irreducible quadratic in $Q(r)$ with roots c_1, c_2. Thus, $E = Q(r)(c_1)$ and $[E : Q(r)] = 2$. Note that $[E : Q] = [E : Q(r)][Q(r) : Q] = 2 \cdot 3 = 6$.*

Exercise 5.3.1 *Prove that $E = Q(r)(c_1)$.*

Example 5.3.3 *Again, let $p(x) = (x^3 + x + 1)$. We construct a splitting field E of $p(x)$, this time regarded as a polynomial in $(Z/2Z)[x]$. Here there is no nice concrete analogue to the field of complex numbers in which we* *can work to construct our splitting field. We will have to use the methods employed in the proof of Theorem 5.3.1.*

Since $p(0) = p(1) = 1$, $p(x)$ is irreducible in $(Z/2Z)[x]$. Therefore, $K = (Z/2Z)[y]/((p(y))$ is a field of dimension 3 over $Z/2Z$ having $2^3 = 8$ elements. Let $e = [(y^3 + y + 1) + y] \in K$. Then, in $K[x]$, we have a factorization $x^3 + x + 1 = (x - e)q(x)$.

Say that $q(x) \in K[x]$ has the form $q(x) = (x^2 + k_1 x + k_2)$. Multiplying and equating coefficients gives us the equations: $-e + k_1 = 0$, $k_2 - ek_1 = 1$, $-ek_2 = 1$. This yields $k_1 = e$, $k_2 = 1 + e^2$. So $q(x) = x^2 + ex + (1 + e^2)$.

The question now is whether or not $q(x)$ is irreducible in $K[x]$. If so, our splitting field will be of the form $E = K[z]/(z^2 + ez + (1 + e^2))$, a field with $8^2 = 64$ elements. If not, then $q(x)$ factors completely in $K[x]$ and K is already our splitting field. This second possibility is actually the case. See Problem 5.3.2.

One naturally would ask, given $p(x) \in F[x]$, to what extent is a splitting field unique? The following theorem provides a nice answer. We provide a commutative diagram which should be helpful in following the proof.

$$
\begin{array}{ccc}
F & \xrightarrow{\phi} & F_1 \\
\downarrow i & & \downarrow i_1 \\
K & \xrightarrow{\phi'} & K_1 \\
\downarrow \bar{\mu} & & \downarrow \bar{\nu} \\
E & \xrightarrow{\Phi} & E_1
\end{array}
$$

Theorem 5.3.2 *Let $\phi : F \cong F_1$ be a field isomorphism and let $p(x) \in F[x]$ be of positive degree. Suppose (E, μ) is a splitting field for $p(x)$ over F. Let (E_1, ν) be a splitting field for $\phi p(x)$ over F_1. Then there exists an isomorphism $\Phi : E \cong E_1$ such that $\Phi \mu = \nu \phi$.*

Proof. Let $p(x)$ have degree n. As in the proof of Theorem 5.3.1, we factor $p(x)$ as a product of irreducible factors in $F[x]$, say $p(x) = p_1(x) \cdot \cdots p_k(x)$, and induct on $n - k$. If $n - k = 0$ then $p(x)$ factors completely in $F[x]$. It follows directly that any splitting field of $p(x)$ over F will be isomorphic to F. (Check this.) Similarly, any splitting field of $\phi p(x)$ over F_1 will be isomorphic to F_1. A moment's thought reveals that our theorem holds in this case.

For the inductive step, say that $p_1(x)$ has degree greater than one. Exactly as in the proof of Theorem 5.3.1, let $K = F[y]/(p_1(y))$ and $i : F \to K$ be the natural embedding. Similarly, put $K_1 = F_1[y]/(\phi p_1(y))$ and let $i_1 : F_1 \to K_1$ be the natural embedding. The isomorphism $\phi : F \to F_1$

extends to an isomorphism $\phi' : K \to K_1$ in the obvious way such that $\phi'i = i_1\phi$. (See Problem 5.3.3.)

We claim that E, E_1 are splitting fields of $ip(x), i_1\phi p(x)$ over K, K_1. First, we need embeddings of K, K_1 into E, E_1. Let $\mu' : F[y] \to E$ be given by $\mu'(\sum_{j=1}^{k} f_j y^j) = \sum_{j=1}^{k} \mu(f_j)a^j$, where $a \in E$ is any root of $p_1(x)$. It is easy to check that μ' is a ring homomorphism and that μ' induces a homomorphism $\bar{\mu} : K \to E$. Since $\bar{\mu}[(p_1(y) + 1] = \mu'(1) = \mu(1) = 1 \neq 0$, the map $\bar{\mu}$ is nonzero. Thus, $\bar{\mu}$ must be a field embedding. In exactly the same manner, using ν, we construct an embedding $\bar{\nu} : K_1 \to E_1$. It is not hard to see that $ip(x)$ and $i_1\phi p(x)$ factor into linear factors in E and E_1. (See Exercise 5.3.2 below.) To finish the proof of the claim we simply note that E, E_1 are generated by the embedded copies of K, K_1 and roots of $ip(x), i_1\phi p(x)$.

To complete the proof of our theorem, first note that, since we have reduced the quantity $n - k$, we can apply the inductive hypothesis to the situation $\phi' : K \cong K_1$ with E, E_1 splitting fields to provide an isomorphism $\Phi : E \to E_1$ with $\Phi\bar{\mu} = \bar{\nu}\phi'$. Following our diagram we have:

$$\Phi\mu = \Phi(\bar{\mu}i) = (\Phi\bar{\mu})i = (\bar{\nu}\phi')i = \bar{\nu}(\phi'i) = \bar{\nu}(i_1\phi) = (\bar{\nu}i_1)\phi = \nu\phi.$$

The proof is complete. ■

Exercise 5.3.2 *Check that* $\mu = \bar{\mu}i$ *and* $\nu = \bar{\nu}i_1$. *Explain why* $i_1\phi p(x)$ *factor into linear factors in* E *and* E_1.

The following corollary gives a strong uniqueness condition for splitting fields.

Corollary 5.3.1 *Let* E, L *be splitting fields for* $p(x)$ *over* F *with associated embeddings* μ, ν. *Then there exists an isomorphism* $\Phi : E \to L$ *such that* $\Phi\mu = \nu$.

We close this section with two interesting and useful results. For $p(x) = p_0 + p_1 x \cdots + p_n x^n$ define $p'(x) = p_1 + 2p_2 x + \cdots + np_n x^{n-1}$. For fairly obvious reasons, $p'(x)$ is called the **derivative** of $p(x)$. A root r in a splitting field E of $p(x)$ over F is called a **multiple root** if there is a factorization in $E[x]$ of the form $ip(x) = (x - r)^k q(x)$ with $k > 1$.

Theorem 5.3.3 *A polynomial* $p(x) \in F[x]$ *has no multiple roots (in any splitting field) if and only if* $p(x)$ *and* $p'(x)$ *are relatively prime in* $F[x]$.

Proof. Suppose $p(x)$ and $p'(x)$ are relatively prime in $F[x]$. Then there exist $u(x), v(x) \in F[x]$ with $(\dagger) : u(x)p(x) + v(x) \, p'(x) = 1$. From equation

(†), it follows that the associated polynomials $ip(x)$ and $ip'(x)$ must remain relatively prime in the ring of polynomials over any extension field (E, i). If $ip(x) = (x-r)^k q(x)$, $k > 1$, in $E[x]$, then $(x-r)$ is a simultaneous divisor of $ip(x)$ and $[ip(x)]' = ip'(x)$ in $E[x]$, a contradiction. Thus, if $p(x)$ and $p'(x)$ are relatively prime in $F[x]$, $p(x)$ can have no multiple roots in any splitting field.

Conversely, let (E, i) be a splitting field for $p(x)$ over F and write $ip(x) = (x-r_1) \cdots (x-r_n)$ in $E[x]$ with the r_j distinct elements of E. Then, by the extended product rule for derivatives, $ip(x)$ and $ip'(x)$ are relatively prime in $E[x]$. Thus, there exist $u(x), v(x) \in E[x]$ with $u(x)ip(x) + v(x)i\,p'(x) = 1$. It follows that $p(x)$ and $p'(x)$ must be relatively prime in $F[x]$. ∎

We can use the previous result, together with our structure theorem for finite abelian groups, to derive a nice fact in field theory.

Theorem 5.3.4 *Any finite subgroup of the multiplicative group of a field is a cyclic group.*

Proof. Let $(F^*, \cdot)$ be the multiplicative group whose elements are the nonzero elements of a field F and let $< 1 > \neq G \leq F^*$ be a finite subgroup. Since $(G, \cdot)$ is a finite abelian group, using the structure theorem for such groups (Theorem 1.11.3), written multiplicatively, $G = G_{p_1} \times \cdots \times G_{p_k}$, where the p_i are distinct primes and each direct factor G_{p_i} is a direct product of cyclic groups of p_i-power order. Say $G_{p_i} = <g_{i1}> \times \cdots \times <g_{it_i}>$, where the cyclic subgroups $<g_{ij}>$ are listed in non-increasing order. Let $G_1 = <g_{11}> \times \cdots \times <g_{1k}>$ be the direct product of the first (and largest) cyclic subgroup in the decomposition of each G_{p_i}. Since the orders of the factors $<g_{1j}>, 1 \leq j \leq k$, are relatively prime, G_1 is cyclic with generator $g = g_{11} \cdots g_{1k}$.

Say that $|g| = m$. In view of the decomposition of G, $|h|$ is a divisor of m for all other elements $h \in G$. (Why?) Put another way, each element of G is a root of the polynomial $p(x) = x^m - 1 \in F[x]$. Note that, if $\operatorname{char} F = p$, then $p \nmid m$. Otherwise, there exists $g \in G$ of order p. But then $(g-1)^p = g^p - 1 = 0$, a contradiction. Thus, $p'(x) = mx^{m-1}$ and $p(x)$ are relatively prime so that $p(x)$ has no multiple roots. It follows that $|G| = m$, so $G = G_1 = <g>$. ∎

Here is an interesting corollary which will be useful later.

Corollary 5.3.2 *Let K be a finite field and F any subfield of K. Then K is a simple algebraic extension of F.*

Proof. Let a be a generator for $(K^*, \cdot)$. Then, plainly, $K = F(a)$. ∎

Thirty-fourth Problem Set

Problem 5.3.1 *Let $F \cong iF \leq K \cong jK \leq E$ and let $f(x) \in F[x]$ be a polynomial of positive degree. Prove that, if (E, j) is a splitting field for $if(x) \in K[x]$ over K, then (E, ji) is a splitting field for $f(x)$ over F.*

Problem 5.3.2 *Using the amusingly strange arithmetic in K, find a factorization of the polynomial $q(x) = x^2 + ex + (1 + e^2)$ (Example 5.3.3) into a product of linear factors in $K[x]$.*

Problem 5.3.3 *Let $\phi : F \to F_1$ be an isomorphism of fields and let $f(x)$ be an irreducible polynomial in $F[x]$. Put $K = F[y]/(f(y))$ and $K_1 = F_1[y]/(\phi f(y))$. Then let $i : F \to K$ and $i_1 : F_1 \to K_1$ be the natural embeddings. Then the isomorphism ϕ extends to an isomorphism $\phi' : K \to K_1$ such that $\phi'i = i_1\phi$.*

Problem 5.3.4 *Let r be the real fifth root of 2 and $\xi = e^{2\pi i/5}$ be a primitive complex fifth root of unity. (a) Explain carefully why $K = Q(\sqrt[5]{2}, \xi)$ is a splitting field for $x^5 - 2$ over Q. (b) Compute $[K : Q]$. Justify your assertion.*

Problem 5.3.5 *Let $p(x) \in F[x]$ be of degree n. Show that the dimension over F of a splitting field of $p(x)$ is at most $n!$.*

Problem 5.3.6 *Let F be a field of characteristic p and $a \in F$. Show that, if $x^p - a$ has a root b in F, then $x^p - a = (x - b)^p$. Hint: Consider the divisibility properties of the binomial coefficients*

5.4   The algebraic closure of a field*

Definition 5.4.1 *A field E is **algebraically closed** if every non-constant polynomial $p(x) \in E[x]$ factors completely into linear factors in $E[x]$.*

In this short section we outline the construction that, given a field F, embeds F into an algebraically closed field A which is generated by F and roots of polynomials in F. This field A, called the **algebraic closure** of F, will be unique up to isomorphism over F. Naturally, we need to use transfinite methods. Zorn's lemma together with an extra construction will do the job. Since knowledge of the details of this construction is not necessary for any of the remaining sections, the instructor might want to skip this section or assign it as an outside assignment.

Without further ado, here is the theorem.

Theorem 5.4.1 *(existence and unicity of the algebraic closure)* Let
F be a field. Then there exists an algebraically closed field A and an embedding $i : F \to A$ such that every element of A is algebraic over iF. If
(A', i') is another such algebraically closed field and embedding, then there
exists an isomorphism $\Psi : A \to A'$ such that $\Psi i = i'$.

Proof. Start with the class $\mathcal{T} = \{(E, i): E$ is a field, $i : F \to E$ is an
embedding and E is algebraic over the subfield $iF\}$. Define an equivalence
relation on $\mathcal{T}$ by saying that $(E, i) \sim (E', i')$ if there is an isomorphism
$\theta : E \to E'$ such that $\theta i = i'$. Let $S = \{(E_\gamma, i_\gamma)\}$, where γ runs through
some index set Γ, consist of one representative of $\mathcal{T}$ chosen from each of its
equivalence classes. Make S into a poset by defining $(E_\alpha, i_\alpha) \leq (E_\beta, i_\beta)$ if
and only if there is a field embedding $\theta_{\alpha\beta} : E_\alpha \to E_\beta$ such that $\theta_{\alpha\beta} i_\alpha = i_\beta$.
It is not difficult to check that $\leq$ is a partial order.

Suppose that $C = \{(E_\gamma, i_\gamma) : \gamma \in \Gamma_0\}$ is a chain in S . Let $P =$
$\bigoplus_{\gamma \in \Gamma_0} E_\gamma$ be the direct sum of the rings E_γ. The "rng" P can be thought
of as the set of finite formal sums $\oplus_{i=1}^k e_{\gamma_i}$, $\{\gamma_1, ... \gamma_k\} \subset \Gamma_0$. The operations
here are componentwise addition and multiplication. If Γ_0 is infinite, then
P lacks an identity-but P satisfies all of other the ring axioms. Now take
I to be the ideal of P generated by the set of elements $\{\theta_{\gamma\delta}(e_\gamma) - e_\gamma :$
$\gamma, \delta \in \Gamma_0, (E_\gamma, i_\gamma) \leq (E_\delta, i_\delta), e_\gamma \in E_\gamma\}$ and put $E = P/I$. Note that E now
is a ring since, for any $\gamma \in \Gamma_0$, the coset $I + 1_\gamma \neq I + 0$ represents the
multiplicative identity of E. The fact that E is a field is straightforward
using the fact that each E_γ is a field. (See Problem 5.4.1 below.)

We next define $j : F \to E$ by $j(f) = I + i_\gamma(f)$, for any $\gamma \in \Gamma_0$. It is
easy to show that j is well-defined embedding. Also, that E is algebraic
over jF. (See Problem 5.4.2 below.) For each $\gamma \in \Gamma_0$ the map $j_\gamma : E_\gamma \to E$
sending $e_\gamma \in E_\gamma$ to $(I + e_\gamma) \in E$ satisfies the relation $j_\gamma i_\gamma = j$. Thus, using
the definition of our partial order, $(E_\gamma, i_\gamma) \leq (E, j)$ for all γ. We have
shown that (E, j) is an upper bound for C.

Applying Zorn's lemma, we conclude that S has a maximal element, say
(A, i). Since $(A, i) \in S$, $i : F \to E$ is a field embedding and A is algebraic
over iF. To complete the existence part of the proof, we claim that A is
algebraically closed. (See Problem 5.4.3 below.)

Next, let $i' : F \to A'$ be a field embedding such that A' is algebraically closed and algebraic over $i'F$. Start with a set S_1 consisting of pairs
(A_δ, Ψ_δ), $\delta \in \Delta$, where A_δ is a field with $iF \leq A_\delta \leq A$ and $\Psi_\delta : A_\delta \to A'$
is an embedding such that $(\Psi_\delta)i = i'$. Make S_1 into a poset by saying
that $(A_\delta, \Psi_\delta) \leq (A_\eta, \Psi_\eta)$ if there is a field embedding $\theta_{\delta\eta} : A_\delta \to A\eta$ with
$(\theta_{\delta\eta})i = i'$. Arguing exactly as in the existence part of the proof, we check
that S_1 satisfies the hypothesis of Zorn's Lemma. Applying Zorn's Lemma,
let (A_0, Ψ) be a maximal element of S_1.

We claim that $A_0 = A$. For proof by contradiction, suppose $a \in A \backslash A_0$. Since a is algebraic over $iF \leq A_0$, a is algebraic over A_0. Let $m(x)$ be the minimal polynomial for a over A_0 and let $a' \in A'$ be a root of $\Psi m(x) \in (\Psi A_0)[x] \leq A'[x]$. Such an a' exists since A' is algebraically closed. Then we can extend Ψ in the natural way to an embedding $\Psi_1 : A_0(a) \to (\Psi A_0)(a') \leq A'$, contradicting the maximality of (A_0, Ψ). Thus, $A = A_0$. We have proven the existence of a field embedding $\Psi : A \to A'$ such that $\Psi i = i'$.

Finally, by a symmetric argument, there is a field embedding $\Omega : A' \to A$ such that $\Omega i' = i$. We have, for each $f \in F$, $(\Omega \Psi) i f = \Omega(\Psi i) f = \Omega i' f = i f$. Hence, the map $\Omega \Psi : A \to A$ is the identity when restricted the subfield iF. By symmetry, $\Psi \Omega : A' \to A'$ is the identity on $i'F$. I'll let you complete the proof in Problem 5.4.4. ∎

Remark 5.4.1 *We've already noted that the field of complex numbers is algebraically closed. To obtain infinitely many other examples, for each prime p, take the algebraic closure $(Z/pZ)^{\sim}$ of the field Z/pZ. The field $Z/pZ^{\sim}$ will be countable (Problem 5.4.5), algebraically closed, of characteristic p.*

Exercise 5.4.1 *Check that $\leq$ is a partial order.*

Exercise 5.4.2 *Check that S_1 satisfies the hypothesis of Zorn's Lemma.*

Exercise 5.4.3 *Find all the places where we used the axiom of choice in this proof.*

Thirty-fifth Problem Set

Problem 5.4.1 *Prove that E is a field.*

Problem 5.4.2 *(a) Show that the map $j(f) = I + i_\gamma(f)$ in the proof of Theorem 5.4.1, is a well defined embedding. (b) Show that E is algebraic over jF. Hint: Γ_0 is a chain.*

Problem 5.4.3 *Prove by contradiction that the field A above is algebraically closed.*

Problem 5.4.4 *Complete the proof above. You need to show that Ψ and Ω are inverse isomorphisms.*

Problem 5.4.5 *Prove that the algebraic closure of a countable field is countable.*

5.5   A structure theorem for finite fields

Let F be a finite field. Then F must be a finite dimensional extension of its prime subfield Z/pZ for some prime p. If $\dim_{Z/pZ} F = n$, then F will contain p^n elements. We classify all finite fields with the following simple theorem.

Theorem 5.5.1 *(structure theorem for finite fields)* *For every prime p and natural number n, there is, up to isomorphism, exactly one field of cardinality p^n. It is a splitting field of $x^{p^n} - x$ over Z/pZ.*

   Proof. Let $p(x) = (x^{p^n} - x) \in (Z/pZ)[x]$. Since $p'(x) = -1$, $p(x)$ and $p'(x)$ are relatively prime. Thus, $x^{p^n} - x$ has p^n distinct roots in any splitting field. Now suppose K is any field with p^n elements. Since the nonzero elements of K form a multiplicative group of order $p^n - 1$, for each $0 \neq k \in K$, $k^{p^n - 1} = 1$. So $k^{p^n} = k$ for all $k \in K$. Thus K is composed entirely of the p^n distinct roots of $x^{p^n} - x$. Plainly, K is a splitting field of $x^{p^n} - x$ over its prime subfield Z/pZ. Since splitting fields are unique up to isomorphism, the proof is complete. ■

   We have shown that, if K is a finite field, $\{|\ K\ |\}$ is a complete set consisting of one isomorphism invariant.

Thirty-sixth Problem Set

Problem 5.5.1 *If $|\ K\ | = p^n$ prove that K contains a subfield of cardinality p^m if and only if m divides n.*

Problem 5.5.2 *If n is a positive integer, prove that there is an irreducible polynomial of degree n in $(Z/pZ)[x]$.*

5.6   The Galois correspondence

In this section we present a deep and central result in algebra, Galois' correspondence between the set of subfields of a certain type of field and the set of subgroups of its automorphism group. This result was published in 1846, fifteen years after his death. For K a field, let $Aut(K)$ be the group consisting of the automorphisms of K (isomorphisms from K onto itself) with binary operation composition.

Definition 5.6.1 *Let $F \leq K$ be fields. Define $Gal_F(K) = \{\alpha \in Aut(K) : \alpha\ |_F = 1_F\}$. If $F \leq L \leq K$, let $\Gamma(L) = \{\alpha \in Gal_F(K) : \alpha\ |_L = 1_L\}$*

If is immediate that $Gal_F(K)$ is a subgroup of $Aut(K)$ and that each $\Gamma(L)$ is a subgroup of $Gal_F(K)$. The subgroup $\Gamma(L)$ is called the **fixing subgroup** of L. The correspondence $L \to \Gamma(L)$ between the set of intermediate subfields between F and K and the set of fixing subgroups of $Gal_F(K)$ is order reversing; that is, if $L \subseteq L'$ then $\Gamma(L') \subseteq \Gamma(L)$.

Exercise 5.6.1 *Verify the claims of the preceding paragraph.*

Definition 5.6.2 *If H is a subgroup of $Gal_F(K)$, put $\mathfrak{F}(H) = \{k \in K : h(k) = k \text{ for all } h \in H\}$.*

Exercise 5.6.2 *Prove that $\mathfrak{F}(H)$ is a subfield of K intermediate between F and K. Prove that if $H \subseteq H'$ are subgroups of $Gal_F(K)$ then $\mathfrak{F}(H') \leq \mathfrak{F}(H)$.*

The subfield $\mathfrak{F}(H)$ is called the **fixed subfield** of H. It follows directly from the definitions that $L \leq \mathfrak{Fb}(L)$ for every intermediate subfield $F \leq L \leq K$ and that $H \leq \Gamma\mathfrak{F}(H)$ for every subgroup $H \leq Gal_F(K)$.

Exercise 5.6.3 *Prove directly from the definitions that $\Gamma(L) = \Gamma\mathfrak{F}\Gamma(L)$ and $\mathfrak{F}(H) = \mathfrak{F}\Gamma\mathfrak{F}(H)$ for all $F \leq L \leq K$ and $H \leq Gal_F(K)$.*

We emphasize that the correspondences $H \to \mathfrak{F}(H)$ and $L \to \Gamma(L)$ take place in the context of some fixed field extension $F \leq K$.

Definition 5.6.3 *A complex number of the form $c = e^{k(2\pi i)/n}$ with $(k, n) = 1$ is called a **primitive n-th root of unity**.*

Exercise 5.6.4 *Show that the multiplicative order of a primitive n-th root of unity is exactly n. Show that any such complex number c will be a generator of the multiplicative group of complex n-th roots of unity.*

Example 5.6.1 *Let $F = Q$ and $K = Q(\sqrt[3]{2}) = \{q_0 + q_1 \sqrt[3]{2} + q_2(\sqrt[3]{2})^2\}$. Every automorphism of K will fix Q so that here $Gal_F(K) = Aut(K)$. An automorphism α of K is determined by its action on $\sqrt[3]{2}$. But $[\alpha(\sqrt[3]{2})]^3 = \alpha[(\sqrt[3]{2})^3] = \alpha(2) = 2$ and the only cube root of 2 in K is $\sqrt[3]{2}$ itself. Hence, for any $\alpha \in Aut(K)$, $\alpha(\sqrt[3]{2}) = \sqrt[3]{2}$. We have shown that $Gal_Q(K) = <1>$. Thus, Q is properly contained in $\mathfrak{F}\Gamma(Q) = \mathfrak{F}(<1>) = Q(\sqrt[3]{2})$.*

Example 5.6.2 *Let $F = Q, K = Q(\rho, c)$, where $\rho = \sqrt[3]{2}$, $c = e^{2\pi i/3}$. Then the three complex cube roots of two are $\rho, c\rho, c^2\rho$. Again, every automorphism of K fixes Q so that $Gal_F(K) = Aut(K)$. An automorphism of K will be determined by its action on ρ and c, hence by its action on $\{\rho, c\rho, c^2\rho\}$, the three cube roots of two. Furthermore, as in the previous example, any* *cube root of two must be mapped to another cube root of two. Therefore, we can regard $Gal_F(K)$ as a subgroup of S_3. Also note that c, a primitive cube root of unity, must be mapped by an automorphism to another primitive cube root of unity, so c maps by any automorphism to either c or c^2. Plainly, the minimal polynomial for c over Q is $(x^3-1)/(x-1) = x^2+x+1$. Hence, $c^2 = -1 - c$.*

Here is the crucial point for this example. *In this case, we claim that we can define an automorphism of K sending ρ to either $\rho, c\rho$ or $c^2\rho$ and, independently, sending c to either c or c^2. Thus, the six possibilities for automorphisms of K will actually occur and $Gal_Q(K)$ will be all of S_3.*

To verify this claim, first note that $\{1, \rho, \rho^2\}$ is a basis for K over its subfield $Q(c)$ and $\{1, c\}$ is a basis for $Q(c)$ over Q. Thus, by Theorem 5.1.2 and Problem 5.1.1, $\{1, c, \rho, c\rho, \rho^2, c\rho^2\}$ is a basis for K over Q. Say, for example, we want to construct an automorphism σ with (1) $\sigma(c) = c^2 = -1 - c$ and (2) $\sigma(\rho) = c\rho$. Then $\sigma^2(\rho) = \sigma(c\rho) = \sigma(c)\sigma(\rho) = c^2 c\rho = \rho$ and $\sigma^2(c) = \sigma(c^2) = \sigma(c)^2 = c^4 = c$. Hence, σ, if it exists, will correspond to a transposition in S_3.

We must also have: (3) $\sigma(c\rho) = \sigma(c)\sigma(\rho) = c^3\rho = \rho$; (4) $\sigma(\rho^2) = [\sigma(\rho)]^2 = c^2\rho^2 = -\rho^2 - c\rho^2$; (5) $\sigma(c\rho^2) = -\rho^2 c^2 (1 + c) = \rho^2 c^4 = c\rho^2$.

Take arbitrary elements $k, k' \in K$, written in the form $k = q_1 1 + q_2 c + q_3\rho + q_4 c\rho + q_5\rho^2 + q_6 c\rho^2$ and $k' = q_1' 1 + q_2' c + q_3'\rho + q_4' c\rho + q_5'\rho^2 + q_6' c\rho^2$. Since σ must be additive and must fix Q, it follows that: $\sigma(k) = q_1 1 - q_2(1 + c) + q_3 c\rho + q_4\rho - q_5(\rho^2 + c\rho^2) + q_6 c\rho^2$. This equation can be rewritten:

$$(s) : \sigma(k) = (q_1 - q_2)1 - q_2 c + q_4\rho + q_3 c\rho - q_5\rho^2 + (q_6 - q_5)c\rho^2$$

with a similar formula for $\sigma(k')$. It should be clear now that $\sigma : K \to K$, defined by formula (s), provides a Q-linear transformation of the rational vector space K. Also, by inspection, $Ker\sigma = 0$. It follows that σ is a bijection on K. So far, so good.

The part that's far from clear is that σ is multiplicative. We need to verify the formula $\sigma(k)\sigma(k') = \sigma(kk')$. This is simply a brute force calculation. Similar arguments verify the existence of the other four non-identity automorphisms. For further elaboration of this example, see Problem 5.6.1.

Before continuing our discussion of fixed fields and fixing subgroups we prove a useful lemma.

Lemma 5.6.1 *Let $\phi : F \cong F'$ and let E and E' be extension fields of F, F' respectively. For simplicity, suppress the embeddings and regard $F \subset E, F' \subset E'$. Suppose $e \in E$ is algebraic over F with minimum polynomial $m(x)$. Then ϕ can be extended to an embedding of $F(e)$ into E' if and only*

*if E' contains a root of $\phi m(x)$. The number of distinct extensions of ϕ is
the number of roots of $\phi m(x)$ in E'.*

Proof. If $e' \in E'$ is a root of $\phi m(x)$, then, exactly as in the proof of
Theorem 5.3.2, we can construct an isomorphism from $F(e)$ to $F'(e')$ with
$\phi(e) = e'$. On the other hand, if $\phi : F(e) \to E'$ is an embedding, then $\phi(e)$
will be a root of $\phi m(x)$. ∎

The following concept will also be of importance.

Definition 5.6.4 *A polynomial $p(x) \in F[x]$ is **separable** if its irreducible
factors have distinct roots (in any splitting field of $p(x)$ over F).*

If $p(x) \in F[x]$ is irreducible, Theorem 5.3.3 shows that the property of
having distinct roots is independent of the splitting field. Also, a moment's
thought shows that, if $p(x)$ is separable considered as a polynomial in $F[x]$,
then $p(x)$ is separable as a polynomial in $K[x]$ for K any extension field of
F.

It's a little hard to come up with an example of a nonseparable polynomial. This is why.

Theorem 5.6.1 *If characteristic $F = 0$ then every $p(x) \in F[x]$ is separable.*

Proof. If $char F = 0$, we show that every irreducible $p(x) \in F[x]$
has distinct roots in any splitting field. This amounts to showing that, if
$p(x)$ is irreducible, then $p(x)$ and $p'(x)$ are relatively prime in $F[x]$. Since
$p(x)$ is irreducible, its only divisors are 1 and $p(x)$ (up to associates). Since
$char F = 0$, $\deg p'(x) = \deg p(x) - 1$ so that $p(x)$ cannot divide $p'(x)$. Hence
p and p' are relatively prime and the proof is complete. ∎

Example 5.6.3 *Let $F = (Z/pZ)(y)$, the field of rational functions with
coefficients in Z/pZ in an indeterminate y. Then $p(x) = x^p - y$ is inseparable in $F[x]$. To see this, let E be a splitting field for $p(x)$ over F and
regard $F \leq E$. If $x - e$ is a linear factor of $p(x)$ in $E[x]$, then $e^p = y$. Since
$char E = p$ and p divides each binomial coefficient $\begin{pmatrix} p \\ k \end{pmatrix}, 1 \leq k < p$, in
$E[x]$ we have the factorization $x^p - y = (x - e)^p$. From this factorization,
it follows that any proper factor of $p(x)$ in $F[x]$ must be an associate of
a polynomial of the form $(x - e)^k$, $1 \leq k < p$. But, a glance at the x^{k-1}
coefficient of $(x - e)^k$ reveals that no such polynomial is in $F[x]$. Hence,
$p(x)$ is irreducible in $F[x]$ and $p(x)$ is nonseparable.*

We continue our investigation of the Galois correspondence with this key theorem.

Theorem 5.6.2 *Let K be the splitting field of a separable polynomial $p(x) \in F[x]$ and regard $F \leq K$. Then $| \, Gal_F(K) \, |= [K : F]$.*

Proof. Let $p_1(x), ..., p_k(x)$ be the distinct irreducible factors of $p(x)$ in $F[x]$. Then K is also the splitting field of $p_1(x) \cdots p_k(x)$ over F. Without loss, suppose that $F \leq K$. Let $K_0 = F$ and, for $1 \leq i \leq k$, let $K_i \leq K$ be a splitting field of $p_i(x)$ over K_{i-1}. We have a chain of fields: $F = K_0 \leq K_1 \leq \cdots \leq K_k = K$. Thus, $[K : F] = [K_1 : K_0][K_2 : K_1] \cdots [K_k : K_{k-1}]$. Each K_i is obtained from K_{i-1} by adjoining the roots of $p_i(x)$. Furthermore, none of the distinct $p_i(x)$ can have common roots in any extension field, since all pairs of distinct irreducible polynomials $p_i(x), p_j(x)$ are relatively prime in $F[x]$. If $\alpha \in Gal_F(K)$, then α can be determined, step by step, by specifying its restriction to the ascending chain of subfields K_i, $i \geq 1$.

Armed with this information, suppose we have determined the restriction of some α to K_{i-1}. Now $K_i = K_{i-1}(r_1, ..., r_t)$, where $t = \deg p_i(x)$ and (by separability) $r_1, ..., r_t$ are the distinct roots of $p_i(x)$. Let $m_1(x)$ be the minimal polynomial for r_1 over K_{i-1}. By Lemma 5.6.1, there are $\deg m_1(x) = [K_{i-1}(r_1) : K_{i-1}]$ distinct extensions of α to $K_{i-1}(r_1)$. Continuing this argument, if $m_j(x)$ is the minimal polynomial for r_j over $K_{i-1}(r_1, ... r_{j-1})$, $j \geq 2$, there are $\deg m_1(x) \deg m_2(x) \cdots \deg m_t(x) = [K_i : K_{i-1}]$ distinct extensions of α from K_{i-1} to K_i. Hence, there are $[K : F]$ extensions of a possible α from F to K. That is, $| \, Gal_F(K) \, |= [K : F]$. ∎

The next result begins with a group of automorphisms and makes claims about its fixed field.

Theorem 5.6.3 *Let G be a finite subgroup of $AutK$ and let F be the fixed field of G. Then $[K : F] \leq | \, G \, |$.*

Proof. Let $| \, G \, |= n$. We show that any $m > n$ elements of K are dependent over F. Let $G = \{g_1 = 1, g_2, ..., g_n\}$ and let $\{e_1, ..., e_m\}$ be a set of m distinct elements of K. Consider the system of n equations in m unknowns:

$$(\dagger) : \sum_{j=1}^{m} g_i(e_j)x_j = 0.$$

Since $m > n$ there is a nonzero solution $(a_1, ... a_m) \in K^m$ to the system. (Refer back to Problem 4.5.6.) Among such solutions choose one, say $(b_1, ... b_m)$, with the least number of nonzero components. Without loss, suppose that $b_1 \neq 0$. Since $b_1^{-1}(b_1, ...b_m)$ is also a solution, we may assume $b_1 = 1$.

The first equation, corresponding to $g_1 = 1$, is $\sum_{j=1}^{m} e_j b_j = 0$. We claim that $\{b_j : j \geq 2\} \subset F$. Then this equation will then exhibit a linear dependence of the elements e_j over F. To prove the claim, suppose that one of the b_j, say b_2, is not in F. Since F is the fixed field of G, that means there is a g_k such that $g_k(b_2) \neq b_2$. If we apply g_k to the system $(\dagger)$ we obtain the system:

$$g_k(\dagger) : \sum_{j=1}^{m} g_k g_i(e_j) x_j = 0.$$

Since left multiplication by g_k permutes the elements of G, the system $g_k(\dagger)$ is just $(\dagger)$ with the equations rearranged. Thus, $(1, b_2, ..., b_m)$ is also a solution to $g_k(\dagger)$. But, since $(1, b_2, ..., b_m)$ is a solution to $(\dagger)$, we can apply g_k to conclude that $(1, g_k(b_2), ..., g_k(b_m))$ is a solution to $g_k(\dagger)$. Hence, $(0, b_2 - g_k(b_2), ..., b_m - g_k(b_m))$ is a solution to $g_k(\dagger)$ and, thus, is a solution to $(\dagger)$. Since $b_2 - g_k(b_2) \neq 0$ and $b_t = 0$ in the original solution implies that $b_t - g_k(b_t) = 0$, we have a contradiction. The proof is complete. ∎

Definition 5.6.5 *An algebraic extension field $K \geq F$ is a **separable extension** of F if the minimal polynomial over F of every element of K is separable.*

If the characteristic of F is zero, every algebraic extension of F is separable. Example 5.6.3 displays an inseparable extension.

The following theorem is most definitely of independent interest. The proof is as in [W].

Theorem 5.6.4 *Let K be a finite dimensional separable extension of F. Then K is a simple algebraic extension; that is $K = F(d)$ for some $d \in K$.*

Proof. Since $[K : F] < \infty$, $K = F(a_1, ..., a_r)$ with $\{a_1, ..., a_r\} \subset K$ some finite set of elements algebraic over F. If F is finite, so is K and we've already noted that any finite field is a simple extension of any of its subfields. Hence, suppose that F is infinite.

It suffices to prove that, if $a, b \in K$ are algebraic over F, then $F(a, b) = F(d)$ for some suitable $d \in K$. Let $f(x), g(x)$ be the minimal polynomials for a, b over F and let L be a splitting field of $f(x)g(x)$ over K. Let $\{a = a_1, ..., a_n\}$, (resp. $\{b = b_1, ..., b_m\}$) be the roots of $f(x)$ (resp. $g(x)$) in L.

Take any element $c \in F$ such that c is unequal to any of the finite set of quotients $\{(a_i - a)/(b - b_k) : 1 \leq i \leq n, 1 < k \leq m\}$. Then let $d = a + bc$. Plainly, $F(d) \subset F(a, b)$. We claim that $F(a, b) = F(d)$. To prove this (at first glance most unlikely) claim, first note that the polynomials $f(d - cx)$

and $g(x)$ have the root b in common. Since $d - cb_k = a_i$ implies that $a + bc - cb_k = a_i$ or $c = (a_i - a)/(b - b_k)$, b is the only root these polynomials have in common.

Thus, the gcd in $L[x]$ of $f(d - cx)$ and $g(x)$ is $(x - b)$. But, since $c \in F$, $f(d - cx)$ and $g(x)$ are in $F(d)[x]$. We cannot have $f(d - cx), g(x)$ relatively prime in $F(d)[x]$, since they share the root b in the larger polynomial ring $L[x]$. It follows that $(x - b)$ must also be the gcd of $f(d - cx), g(x)$ in $F(d)[x]$. But then $b \in F(d)$. Hence, $a \in F(d)$ as well and $F(a, b) = F(d)$. The proof is complete. ∎

I've always found this next corollary kind of amazing.

Corollary 5.6.1 *Let F be a field with $Q \subset F \subset C$ such that $[F : Q] < \infty$. (Such a field is called an **algebraic number field**.) Then $F = Q(d)$.*

Definition 5.6.6 *An algebraic extension $K \geq F$ is **normal** if every irreducible polynomial in $F[x]$ that has a root in K factors completely in K.*

Put another way, K is normal over F if and only if, whenever K contains a root of an irreducible polynomial $p(x) \in F[x]$, K contains a splitting field for $p(x)$.

We have one more preliminary theorem, again interesting in its own right, before our main result.

Theorem 5.6.5 *Let $K \geq F$. Then the following conditions are equivalent:*

(1) K is a splitting field of a separable polynomial $p(x) \in F[x]$;
(2) the group $G = Gal_F(K)$ is finite and $F = \mathfrak{F}(G)$;
(3) K is finite dimensional and normal and separable over F.

Moreover, under the assumption of conditions (1)-(3), if $F = \mathfrak{F}(H)$ for some subgroup $H \leq Aut(K)$, then $H = G$.

Proof. (1) → (2) Assume (1) and let $G = Gal_F(K)$. By Theorem 5.6.2, $|G| = [K : F]$, so G is finite. Now let $F' = \mathfrak{F}(G)$. Plainly, $F \leq F' \leq K$, so K is also a splitting field for $p(x)$ over F'. Thus, we can apply Theorem 5.6.2 again to conclude that $|G'| = [K : F']$, where $G' = Gal_{F'}(K)$. Note that $G' \leq G$.

In the context of the subgroup-subfield correspondence induced by the original extension $F \leq K$, we can regard $G' = \Gamma F' = \Gamma \mathfrak{F}(G)$. But, directly from the definitions, $G \leq \Gamma \mathfrak{F}(G)$. Hence, $G \leq G'$ and $G = G'$. Then, using the dimension relations of the previous paragraph, we must have $F = F'$ as well. Thus, $F = F' = \mathfrak{F}(G)$.

(2)$\rightarrow$ (3) If (2) holds then, by Theorem 5.6.3, $[K : F] \leq | G | < \infty$. We show that K is a normal and separable extension of F.

Let $f(x) \in F[x]$ be irreducible and suppose that $r \in K$ with $f(r) = 0$. Let $S = \{r = k_1, ..., k_j\}$ be the set of distinct elements of the form $gr, g \in G$. Since the elements of G are automorphisms of K, $S \subset K$. Moreover, G fixes the coefficients of any polynomial in $F[x]$. So, if we apply elements of G to the equation $f(r) = 0$ we obtain the equations $f(k_i) = 0, 1 \leq i \leq j$. Thus, $f(x)$ is divisible by each $x - k_i$ in $K[x]$. Since the k_i are distinct, $f(x)$ is divisible by their product $q(x) = (x - k_1)(x - k_2) \cdots (x - k_j)$ in $K[x]$.

If $h \in G$, then left multiplication by h permutes the elements of S. Thus, $hq(x) = h \prod_{i=1}^{j}(x - k_i) = \prod_{i=1}^{j}(x - h(k_i)) = q(x)$ for all $h \in G$. It follows that the **coefficients** of $q(x)$ are in $\mathfrak{F}(G) = F$. (Using the fact that, if each $h \in G$ permutes the roots of a polynomial $q(x)$ in some splitting field, then the coefficients of $q(x)$ must lie in $\mathfrak{F}(G)$ is a standard Galois theory trick. We'll see it used again.)

Now, since $q(x), f(x)$ are both polynomials in $F[x]$ and $q(x) \mid f(x)$ in $K[x]$, we have that $q(x)$ actually divides $f(x)$ in $F[x]$. (Think about this for a moment.) But $f(x)$ is irreducible in $F[x]$. Hence, $f(x) = cq(x)$ for $0 \neq c \in F$. This shows that, if an irreducible polynomial $f(x) \in F[x]$ has a root in K, then $f(x)$ factors completely in $K[x]$. In other words, K is a normal extension of F.

In particular, the minimal polynomial $m(x)$ of any $k \in K$ over F factors completely into distinct linear factors in $K[x]$. Hence, K is also separable over F.

(3) $\rightarrow$ (1) Using (3), it is easy to write $K = F(k_1, ..., k_r)$ where $\{k_i : 1 \leq i \leq r\}$ is the set of distinct roots of a separable polynomial $p(x) \in F[x]$. See Problem 5.6.7.

We complete the proof by establishing the final claim. Suppose that $F \leq K$ with G, F, K as in (2). Let H be an arbitrary subgroup of $Aut(K)$ such that $F = \mathfrak{F}(H)$. Then, under the subgroup-subfield correspondence associated with the extension $F \leq K$, we have $H \leq \Gamma \mathfrak{F}(H) = \Gamma F = G$. Moreover, by (1), K is the splitting field of a separable polynomial $p(x) \in F[x]$. Since F is the fixed field of both G and H, using Theorem 5.6.2, $| H | = [K : F] = | G |$. Hence, $H = G$ and the proof is complete. ∎

Exercise 5.6.5 *Explain in the proof* (2) $\rightarrow$ (3) *why* $q(x) \mid p(x)$ *in* $F[x]$.

An extension K of F satisfying any of the equivalent conditions of Theorem 5.6.5 is called a **Galois extension** of F. If K is a Galois extension of F, we call the group $Gal_F(K)$ the **Galois group** of K over F. If $p(x) \in F[x]$ is separable, by the **Galois group of** $p(x)$ we mean the Galois group of a splitting field K of $p(x)$ over F. Since any two splitting fields of a polynomial $p(x) \in F[x]$ are isomorphic respecting the embeddings of F, the Galois group of $p(x)$ has no ambiguity.

We close this section with our main result, providing precise details on the subgroup-subfield correspondence for Galois extensions.

Theorem 5.6.6 *(Galois Correspondence) Let K be a Galois extension of F and let $G = Gal_F(K)$. Then the associated maps $\mathfrak{F} : H \to \mathfrak{F}(H)$ and $\Gamma : E \to \Gamma(E)$ are inverse bijections between the set of subgroups $H \leq G$ and the set of intermediate subfields $F \leq E \leq K$. Moreover, if $H \leq G$, then $\mid H \mid = [K : \mathfrak{F}(H)]$ and $[G : H] = [\mathfrak{F}(H) : F]$. Finally, $H \lhd G$ if and only if $\mathfrak{F}(H)$ is a normal extension of F. In this case $Gal_F(\mathfrak{F}(H)) \cong G/H$.*

Proof. Suppose K is a Galois extension of F with Galois group G. In particular, K is the splitting field of a separable polynomial $p(x) \in F[x]$. Let $H \leq G$ and put $E = \mathfrak{F}(H)$. Then $F \leq E \leq K$. Since K is the splitting field of $p(x) \in F[x] \leq E[x]$, we can apply the final claim of Theorem 5.6.5 to H and E. We conclude that $H = Gal_E(K)$. Then, by Theorem 5.6.2, we have $\mid H \mid = [K : E]$. Thus,

$$[G : H] \mid H \mid = \mid G \mid = [K : F] = [K : E][E : F] = \mid H \mid [E : F].$$

It follows that $[G : H] = [E : F]$. Moreover, using the definitions,

$$\Gamma\mathfrak{F}(H) = \Gamma(E) = \{\alpha \in Aut(K) : \alpha \mid_E = 1\} = Gal_E(K) = H.$$

Now let E be an arbitrary intermediate subfield, $F \leq E \leq K$. Using condition (1) of Theorem 5.6.5, we see that K is a Galois extension of E. Thus, by condition (2) of that theorem, E is the fixed of $Gal_E(K)$. But $Gal_E(K)$ corresponds with $\Gamma(E)$. Hence, $\mathfrak{F}\mathfrak{b}(E) = E$.

At this point, we have shown that the correspondences $\mathfrak{F}$ and Γ are inverse bijections and that the desired dimension relations hold.

Next, suppose that $F \leq E \leq K$ and that E is a normal field extension of F. Since K is finite dimensional separable over F, E is also finite dimensional separable over F. Then, by Theorem 5.6.4 and the fact that E is normal over F, E is the splitting field of a separable polynomial $p(x) \in F[x]$. Say $\deg p(x) = t$. Any automorphism $g \in G$ will fix F, thus will permute the (distinct) roots $\{r_1, ..., r_t\}$ of $p(x)$. Since $E = F(r_1, ..., r_t)$, the restriction homomorphism $\rho : g \to g \mid_E$ will map G into the Galois group of E over F.

Put $H = \Gamma(E)$. Then $H = \{g \in G : g \mid_E = 1\} = \ker \rho$. Hence, H is a normal subgroup of G. Now K is also the splitting field of $p(x)$ over E. Therefore, we can use Theorem 5.3.2 to conclude that every automorphism of E extends to an automorphism of K. That is, the restriction map ρ is onto. By the first group isomorphism theorem, $G/H \cong Gal_F(E)$.

Finally, suppose that H is an arbitrary normal subgroup of G and put $E = \mathfrak{F}(H)$. If $g \in G$, it is a simple exercise to show that $\mathfrak{F}(gHg^{-1}) = gE$. Since $gHg^{-1} = H$ for all $g \in G$, $gE = E$ for all $g \in G$. We claim that this fact implies that E is a normal extension of F. To verify the claim, let $q(x) \in F[x]$ be irreducible with a root $e \in E$. If $q(x)$ had a root $r \notin E$, then the isomorphism $F(e) \cong F(r)$ could be extended to an automorphism g of K fixing F. Then, for this $g \in G$ we would have $g(E) \nsubseteq E$, a contradiction. The proof is complete! ∎

Thirty-seventh Problem Set

Problem 5.6.1 *Refer back to Example 5.6.2. (a) Find a formula like the one displayed as (s) for an automorphism τ of order three. Note that then we can regard S_3 as $S_3 = <\sigma, \tau>$. (b) Find the fixed subfield of $K = Q(\rho, c)$ corresponding to each subgroup of S_3. The easiest way to do this is to write an arbitrary element of K as a rational combination of basis vectors, then solve for the necessary conditions for that element to be fixed by all the members of a specified subgroup. The subgroups of S_3 can be written in terms of the generators σ, τ.*

Problem 5.6.2 *Let $F = (Z/pZ)(y)$ as in Example 5.6.3. Let $K = F(e)$, where $e^p = y$. Show that $| Gal_F(K) | \neq [K : F]$. What goes wrong in the proof of Theorem 5.6.2 for this example?*

Problem 5.6.3 *Determine the Galois group of $p(x) = (x^2 - 2)(x^2 - 3)$ over Q. Find all intermediate subfields between Q and the splitting field $K \subset \mathcal{R}$. Why do you know you have them all?*

Problem 5.6.4 *Let $K = (Z/pZ)(y)$ where y is transcendental over Z/pZ. Let G be the cyclic subgroup of $Aut(K)$ generated by the automorphism $\alpha : y \to y + 1$. Find $F = \mathfrak{F}(G)$ and $[K : F]$.*

Problem 5.6.5 *Prove that the Galois group of $x^p - 1$ over Q is cyclic of order $p - 1$.*

Problem 5.6.6 *Find an example of fields $F \leq K \leq L$ such that K is a normal extension of F and L is a normal extension of K, but L is not a normal extension of F.*

Problem 5.6.7 *Finish the proof of $(3) \to (1)$ in Theorem 5.6.5.*

Problem 5.6.8 *Find a complex number d such that $Q(i, \sqrt{2}) = Q(d)$.*

5.7 Galois criterion for solvability by radicals

The quadratic formula $x = (-b \pm \sqrt{b^2 - 4ac})/2a$, giving the roots of the quadratic $ax^2 + bx + c$ in terms of its coefficients, has been known since antiquity. During the period of the Italian Renaissance similar but much more complex formulas were obtained, giving the roots of general cubic and quartic polynomials in terms of rational operations and root extractions involving the coefficients of these polynomials. This procedure is called *solving the equation by radicals*. Up until the mid 1800's a number of attempts were made to generalize these results to polynomials of degree 5 and higher. These attempts were doomed to failure. The theorem of Abel-Ruffini, published in 1827, states that the general equation of n-th degree, $n \geq 5$, is not solvable by radicals. Before his untimely death at age 21, Galois gave a complete version of the proof of Abel-Ruffini's Theorem, together with a criterion for solvability by radicals and the subgroup-subfield correspondence of the previous section. It's fair to say that Galois' work was and remains the most profound work in algebra done by a single individual. We start by saying exactly what we mean when we say $p(x) \in F[x]$ is solvable by radicals.

Definition 5.7.1 *A polynomial $p(x) \in F[x]$ of positive degree is **solvable by radicals over** F if there is a tower of subfields $F = F_0 < F_1 < \cdots < F_n$ such that $F_{i+1} = F_i(d_i)$ with $d_i^{k_i} = a_i \in F_i$, $0 \leq i \leq n-1$, and such that F_n contains a splitting field of $p(x)$ over F.*

This means that every root of $p(x)$ can be obtained starting from a finite set of elements from F and performing a finite series of operations involving rational combinations and k_i-th roots.

Exercise 5.7.1 *Give a simple proof of the statement immediately above, by induction on the length of the field tower.*

We are ready to embark on the proof of Galois' magnificent theorem: Let F be a field of characteristic zero and let $p(x) \in F[x]$. Then the equation $p(x) = 0$ is solvable by radicals over F if and only if the Galois group of $p(x)$ over F is a solvable group. In this part we mostly follow the treatment in [W]. A series of preliminary facts will lead to the main result.

Definition 5.7.2 *Let C_n denote the splitting field of $x^n - 1$ over F. The field C_n is called the **cyclotomic field of order** n over F.*

Theorem 5.7.1 *The Galois group of C_n over Q is isomorphic to $U(n)$, the multiplicative group of units of Z/nZ. In particular this Galois group is abelian.*

Proof. Note that C_n is a Galois extension of Q. Let $c = e^{2\pi i/n}$ be the standard complex primitive n-th root of unity. Since the multiplicative order of c is exactly n, c will be a generator for the multiplicative subgroup of all complex n-th roots of unity. Thus, our splitting field $C_n = Q(c)$. Plainly, each automorphism α of $Q(c)$ is determined by $\alpha(c)$ and $\alpha(c)$ must be a primitive n-th root of unity. Thus, we can label each $\alpha \in G = Gal_Q(C_n)$ by $\alpha = \alpha_t$, where $\alpha_t(c) = c^t$ for some $t \in U(n)$. The map $\theta : \alpha_t \to t$ certainly is a multiplicative monomorphism from G into $U(n)$. We must check that θ is epic; that is, for each $u \in U(n)$, there is a corresponding automorphism $\alpha_u \in G$.

To check this, consider the n-th **cyclotomic polynomial**, denoted $\phi_n(x)$, given by the formula $\phi_n(x) = \prod_{u \in U(n)} (x - c^u)$. For each $\alpha_t \in G$, we have: $\alpha_t \phi_n(x) = \prod_{u \in U(n)} [x - \alpha_t(c^u)] = \prod_{u \in U(n)} [x - c^{tu}] = \phi_n(x)$. The last equality comes from the fact that, for fixed $t \in U(n)$, the map $u \to tu$ is a bijection of the set $U(n)$. Using my favorite Galois theory trick, since the coefficients of $\phi_n(x)$ are fixed by all elements of the Galois group G, we can conclude that $\phi_n(x) \in Q[x]$.

Plainly, $\phi_n(c^u) = 0$ for each $u \in U(n)$. Thus, if we can show that $\phi_n(x)$ is irreducible in $Q[x]$, then $\phi_n(x)$ will be the minimal polynomial for each c^u. But then, by Lemma 5.6.1, for all $u \in U(n)$, we will have our desired automorphism of C_n mapping $c \to c^u$.

It remains to show that $\phi_n(x)$ is irreducible. First, since $\phi_n(x) \in Q[x]$ is a product of one linear factor for each primitive n-th root of unity, there is a factorization in $Q[x]$ of the form $x^n - 1 = \phi_n(x)a(x)$. By the proof of Theorem 2.4.5, since $(x^n - 1) \in Z[x]$, it follows that $\phi_n(x) \in Z[x]$. (Think about this point again for a minute.)

The remainder of the proof is from [W]. Let $f(x)$ be the minimum polynomial for c over Q. As above, since $f(x) \mid x^n - 1$, we have $f(x) \in Z[x]$. Now let p be a prime not dividing n and let $g(x) \in Z[x]$ be the minimum polynomial over Q of the primitive n-th root c^p. Since both $f(x)$ and $g(x)$ are monic and irreducible in $Q[x]$, either $f(x) = g(x)$ or $f(x)$ and $g(x)$ are relatively prime.

Suppose that $f(x)$ and $g(x)$ are relatively prime. We will obtain a contradiction. In this case $f(x)$ and $g(x)$ can have no roots in common. Since both polynomials divide $x^n - 1$ in $Z[x]$, there must be a factorization in $Z[x]$ of the form:

$$(1) : x^n - 1 = f(x)g(x)h(x).$$

Moreover, the polynomial $g(x^p)$ has c as a root, whence, there is a factorization in $Z[x]$:

$$(2) : g(x^p) = f(x)k(x).$$

View equations (1) and (2) modulo p, that is regard all polynomials as being in $(Z/pZ)[x]$. Operating mod p, we have:

$$(2)_p : g(x^p) = g(x)^p = f(x)k(x).$$

By $(2)_p$ any irreducible factor $q(x)$ dividing $f(x)$ in $(Z/pZ)[x]$ must also divide $g(x)$. Therefore, by $(1)_p$, $q(x)^2$ must divide $x^n - 1 = f(x)g(x)h(x)$ in $(Z/pZ)[x]$. But, since $p \nmid n$, $x^n - 1$ is relatively prime to its derivative nx^{n-1} in $(Z/pZ)[x]$. Hence $x^n - 1$ has no multiple roots in (Z/pZ)-a contradiction.

It follows that, for each $p \nmid n$, $f(x) = g(x)$ is the minimal polynomial for c^p over Q. Now let c^u be any primitive n-th root of unity. Note that $u \in U(n)$ is of the form $u = p_1 p_2 \cdots p_k$ with the p_i primes and prime to n. By what we just did, the minimal polynomial of c^{p_1} over Q is $f(x)$. But then, using the same argument, the minimal polynomial of $(c^{p_1})^{p_2}$ over Q is also $f(x)$. It follows that $f(x)$ is the minimal polynomial over Q for all the primitive n-th roots of unity. Thus, $f(x)$ divides $\phi_n(x)$ in $Z[x]$. But each primitive n-th root $c^u, u \in U(n)$, is a root of $f(x)$, so $\phi_n(x) = f(x)$. Since $f(x)$ is irreducible in $Q[x]$, so is $\phi_n(x)$. Hence, we can write $G = \{\alpha_u : u \in U(n)\}$ as desired. The proof is complete. ∎

We note that, if F is **any** field of characteristic zero, the polynomial $x^n - 1 \in F[x]$ is separable. Hence, if K is a splitting field for $x^n - 1$, the multiplicative subgroup $U < K$ of n-th roots of unity has order n. By Theorem 5.3.4, U is cyclic. In this more general situation a **primitive n-th root of unity** can be taken to be any cyclic generator c for U.

In view of the above remark, we can define a cyclotomic polynomial $\phi_n(x) = \prod_{u \in U(n)} (x - c^u)$ for an arbitrary field F of characteristic zero. Let G be the Galois group of the splitting field F_n of $x^n - 1$ over F. If we try to repeat the proof of Theorem 5.7.1 in this more general situation, the only problem will be that $\phi_n(x)$ may not be irreducible over F. If $\phi_n(x) = f(x)g(x)$ in $F[x]$, each primitive n-th root of unity will be a root of either $f(x)$ or $g(x)$. Say that $f(c) = 0$. Then any automorphism $\alpha \in G$ is determined by $\alpha(c) = c^t, t \in U(n)$, but c^t must remain a root of $f(x)$. Hence, the correspondence $\alpha \to t$ will not be onto. But, in any event, the Galois group of F_n over F will still be isomorphic to a subgroup of $U(n)$.

Here is a similar result, whose proof is a good bit easier.

Theorem 5.7.2 *Let F be a field of characteristic zero that contains all the n-th roots of 1, and let $a \in F$. Then the Galois group G of $x^n - a$ over F is isomorphic to a subgroup of the additive cyclic group $Z(n)$. In particular, G is cyclic. If $x^n - a$ is irreducible in $F[x]$, then $G \cong Z(n)$.*

Proof. Let K be a splitting field of $x^n - a$ over F. If $K \neq F$, let $b \in K, b \notin F$ with $b^n = a$. If c is a primitive n-th root of 1, then the roots of $x^n - a$ are $\{b, cb, ..., c^{n-1}b\}$. Since F contains all n-th roots of 1, $K = F(b)$. Each $\alpha \in G = Gal_F(K)$ is determined by $\alpha(b)$ and $\alpha(b) \in \{b, cb, ..., c^{n-1}b\}$. Consider the map $\sigma : \alpha \to t$ from G to $Z(n)$ defined by the condition $\sigma(\alpha) = t$ if $\alpha(b) = c^t b$. If $\alpha, \beta \in G$ with $\alpha(b) = c^t b, \beta(b) = c^l b$, then $\alpha\beta(b) = \alpha(c^l b) = \alpha(c^l)\alpha(b) = c^l c^t b = c^{l+t} b$. Hence σ is a homomorphism from G into the additive group $Z(n)$. It is immediate that $Ker\ \sigma = \{1\}$ so that G is isomorphic to a subgroup of $Z(n)$. If $x^n - a$ is irreducible in $F[x]$, then the minimal polynomial for each $c^t b$ must be $x^n - a$. Thus, for each t, there will be an automorphism α with $\alpha(b) = c^t b$. In this case σ is onto and $G \cong Z(n)$. ∎

Since the group $Z(p)$ has no proper subgroups, we have the following.

Corollary 5.7.1 *Let F be a field of characteristic zero containing all the p-th roots of 1, p a prime, and let $a \in F$. Then $x^p - a$ is either irreducible or splits completely in $F[x]$. If $x^p - a$ is irreducible in $F[x]$, then its Galois group over F is isomorphic to $Z(p)$.*

We need two technical lemmas before presenting our main result.

Lemma 5.7.1 *Let $p(x) \in F[x]$ be solvable by radicals over F, where F has characteristic zero. Then there is a tower of fields $F < K_0 < K_1 < \cdots < K_n$ such that: (1) $K_0 = F(c)$ with c a primitive n-th root of unity and K_0 is a normal separable extension of F; (2) for $0 \le i \le n - 1$, each $K_{i+1} = K_i(b_i)$ is a normal separable extension of K_i with $b_i^{k_i} = a_i \in K_i$ and (3) K_n contains a splitting field of $p(x)$ over F.*

Proof. Since $p(x)$ is solvable by radicals over F, there is a tower of fields $F = F_0 < F_1 < \cdots < F_n$ such that $F_{i+1} = F_i(d_i)$ with $d_i^{k_i} = a_i \in F_i$, $0 \le i \le n - 1$, and such that F_n contains a splitting field of $p(x)$ over F. If $k_i = lm$, then we can refine the extension $F_i < F_i(d_i)$ to $F_i < F_i(d_i^l) < F_i(d_i)$. Thus, we may assume that all the positive integers k_i are prime.

Let m be the least common multiple of the k_i's. Take K_0 to be a splitting field of $x^m - 1$ over F. Since $char F = 0$, we can apply Theorem 5.6.5 to see that K_0 is a normal separable extension of F. Moreover, $K_0 = F(c)$ for c a primitive n-th root of unity. For $i \ge 0$, consider the separable polynomial $(x^{k_i} - a_i) \in K_i[x]$. By Corollary 5.7.1, there are two possibilities. If $(x^{k_i} - a_i)$ splits completely in $K_i[x]$, put $K_{i+1} = K_i$ and renumber the fields K_j from this point on. Otherwise, $(x^{k_i} - a_i)$ is irreducible in $K_i[x]$ and $K_{i+1} = K_i(b_i)$ with $b_i^{k_i} = a_i \in K_i$. Furthermore, K_{i+1} is a splitting field of $x^{k_i} - a$ over K_i. Again by Theorem 5.6.5, K_{i+1} is a normal separable extension of K_i.

Our final claim is that we can embed each field F_i into the corresponding field $K_i, 0 \leq i \leq n$. (See Problem 5.7.5.) Hence, K_n will contain a splitting field of $p(x)$ over F. ■

Lemma 5.7.2 *Let K be a normal extension of F of degree p and suppose that F has characteristic zero and contains all the p-th roots of unity. Then $K = F(d)$ with $d^p \in F$.*

Proof. Since K is a finite dimensional extension of F and $char F = 0$, by Theorem 5.6.4 K is a simple algebraic extension. Say that $K = F(a)$. Since K is a finite dimensional normal separable extension of F, K is a Galois extension of F. Let σ be a generator of the cyclic p-group $Gal_F(K)$.

Now let $\omega \in F$ be a p-th root of unity, $\omega \neq 1$. For $0 \leq i < p$ consider the quantities

$$d_i = a + \omega^i \sigma(a) + (\omega^2)^i \sigma^2(a) + \cdots + (\omega^{p-1})^i \sigma^{p-1}(a).$$

A simple calculation shows that $\sigma(d_i) = \omega^{-i} d_i$ for each i. Therefore, $\sigma(d_i^p) = (\omega^{-i})^p d_i^p = d_i^p$. It follows that $d_i^p \in F$. Since each ω^i satisfies the cyclotomic polynomial of degree $p-1$, another simple calculation shows that $\sum_{j=0}^{p-1} d_j = pa$. Since a is not in F, a field of characteristic zero, neither is pa. Thus, at least one d_j is not in F. Put $d = d_j$. Then $K = F(d)$ and the proof is complete. ■

Exercise 5.7.2 *Check out the details of the above proof, namely that $\sigma(d_i) = \omega^{-i} d_i$ for each i and $\sum_{j=0}^{p-1} d_j = pa$.*

We are now ready for our main result.

Theorem 5.7.3 *Let F be a field of characteristic zero and let $p(x) \in F[x]$. Then $p(x)$ is solvable by radicals over F if and only if its Galois group over F is solvable.*

Proof. Suppose $p(x)$ is solvable by radicals over F. Let $F < K_0 < K_1 < \cdots < K_n$ be the tower of fields produced by Lemma 5.7.1. Consider the normal series of subgroups $< e >= Gal_{K_n}(K_n) \lhd Gal_{K_{n-1}}(K_n) \lhd \cdots \lhd Gal_{K_1}(K_n) \lhd Gal_{K_0}(K_n) \lhd Gal_F(K_n)$. Each group is normal in the next one since each subfield is a normal extension of the previous one. By the fundamental subgroup-subfield correspondence theorem, $Gal_{K_i}(K_n)/Gal_{K_{i+1}}(K_n)$ is isomorphic to $Gal_{K_i}(K_{i+1})$ for all $i \geq 0$. Furthermore, by the dimension relations, each Galois group $Gal_{K_i}(K_{i+1})$ is cyclic of prime order. Finally, $Gal_F(K_n)/G_{K_0}(K_n) \cong Gal_F(K_0)$, which is abelian by Theorem 5.7.1. We have proved that $Gal_F(K_n)$ is solvable.

To complete this direction of the proof, let $K \leq K_n$ be a splitting field for $p(x)$ over F. Then $Gal_F(K)$, the Galois group of $p(x)$ over F, is isomorphic to $Gal_F(K_n)/G_K(K_n)$. Hence, $Gal_F(K)$ is also solvable.

Conversely, let K be a splitting field for $p(x)$ over F and suppose that $Gal_F(K)$ is solvable. Say that $[K : F] = n$. Let N be a splitting field of $x^n - 1$ over K. We have $F \leq K \leq N$, $G_K(N) \lhd G_F(N)$ (since K is a splitting field over F) and $Gal_F(K)$ is solvable. Thus, the factor group $Gal_F(N)/Gal_K(N) \cong Gal_F(K)$ is solvable. Since, by Theorem 5.7.1, $Gal_K(N)$ is abelian, it follows that $Gal_F(N)$ is solvable.

Since $Gal_F(K)$ is solvable, arguing in the converse direction to the proof in the first paragraph, we get a tower of subfields $F = F_0 < F_1 < \cdots < F_r = K$ such that each F_{i+1} is a separable normal extension of degree p_i over F_i. Say that F_{i+1} is the splitting field of the separable polynomial $f_i(x) \in F_i[x]$. Let K_0 be the splitting field of $x^n - 1$ over F taken inside the field N. For $i \geq 0$ let K_{i+1} be the splitting field of $f_i(x)$ over K_i, constructed as a subfield of N. Note that $K_r \geq F_r = K$.

The restriction of an element of $Gal_{K_i}(K_{i+1})$ to F_{i+1} induces a homomorphism from $Gal_{K_i}(K_{i+1})$ to $Gal_{F_i}(F_{i+1})$. An element in the kernel of this homomorphism fixes the roots of $f_i(x)$ and, hence, fixes F_{i+1}. Thus, the restriction map is monic and each $Gal_{K_i}(K_{i+1})$ is isomorphic to a subgroup of $Gal_{F_i}(F_{i+1}) \cong Z(p_i)$. It follows that, for $i \geq 0$, $[K_{i+1} : K_i] = 1$ or p_i. By Lemma 5.7.2, $K_{i+1} = K_i$ or $K_{i+1} = K_i(d_i)$ with $d_i^{p_i} = a_i \in K_i$.

Moreover, $Gal_F(K_0)$, a finite abelian group, can be written as the union of an ascending chain of subgroups, each of prime index in the next. This fact, combined with the final claim of the previous paragraph, shows that we can find a tower of subfields $F = L_0 < L_1 < \cdots < L_t$ such that, for each i, $L_{i+1} = L_i(d_i)$ with $d_i^{p_i} = l_i \in L_i$ and such that the splitting field K is a subfield of the top field L_t. We have shown that $p(x)$ is solvable by radicals over F. The proof is complete. ∎

Thirty-eighth Problem Set

Problem 5.7.1 *Let K be a Galois extension of F and let L, M be intermediate subfields. Show that $\alpha(L) = M$ for some $\alpha \in Gal_F(K)$ if and only if $\alpha Gal_L(K)\alpha^{-1} = Gal_M(K)$.*

Problem 5.7.2 *Find the Galois group of $x^4 - 2$ over Q. Is $x^4 - 2$ solvable by radicals over Q?*

Problem 5.7.3 *Do the same problem for $x^4 + x^2 + 1$.*

Problem 5.7.4 *Do the same problem for $x^5 - 1$.*

Problem 5.7.5 *In the proof of Lemma 5.7.2, verify that each field F_i can be embedded in the corresponding field K_i.*

5.8 The general equation of degree n

Let's stop to think what we mean when we say that the quadratic formula $x = -b \pm \sqrt{b^2 - 4ac}/2a$ gives us the roots of the general quadratic equation $ax^2 + bx + c = 0$ over some field F of characteristic zero. What we mean is that the formula works for $a, b, c \in F$ unrestricted coefficients, that is, subject to no algebraic relations at all. Put another way, we are really solving the quadratic over $F(a, b, c)$, where a, b, c are transcendental variables over F. Additionally, in considering the general equation of degree n, it plainly does no harm to assume that the equation is monic. These considerations motivate the following definition.

Definition 5.8.1 *Let F be a field of characteristic zero. The **general polynomial equation of degree** n **over** F is $p_n(x) = \sum_{i=0}^{n} a_i x^i = 0$, where $a_n = 1$ and $a_0, ..., a_{n-1}$ are commuting transcendental variables over F. We say the general polynomial equation of degree n is solvable by radicals over F if $p_n(x) = 0$ is solvable by radicals over the field $F(a_0, ..., a_{n-1})$.*

Here is the key fact.

Theorem 5.8.1 *For every field F of characteristic zero and every positive integer n, the Galois group of $p_n(x)$ over $F(a_0, ..., a_{n-1})$ is isomorphic to S_n.*

Proof. Let F be a field of characteristic zero, $p_n(x)$ be the general polynomial of degree n over F and put $L = F(a_0, ..., a_{n-1})$. Let $K = L(r_1, ..., r_n)$ be a splitting field of $p_n(x)$ over L, where the r_i are the distinct roots of $p_n(x)$. In $K[x]$, we have $p_n(x) = (x - r_1) \cdots (x - r_n)$. Expanding this product and comparing coefficients shows that the elements a_i are the **elementary symmetric polynomials** in the r_i; that is, $a_{n-1} = -\sum_{1 \le i \le n} r_i$, $a_{n-2} = \sum_{1 \le i < j \le n} r_i r_j, ..., a_0 = \pm r_1 \cdots r_n$. Thus, we can regard $L \le K$ as the subfield obtained by adjoining the elementary symmetric polynomials in the r_i to F.

As always, we can identify $Gal_L(K)$ with a subgroup of S_n, the subgroup of all legal permutations of the roots of the polynomial under consideration. But here **any** permutation of the r_i induces an automorphism of K which leaves L fixed. Hence, $Gal_L(K) = S_n$.

We can obtain a little more information. The subfield L, being the fixed field of its Galois group S_n, contains all **symmetric rational functions**, that is rational functions in the variables $r_1, ..., r_n$ invariant under any permutation of these variables. Hence, any symmetric rational function can be written as a rational combination of the elementary symmetric polynomials. ■

Since S_n is solvable if and only if $n < 5$, we have the following corollary.

Corollary 5.8.1 *The general polynomial equation $\sum_{i=0}^{n} a_i x^i = 0$ is solvable by radicals over an arbitrary field of characteristic zero if and only if $n < 5$.*

We close the Chapter by presenting an example from [W] of a specific $p(x) \in Q[x]$ whose Galois group over Q is S_5.

Example 5.8.1 *Let $p(x) = x^5 - 4x + 2 \in Q[x]$. By Eisenstein's criterion, $p(x)$ is irreducible in $Q[x]$. Using the intermediate value theorem, we can see that $p(x)$ has three real roots. Let $K \le C$ be a splitting field for $p(x)$ and $r, s \in K$ be roots of $p(x)$. Since $p(x)$ is irreducible, it is the minimal polynomial over Q for both r and s. Thus, there is an $\alpha \in G = Gal_Q(K)$ such that $\alpha(r) = s$. Moreover, complex conjugation permutes the two complex roots of $p(x)$ and fixes the real roots. Hence, if we regard $G \le S_5$, we have that G acts transitively on $\{1, ..., 5\}$ and that G contains a transposition. The only such subgroup of S_5 is S_5 itself so that $G = S_5$.*

Thirty-ninth Problem Set

Problem 5.8.1 *Find $p(x) \in Q[x]$ whose Galois group over Q is S_3.*

Further reading: For me [J2] is the best thing to look at next for more advanced topics in field theory. Also, the first part of [Ka2] contains a compact, nicely presented version of the basic material.

Chapter 6

Topics in Noncommutative Rings

6.1 Introduction

In this chapter we present some of the basics of the classical theory of non-commutative rings. In my mind no one can improve on the elegant exposition in Herstein's monograph Noncommutative Rings [H]. So this chapter follows closely the exposition of parts of this monograph. Throughout, the symbol $_RM$ will denote the fact that R is a (not necessarily commutative) ring and that M is a left R-module. All R-modules will be left R-modules, but we will show that the theory we develop is right-left symmetric. Recall that $(0 : M)$, the annihilator of M, is the ideal consisting of all $r \in R$ such that $rM =< 0 >$. For $m \in M$, also recall that $(0 : m)$, the annihilator of m, is the left ideal consisting of all $r \in R$ such that $rm = 0$.

Exercise 6.1.1 *Prove that $(0 : M)$ is an ideal of R.*

Definition 6.1.1 *A module $_RM$ is called **faithful** if $rM =< 0 >$ only if $r = 0$.*

Exercise 6.1.2 *Prove that $_RM$ can be naturally made into a faithful $R/(0 : M)$ module.*

Exercise 6.1.3 *Prove that the factor ring $R/(0 : M)$ is isomorphic to a subring of $E(M, +)$. Here $E(M, +)$ is the set of endomorphisms of the additive abelian group $(M, +)$, made into a ring via the operations of pointwise* *addition of endomorphisms as addition and composition of endomorphisms as multiplication. The proof is not at all difficult, but you'll need to use all of the module properties.*

A ring homomorphism from a ring R into the endomorphism ring $E(A, +)$, for some abelian group A, is called a **representation** of R.

Exercise 6.1.4 *Think for a few minutes to convince yourself that there is a natural correspondence between left R-modules and representations of R.*

6.2   Simple modules

Definition 6.2.1 *A module $_RM$ is **simple** if $M \neq\; <0>$ and if $<0>$ and M are the only submodules of M.*

Definition 6.2.2 *Let M be a left R-module. The **commuting ring** $C(M)$ of $_RM$ is the ring $Hom_R(M, M)$.*

Exercise 6.2.1 *Prove that $C(M)$ is a subring of $E(M, +)$.*

Lemma 6.2.1 *(Schur's Lemma) If M is simple, then $C(M)$ is a division ring.*

Proof. We need to show that, if $0 \neq \lambda \in C(M)$, then $\lambda^{-1} \in C(M)$. But if $\lambda \in C(M)$ both $Ker\,\lambda$ and $Im\,\lambda$ are submodules of M. Since M is simple and $0 \neq \lambda$, we must have $Ker\,\lambda =<0>$ and $Im\,\lambda = M$. That is, λ is a bijection on the set M. We claim that the inverse bijection λ^{-1} is also an element of $C(M)$, that is $\lambda^{-1} \in E(M, +)$ and, for all $r \in R$, $\lambda^{-1}r = r\lambda^{-1}$ as endomorphisms of M. We have, for all $m, m' \in M$, $\lambda(\lambda^{-1}m + \lambda^{-1}m') = \lambda\lambda^{-1}m + \lambda\lambda^{-1}m' = m + m' = \lambda\lambda^{-1}(m + m')$. Since λ is monic, $\lambda^{-1}m + \lambda^{-1}m' = \lambda^{-1}(m+m')$ and $\lambda^{-1} \in E(M, +)$. Similarly, for all $m \in M, r \in R$, $\lambda(\lambda^{-1}rm) = rm = r(\lambda\lambda^{-1}m) = \lambda(r\lambda^{-1}m)$. Hence, $\lambda^{-1}rm = r\lambda^{-1}m$ and the proof is complete. ∎

We close this section with a simple theorem (sorry about that). Recall that, if L is a left ideal of R, we let R/L be the abelian factor group $R/L = \{L + r : r \in R\}$. Then R/L becomes a natural left R-module by defining $r'(L + r) = L + r'r$.

Theorem 6.2.1 *Every simple R-module is isomorphic to R/L for some maximal left ideal $L \subset R$.*

Proof. By the correspondence theorem for left ideals, if $L \subset R$ is a maximal left ideal, then R/L will be a simple R-module. Conversely, suppose that $_RM$ is simple and let $0 \neq m \in M$. Since $m = 1m \in Rm \leq M$, then $Rm = M$. Thus, the map $r \to rm$ is an R-module epimorphism from $_RR$ to $_RM$ with kernel $L = (0 : m)$. By the first module isomorphism theorem, $R/L \cong M$. By the submodule correspondence theorem, L is a maximal submodule of $_RR$. Equivalently, L is a maximal left ideal of R. ∎

Fortieth Problem Set

Problem 6.2.1 *Show that the simple Z-modules are just the cyclic groups of the form $Z(p)$ for p a prime.*

Problem 6.2.2 *Let V be a finite dimensional vector space over a field F and let $R = L_F(V)$ be the ring of linear transformations from V to V. Prove that V is a simple R-module.*

Problem 6.2.3 *Let $M = V$ and $R = L_F(V)$ be as above. Show that $C(M) = F_l$, the set of scalar multiplications by elements of F.*

Problem 6.2.4 *Let $M = \mathcal{R}^2$ be an $\mathcal{R}[x]$-module via the action $p(x)v = p(T)v$, where $T : \mathcal{R}^2 \to \mathcal{R}^2$ has matrix $\begin{bmatrix} 0 & -1 \\ 1 & 0 \end{bmatrix}$ with respect to the standard basis. Prove that $_{\mathcal{R}[x]}M$ is simple. Compute and identify $C(M)$.*

6.3 The Jacobson Radical

Definition 6.3.1 *Let S be the class of all simple left R-modules. We define the **Jacobson Radical**, $J(R)$, of a ring R by $J(R) = \cap_{M \in S}(0 : M)$.*

Since each annihilator $(0 : M)$ is an ideal of R so is $J(R)$.

Remark 6.3.1 *Since we're intersecting ideals of R, there's no logical difficulty in taking the intersection over the class S instead of over a set. Alternately, note that, if $_RM \cong_R M'$, then $(0 : M) = (0 : M')$. In view of this fact and Theorem 6.2.1, we could define $J(R) = \cap_{L \in \max(R)}(0 : R/L)$, where $\max(R)$ is the set of all maximal left ideals of a ring R.*

The following theorem gives another intersection characterization for $J(R)$.

Theorem 6.3.1 *We have $J(R) = \cap_{L \in \max(R)} L$.*

Proof. Each $(0 : M) = \cap_{0 \neq m \in M}(0 : m)$. If M is simple, $M = Rm \cong R/(0 : m)$ for each nonzero $m \in M$. By the submodule correspondence theorem, each $(0 : m)$ is a maximal left ideal of R. Thus, $J(R) = \cap_{M \in \mathcal{S}}(0 : M) = \cap_{M \in \mathcal{S}, 0 \neq m \in M}(0 : m) \subseteq \cap_{L \in \max(R)}L$. On the other hand, if L is any maximal left ideal of R, then $R/L \in \mathcal{S}$ with $L = (0 : L+1) \supseteq (0 : R/L)$. Hence, $\cap_{L \in \max(R)}L \supseteq J(R)$. $\blacksquare$

We next would like to characterize $J(R)$ as the largest ideal of R satisfying a property which depends only on the internal ring structure of R. To do this we need a definition.

Definition 6.3.2 *An element $r \in R$ is* **left quasi-regular** *(lqr) if $1 + r$ has a left inverse r' in R. The element r' is called the* **left quasi-inverse** *of r. A (left) ideal is a* **lqr (left) ideal** *if all of its elements are left quasi-regular.*

Theorem 6.3.2 *The Jacobson Radical is a lqr ideal of R that contains all lqr left ideals of R. Thus, $J(R)$ is the unique maximal lqr left or two-sided ideal of R.*

Proof. Let $j \in J(R)$ and suppose that $1 + j$ has no left inverse in R. We'll derive a contradiction. We have that $R(1 + j)$ is a proper left ideal of R. Consider the set $\mathcal{L}$ of proper left ideals of R which contain $R(1 + j)$. Partially order $\mathcal{L}$ by set inclusion. We can then apply Zorn's Lemma (check this) to obtain a maximal element $L \in \mathcal{L}$. Note that L is also maximal in the set of proper left ideals of R. Thus, by Theorem 6.3.1, $j \in L$. But then $1 = (1 + j) - j \in L$, a contradiction. We've shown that each element of $J(R)$ is left quasi-regular.

Now let L be a left quasi-regular left ideal of R. To show $L \subset J(R)$, we show that $LM =< 0 >$ for M any simple R-module. The proof is again by contradiction. Suppose that $Lm \neq< 0 >$ for some element m of a simple module M. Thus, the submodule $Lm = M$, so there exists $l \in L$ such that $lm = -m$. Hence, $(1 + l)m = 0$. But $(1 + l)$ is left invertible and $m \neq 0$, a contradiction. The proof is complete. $\blacksquare$

Suppose $j \in J(R)$ has left quasi-inverse $j' = 1 + j_1$. By definition, $(1 + j_1)(1 + j) = 1$ so that $j_1 = -(j + j_1 j) \in J(R)$. Thus, $(1 + j_1)$ has a left multiplicative inverse in R. Since $(1 + j)$ is a right inverse for $(1 + j_1)$, it follows that $(1 + j)$ is invertible in R with two-sided inverse $(1 + j_1)$. (Why?) We have shown that $J(R)$ is actually a **quasi-regular ideal**; that is $(1 + j)$ is invertible in R for each $j \in J(R)$.

Now imagine that we had developed our theory starting with simple right R-modules. We would have constructed a "right" Jacobson Radical $J(R)_r = \cap_{M \in \mathcal{S}_r}(0 : M)$, where $\mathcal{S}_r$ is the class of simple right R-modules.

Following the development above, we would show that $J(R)_r$ is a quasi-regular ideal and is the unique maximal right quasi-regular right ideal of R. By symmetry, we see that $J(R)_r = J(R)$ (although the collections $\mathcal{S}$ and $\mathcal{S}_r$ may be quite different).

Definition 6.3.3 *A one sided ideal $I \leq R$ is called **nil** if $x \in I \to x^{n(x)} = 0$ for some positive integer $n(x)$.*

Theorem 6.3.3 *Every nil left or right ideal $I \leq R$ is contained in $J(R)$.*

Proof. Let $x \in I$ and suppose that $x^n = 0$. Then $1 + x$ is invertible with inverse $(1 - x + x^2 - \cdots \pm x^{n-1})$. Hence I is a quasi-regular one sided ideal so that, by our previous remarks, $I \leq J(R)$. ∎

Definition 6.3.4 *A ring R is called **Artinian**, if, whenever $L_1 \supseteq L_2 \supseteq \cdots \supseteq L_n \supseteq \cdots$ is a descending chain of left ideals of R, then there exists an integer k such that $L_j = L_k$ for all $j \geq k$. In other words, there is no infinite properly descending chain of left ideals of R.*

Example 6.3.1 *The ring of integers Z is not Artinian since $Z \supset 2Z \supset 4Z \supset \cdots$ is an infinite properly descending chain of ideals.*

Example 6.3.2 *Any finite dimensional algebra A over a field F is Artinian. This follows since, if $L \subset A$ is a left ideal and $c \in F$, then $cL = c(1_A)L = (c1_A)L \subset L$. Hence, L is also an F-subspace. Since $\dim_F(A) < \infty$, A is Artinian.*

Our version of Artinian is left Artinian. There is an obvious notion of right Artinian. In general, these two concepts are not related. See Problem 6.3.8.

We close this section with a partial converse to Theorem 6.3.3. For a one or two-sided ideal I and a positive integer n, let I^n be the ideal consisting of all finite sums of products of n (not necessarily distinct) terms taken from I. Note that we have a descending chain of ideals $I \geq I^2 \geq \cdots \geq I^n \geq \cdots$. An ideal I is **nilpotent** if there exists a positive integer n such that $I^n = 0$. Every nilpotent ideal is automatically nil, but, in general, nil ideals need not be nilpotent. See Problem 6.3.9. In any event, all nilpotent one or two-sided ideals are contained in $J(R)$. In the Artinian case, we can say much more.

Theorem 6.3.4 *Let R be an Artinian ring and $J = J(R)$. Then J is nilpotent and, thus, is the unique maximal nilpotent one or two-sided ideal of R.*

Proof. Consider the descending chain $J \geq J^2 \geq \cdots \geq J^n \geq \cdots$. Since R is Artinian, there exists n such that $J^n = J^{n+1} = \cdots = J^{2n}$. Hence, $J^{2n}x = 0$ implies $J^n x = 0$.

Suppose $J^n \neq 0$. We'll derive a contradiction. Let $W = \{x : J^n x = 0\}$. Then W is an ideal of R. (Check this.) We cannot have $J^n \leq W$ for then $J^n J^n = J^{2n} = 0$ so that $J^n = 0$. So in the factor ring $\bar{R} = R/W$ the ideal $\bar{J}^n = (J^n + W)/W$ is nonzero. Furthermore, suppose $\bar{J}^n \bar{x} = 0$ for some $\bar{x} = W + x \in \bar{R}$. Then $J^n x \leq W$ so that $J^{2n}x = 0$. But then $J^n x = 0$, that is $x \in W$. It follows that $\bar{x} = 0$.

Since $\bar{J}^n$ is a nonzero ideal of $\bar{R}$ and $\bar{R}$ is an Artinian ring (why?), $\bar{J}^n$ contains a minimal nonzero left ideal $\bar{L}$. Under the natural action of left multiplication, $\bar{L}$ will be a simple left R-module. (Think about this for a minute.) Thus, by definition of J, $J\bar{L} = 0$. So, plainly, $\bar{J}^n \bar{L} = 0$ as well. But then, by the previous paragraph, $\bar{L} = 0$, our desired contradiction. It follows that $J^n = 0$ and the proof is complete. ∎

Corollary 6.3.1 *In an Artinian ring any nil or quasi-regular one or two-sided ideal is nilpotent.*

Proof. Any such ideal is contained in $J(R)$. ∎

Forty-first Problem Set

Problem 6.3.1 *Prove that, if $r \in R$ has a quasi-inverse r', then r' is unique.*

In the next four problems find $J(R)$ for the given ring R.

Problem 6.3.2 $R = Z$.

Problem 6.3.3 $R = Z_p$. *Recall that, for a prime p, Z_p is the subring of Q consisting of all rationals r/s in reduced form such that $p \nmid s$.*

Problem 6.3.4 $R = Z/(12Z)$.

Problem 6.3.5 *R is the ring of $n \times n$ lower triangular matrices.*

Problem 6.3.6 *Prove that $J[R/J(R)] = <0>$ for any ring R.*

Problem 6.3.7 *Find a ring R with quasi-regular elements but no quasi-regular ideals. The radical of this ring will be zero, even though the ring has quasi-regular elements. Hint: try a well-known ring.*

Problem 6.3.8 *Let R be the subring of 2×2 matrices of the form $\left\{ \begin{bmatrix} z & 0 \\ r & s \end{bmatrix} \right.$: $z \in Z$, $r, s \in \mathcal{R}\}$. Show that R is Artinian but not right Artinian. Now give an example of a ring which is right Artinian but not Artinian.*

Problem 6.3.9 *Let R be the ring direct product $R = \prod_{i=2}^{\infty} Z/2^i Z$ and let $I = \bigoplus_{i=2}^{\infty} (2Z/2^i Z)$. Here $\bigoplus_{i=2}^{\infty} (2Z/2^i Z)$ denotes the following subset of R : $\{(r_i) : r_i \in (2Z/2^i Z) \text{ and } r_i = 0 \text{ for all but finitely many } i\}$. Prove that I is a nil, but not nilpotent, ideal of R.*

6.4 The Jacobson Density Theorem

In this chapter we are going to expand our original definition of vector space to include modules over division rings, not just modules over fields. Almost all the standard vector space theorems work in this new setting; we just have to be a little careful since the ring of scalars need not be commutative. In particular, if V is a left vector space over a division ring D, left multiplication by an element in D but not in the center of D will not be a linear transformation. Also, in this context, we must distinguish between left and right vector spaces.

Let $_R M$ be simple with centralizer $C(M)$. By Schur's Lemma, $C(M)$ is a division ring. Since $C(M)$ is a subring of $E(M, +)$, M can be regarded in the natural way as a left vector space over $C(M)$.

We are going to do something that will look odd; we'll see the advantage of this viewpoint when we prove the Wedderburn-Artin Theorem. Let Γ be the **opposite ring** of $C(M)$; that is, $(\Gamma, +)$ and $(C(M), +)$ are isomorphic abelian groups, but the product xy of two elements $x, y \in \Gamma$ is defined as $yx \in C(M)$. Put simply, we have "turned the multiplication around" to go from the division ring $C(M)$ to the division ring Γ.

Exercise 6.4.1 *Prove that M as above can be regarded as a right vector space over Γ.*

We call Γ the **centralizer** of M. We are now ready for the Jacobson Density Theorem. We present a slightly reworded version of the proof of Theorem 2.1.2 in [H].

Theorem 6.4.1 *(Jacobson Density Theorem)* *Let M be a simple left R-module with centralizer Γ. Regard M as a right vector space over Γ. Suppose that $\{m_1, ..., m_k\} \subset M$ is a finite Γ-independent set and $\{n_1, ..., n_k\} \subset M$ is an arbitrary subset. Then there exists $r \in R$ such that $rm_i = n_i, 1 \leq i \leq k$.*

Proof. If $k = 1$, the theorem follows by simplicity of $_R M$. Henceforth, suppose $k \geq 2$. To prove the theorem, it suffices to prove the following claim: Given V, an arbitrary subspace of M of finite positive dimension, and a vector $m \notin V$, there exists $r \in R$ such that $rV =< 0 >$ but $rm \neq 0$.

We first verify that our claim suffices to prove the theorem. Suppose that, given V, m, we can always find such an r. Then $Rrm \neq< 0 >$ so that $Rrm = M$. We can thus find $s \in R$ such that srm is arbitrary and $srV =< 0 >$. Given $\{m_1, ..., m_k\}, k \geq 2$, a Γ-independent subset of M and $\{n_1, ..., n_k\}$ an arbitrary subset of M, let V_i, $1 \leq i \leq k$, be the subspace spanned by $\{m_j : j \neq i\}$. For each i, since $m_i \notin V_i$ and $\dim_\Gamma V_i = k - 1$, we can invoke our claim to find $r_i \in R$ such that $r_i m_i = n_i$ and $r_i V_i =< 0 >$. Then the element $r = \sum_{i=1}^{k} r_i$ will be such that $rm_i = n_i, 1 \leq i \leq k$.

Now we prove our claim by induction on d, the dimension of V over Γ. If Inductively, suppose our claim holds for all subspaces of dimension less than d. Write V as a direct sum of subspaces $V = V_0 \oplus w\Gamma$, where $\dim_\Gamma V_0 = (d - 1)$. By our inductive hypothesis, for $m \notin V_0$, there exists $r \in (0 : V_0)$ such that $rm \neq 0$. (Note that, if $d = 1$, then $V_0 =< 0 >$ and $(0 : V_0) = R$. So this case is covered in the rest of the proof.)

Suppose that our claim is false for V. Then there exists $m \in M, m \notin V$ such that, whenever $rV =< 0 >$, then $rm = 0$. We will obtain a contradiction. By our inductive hypothesis, since $w \notin V_0$, $(0 : V_0)w \neq< 0 >$. But $(0 : V_0)$ is a left ideal of R so that $(0 : V_0)w$ is a submodule of M. By simplicity of M, we must have $(0 : V_0)w = M$.

Define $\theta : M \rightarrow M$ by $\theta(x) = am$, where $a \in (0 : V_0)$ is such that $aw = x$. Note that θ is well defined since, if $x = aw = a'w$ for $a, a' \in (0 : V_0)$, then $(a - a') \in (0 : V_0) \cap (0 : w) = (0 : V)$. By the choice of m, $(a - a')m = 0$.

It's easy to check that $\theta \in E(M, +)$. (Check it.) Additionally, if $x \in M$ and $x = aw$ for $a \in (0 : V_0)$, then, for $r \in R$, $r\theta(x) = r(am) = (ra)m = \theta(rx)$. Thus, $\theta \in C(M)$.

Let $\gamma \in \Gamma$ be the image of θ under the ring anti-isomorphism $\alpha \rightarrow \alpha^0$ from $C(M)$ to Γ. We have, for all $a \in (0 : V_0)$, $am = \theta(aw) = (aw)\gamma = a(w\gamma)$. This equality can be written $(0 : V_0)(m - w\gamma) =< 0 >$. By the inductive hypothesis, the element $(m - w\gamma)$ must lie in V_0. Thus, $m \in V_0 \oplus w\Gamma = V$, our desired contradiction. The proof is complete. ∎

Definition 6.4.1 *Let $_R M_\Gamma$ be a left R-module and a right vector space over a division ring Γ. If R satisfies the conclusion of the previous theorem, we say that R acts **densely** on M_Γ.*

The reason for the terminology is that we can make M_Γ into a topological vector space by defining a neighborhood base for $0 \in M_\Gamma$ to be the set

$\{A_i : i \in I\}$ of all annihilators of finite dimensional subspaces of M_Γ . A neighborhood base for an arbitrary $m \in M_\Gamma$ is then defined to be the set $\{A_i + m\}$. A ring R that acts densely on M_Γ according to the definition above acts densely in the topological sense with respect to the topology just defined.

6.5 Semisimple Artinian rings

Definition 6.5.1 *A ring R is called **semisimple** if $J(R) = <0>$.*

Definition 6.5.2 *An **algebra over a division ring** D is a ring R that is also a right vector space over a D. An associative axiom holds, connecting the ring and vector space structures: $(rs)d = r(sd)$ for all $r, s \in R, d \in D$.*

Note that our definition is consistent with our previous definition of an algebra over a field. If F is a field then an F-algebra R can be regarded interchangeably as a left or right F-vector space.

As in the case of finite dimensional algebras over a field (see Example 6.3.2), each finite dimensional algebra R_D over a division ring D is both Artinian and right Artinian. This follows since R is also naturally a left vector space over the division ring C, the opposite ring of D. In this context, plainly, $\dim_C(_C R) = \dim_D(R_D)$. Thus, any properly descending chain of either left ideals or of right ideals can contain no more terms than this common dimension.

Definition 6.5.3 *A **basis** β for V_D, a right vector space over the division ring D, is a maximal D-independent subset of V.*

As for vector spaces over fields, we can use Zorn's Lemma to show that any right vector space $V = V_D$ has a basis. Moreover, with only a slight adjustment of the proofs of Theorems 4.3.3 and 4.3.4, we can show that the cardinality of a D-basis β is an invariant of V_D and that this cardinal determines V up to vector space isomorphism. Additionally, the proof of 4.3.2 works verbatim to establish the following result.

Theorem 6.5.1 *Let M be a right vector space over the division ring D. Then the following are equivalent: (1) β is a basis for M; (2) β is a minimal spanning set for M over D; (3) Each $m \in M$ can be uniquely expressed as a finite linear combination $m = \sum_{i=1}^{k} b_i d_i$ with the b_i in β and the d_i in D.*

Let M_D be finite dimensional with basis $\beta = \{b_1, ... b_k\}$. For $m \in M$ we define the **coordinates of m with respect to** β to be the (unique) $1 \times k$ column matrix $[m]_\beta = \begin{bmatrix} d_1 & d_2 & \cdot & \cdot & d_k \end{bmatrix}^t$ of elements in D such that $m = \sum_{i=1}^{k} b_i d_i$.

Definition 6.5.4 *Let M be a right D-vector space of dimension k, $\beta = \{b_1, ..., b_k\}$ an ordered basis for M, and T a linear operator on M. Then the **matrix of T with respect to** β, $[T]_\beta$, is the $k \times k$ matrix $[d_{ij}]$ with entries in D defined by the equations: $T(b_j) = \sum_{i=1}^{k} b_i d_{ij}$, $1 \le j \le k$. That is, the j-th column of the matrix $[T]_\beta$ is the coordinate column of the vector $T(b_j)$.*

Note that, even though M is a right vector space, we retain our custom of writing linear operators on the left. The fact that T respects scalar multiplication now becomes as associative law: $(Tm)d = T(md)$ for all $m \in M, d \in D$.

We now have the division ring analogue of Theorem 4.5.1.

Theorem 6.5.2 *Let T be a linear operator on the finite dimension right vector space M_D and let β be an ordered basis. Then, for each $m \in M$, $[T]_\beta[m]_\beta = [T(m)]_\beta$.*

Proof. See Problem 6.5.1 below. ∎

You should also work through Problem 6.5.2 to see why we choose to make M a right vector space rather than a left vector space.

The proof of our next theorem comes more or less verbatim from the proof of Theorem 4.5.2. Most of the verification is straightforward. Proving the equality $[T]_\beta[S]_\beta = [TS]_\beta$ requires a little care. If you write out the proof you again will see the point of using the set-up of a right vector space.

Theorem 6.5.3 *Let M be a right vector space of dimension n over the division ring D. Then the mapping $T \to [T]_\beta$ is a D-algebra isomorphism from the algebra of D-linear operators on M_D to the algebra of $n \times n$ matrices over D.*

Remark 6.5.1 *All of the linear algebra above can be generalized to infinite dimensional vector spaces over division rings. Since this more general formulation is not necessary for the proof of the Wedderburn-Artin Theorem, we stay with the finite dimensional case.*

We are now ready to prove the celebrated Wedderburn-Artin Theorem, certainly the most significant theorem in noncommutative ring theory.

Theorem 6.5.4 *(**Wedderburn-Artin**) A ring R is semisimple Artinian if and only if R is isomorphic to a ring direct product $\prod_{i=1}^{k} M_{n_i}(\Gamma_i)$. Here $M_{n_i}(\Gamma_i)$ is the ring of $n_i \times n_i$ matrices over the division ring Γ_i.*

Proof. We'll do the "if" part first, since it's easier. The properties of being semisimple and of being Artinian are each invariant under ring isomorphism, so, without loss, put $R = \prod_{i=1}^{k} M_{n_i}(\Gamma_i)$. For each fixed i, $1 \leq i \leq k$, let $e_i \in \prod_{i=1}^{k} M_{n_i}(\Gamma_i)$ be the element which has i-th component the identity matrix $I_{n_i} \in M_{n_i}(\Gamma_i)$ and zero component elsewhere. The e_i are orthogonal idempotents in the center of R with $1_R = \sum_{i=1}^{k} e_i$. It follows that each left or two-sided ideal $L \subset R$ is of the form $L = \sum_{i=1}^{k} e_i L = \sum_{i=1}^{k} L e_i$. Let J be the Jacobson radical of R. If $J \neq < 0 >$, then at least one $e_i J$ is a nonzero quasi-regular ideal of $M_{n_i}(\Gamma_i)$. But $M_{n_i}(\Gamma_i)$ has no proper ideals. (See Problem 6.5.5 below for an outline of the proof.) Hence, $M_{n_i}(\Gamma_i) = e_i J$ is quasi-regular. This is nonsense, since the identity element of any ring cannot have a quasi-inverse. Thus, R is semisimple.

Let R be a ring direct product, $R = \prod_{i=1}^{k} R_i$, where each ring R_i is Artinian. We show that R is Artinian. Let $L_1 \geq L_2 \geq \cdots \geq L_j \geq \cdots$ be a descending chain of left ideals of R. Then, for each i, $L_1 e_i \geq L_2 e_i \geq \cdots \geq L_j e_i \geq \cdots$ is a descending chain of left ideals of the ring $Re_i = R_i$. Since each R_i is Artinian, each of these chains must stabilize. If n is such that $L_n e_i = L_{n+t} e_i$ for $1 \leq i \leq k$ and all $t \geq 0$, then $L_n = L_{n+t}$ for all $t \geq 0$. It follows that R is Artinian.

Thus, to prove that the ring R of our theorem is Artinian, it suffices to prove that each $M_i = M_{n_i}(\Gamma_i)$ is Artinian. We simply note that M_i is a left vector space over the division ring C_i, the opposite ring of Γ_i, and $\dim_{C_i}({}_{C_i} M_i) = \dim_{\Gamma_i}(M_i)_{\Gamma_i} = n_i^2$. Problem 6.5.4 outlines a different proof that each matrix algebra M_i is Artinian.

Conversely, suppose that R is semisimple Artinian. Since R is Artinian, we can choose a finite set of simple R-modules $\{S_1, ..., S_k\}$ such that $< 0 >= J(R) = \cap_{i=1}^{k}(0 : S_i)$. Otherwise, we could construct a proper infinite descending chain of intersections of annihilators of simple left R-modules, contrary to the assumption that R is Artinian. It does no harm to assume that our intersection is **irredundant**, that is if we delete any annihilator from the intersection, the intersection of the remaining ones is nonzero.

Then, it is easy to check that the natural map $\alpha : R \to \prod_{i=1}^{k} R/(0 : S_i)$ given by $\alpha(r) = ((0 : S_1) + r, (0 : S_2) + r, ..., (0 : S_k) + r)$ is a ring embedding. For each i, by the Jacobson density theorem, $R/(0 : S_i)$ is isomorphic to a dense subring of $M_{n_i}(\Gamma_i)$, where Γ_i is the centralizer of S_i and $\dim_{\Gamma_i}(S_i) = n_i$. Here we are identifying $M_{n_i}(\Gamma_i)$ with the ring of linear operators on the right Γ_i-vector space S_i. (Note that we cannot have $\dim_{\Gamma_i}(S_i) = \infty$ for any i, lest we use the Jacobson density theorem to construct an infinite properly descending chain of left ideals of R. (See Problem 6.5.7.) But, since $\dim_{\Gamma_i}(S_i) = n_i < \infty$, the only dense subring of $\mathcal{L}_{\Gamma_i}(S_i) \cong M_{n_i}(\Gamma_i)$ is $M_{n_i}(\Gamma_i)$ itself. Thus we have a ring embedding

$\beta : R \to \prod_{i=1}^{k} M_{n_i}(\Gamma_i)$.

Since $\cap_{i=1}^{k}(0 : S_i)$ is irredundant, for each fixed j, $\cap_{i=1, i \neq j}^{k}(0 : S_i)$ maps via the embedding β onto a nonzero ideal of $M_{n_j}(\Gamma_j)$. But, the matrix ring $M_{n_j}(\Gamma_j)$ has no proper ideals so that $\beta[\cap_{i=1, i \neq j}^{k}(0 : S_i)] = M_{n_j}(\Gamma_j)$. It follows that β is epic and the proof is complete. ∎

We note for future reference that the simple R-modules $\{S_1, ..., S_k\}$ used in the proof are pairwise nonisomorphic. This follows from the fact that, if $_RS_i \cong_R S_j$ with $i \neq j$, then $(0 : S_i) = (0 : S_j)$, contradicting the fact that the intersection $\cap_{i=1}^{k}(0 : S_i)$ was chosen to be irredundant. The following more precise result is of interest.

Corollary 6.5.1 *Each simple R-module is isomorphic to (exactly) one of the modules S_i above.*

Proof. Let $_RS$ be simple. Since each idempotent e_i of the proof above is central, either $e_iS = S$ or $e_iS = 0$. Note that $e_iS = S$ implies that $e_jS = 0$ for $j \neq i$ so that S is naturally a $e_iR = M_{n_i}(\Gamma_i)$ module.

Fix the unique i with $e_iS = S$. For $1 \leq j \leq n_i$, let L_j be the set of all matrices in e_iR with arbitrary elements in the j-th column and zero's elsewhere. In Problem 6.5.3, you are asked to prove that each L_j is a simple e_iR-module and that $L_j \cong_{e_iR} L_{j'}$ for all $1 \leq j, j' \leq n_i$. Note that, since R is the ring direct product, $R = \prod_{t=1}^{k} e_tR$, these e_iR-isomorphisms can be regarded just as well as R-isomorphisms. Let $0 \neq s \in S$. Since $e_is = s$ and e_iR is a sum of its left ideals L_j, there exists some L_j with $L_js \neq 0$. Then $L_j \to L_js$ provides an R-isomorphism from the simple left R-module L_j onto the simple module S.

Finally, note that the R-module S_i from our original list of nonisomorphic simple left R-modules has $e_iS = S_i$. Hence, S_i is isomorphic to any of the minimal left ideals L_j above. We have shown that any simple module S is isomorphic to one of the modules $\{S_1, ..., S_k\}$. ∎

As fine as it is, we hesitate to call the Wedderburn-Artin Theorem a structure theorem, since the structural description involves division rings, rather than cardinal numbers or some other fairly simple mathematical entities. Many mathematicians have spent a good part of their careers studying division rings.

Forty-second Problem Set

Problem 6.5.1 *Prove Theorem 6.5.2, using the proof of Theorem 4.5.1 as a guide.*

Problem 6.5.2 *Suppose that M is a left vector space of dimension two over a division ring D and let T be a D-linear transformation of M. Show that there is no reasonable way to construct a 2×2 matrix A with entries in D such that $A[m]_\beta = [T(m)]_\beta$ for all $m \in M$ for any D-basis β.*

Problem 6.5.3 *Let $R = M_n(D)$ and, for $1 \le j \le n$, let L_j be the set of all matrices with arbitrary elements in the j-th column and zero's elsewhere. Prove that each L_j is a simple R-module and that $L_j \cong_R L_{j'}$ for all j, j'.*

Problem 6.5.4 *Let $R = M_n(D)$, where D is a division ring. Prove that, if L is a nonzero left ideal of R, then, as a left R-module, $L = \bigoplus_{j=1}^{m \le n} N_j$, where each N_j is R-isomorphic to one (hence all) of the simple R-modules L_j of the previous problem. Deduce that R is Artinian and that all minimal left ideals of R are R-isomorphic.*

Problem 6.5.5 *Show that, if $I \ne (0)$ is an ideal of $M_n(D)$, then $I = M_n(D)$. Hint: Let A be a nonzero matrix in I. Use left and right multiplications by appropriate elements of $M_n(D)$ to obtain the matrix*

$$
B = \begin{bmatrix}
1 & 0 & \cdot & \cdot & 0 \\
0 & \cdot & \cdot & \cdot & 0 \\
\cdot & \cdot & \cdot & \cdot & \cdot \\
\cdot & \cdot & \cdot & \cdot & \cdot \\
0 & \cdot & \cdot & \cdot & 0
\end{bmatrix} \in I.
$$

Then prove that I contains the identity matrix, so that $I = M_n(D)$.

Problem 6.5.6 *Show that, if $\dim_{\Gamma_i}(S_i) = \infty$ for some i, we can use the Jacobson density theorem to construct an infinite properly descending chain of left ideals of R.*

6.6 Structure of complex group algebras

In this section we outline the application of the Wedderburn-Artin Theorem to the study of certain group algebras.

Definition 6.6.1 *Let F be a field and $G = \{g_1, ..., g_n\}$ be a multiplicative group. The **group algebra** FG is the F-vector space on G, $\bigoplus_{i=1}^{n} Fg_i$, with ring product naturally induced by the multiplication in G.*

We use summation notation to write each element of $y \in FG$ in the form $y = \sum_{i=1}^{n} f_i g_i : \{f_1, ... f_n\} \subset F$.

Example 6.6.1 *Let* $G = < x >$, *where* x *is an element of order three. Then, in* FG *for any field* F, $(c_0e + c_1x + c_2x^2)(d_0e + d_1x + d_2x^2) = (c_0d_0 + c_1d_2 + c_2d_1)e + (c_0d_1 + c_1d_0 + c_2d_2)x + (c_0d_2 + c_1d_1 + c_2d_0)x^2$.

Exercise 6.6.1 *Compute the formula for the product of two arbitrary elements of* FG, *where* G *is the Klein 4-group. Recall that* $G = \{e, a, b, c\}$ *such that* e *is the identity,* $a^2 = b^2 = c^2 = e$, *and the product in either order of any two distinct elements from* $\{a, b, c\}$ *is the third.*

For the remainder of the chapter, G will be a multiplicative group of order n and $F = C$, the field of complex numbers. Then CG is an n-dimensional C-algebra; hence, CG will be Artinian. The following result, whose proof is due to Maschke, implies that CG is semisimple as well. We state Maschke's classical theorem, but rather than proving the general result, we will be content with giving a simpler argument from [H] designed specifically for CG.

Theorem 6.6.1 *Let* G *be a finite multiplicative group and* F *be a field such that either* $\operatorname{char} F = 0$ *or* $\operatorname{char} F = p$, *where* p *does not divide the order of* G. *Then the group algebra* FG *is semisimple.*

Proof. Let $F = C$. Define a bijection $x \to x^*$ from CG to itself as follows. If $x = \sum c_i g_i$, then $x^* = \sum \bar{c}_i g_i^{-1}$, where $\bar{c}_i$ is the complex conjugate of c_i. (Check that $x \to x^*$ is a bijection.) The following properties are not difficult to ascertain for all $x, y \in CG$, $\alpha, \beta, \in C$: 1) $x^{**} = x$; 2) $(\alpha x + \beta y)^* = \bar{\alpha} x^* + \bar{\beta} y^*$; 3) $(xy)^* = y^* x^*$.

Note that $xx^* = \sum |c_i|^2 e + \sum_{g_i \neq e} d_i g_i$ for suitable complex coefficients d_i. (Check this as an exercise.) Hence, if $xx^* = 0$, then $\sum |c_i|^2 = 0$ so that $x = 0$.

Let J be the Jacobson radical of CG. Since CG is Artinian, $J^k = 0$ for some positive integer k. If $0 \neq j \in J$, put $v = jj^*$. Then $v \in J$ and, by our previous note, $v \neq 0$. Thus, we can choose a positive even integer $2s$ such that $v^{2s} = 0$ but $v^s \neq 0$.

By properties 1) and 3), $v^* = (jj^*)^* = jj^* = v$. But then, $0 = v^{2s} = v^s(v^*)^s = v^s(v^s)^*$. It follows that $v^s = 0$, contradicting the choice of s. Hence, CG is semisimple, as desired. ∎

Applying the Wedderburn-Artin Theorem we have:

Corollary 6.6.1 *The group algebra* CG *is isomorphic to a ring direct product* $\prod_{i=1}^{k} M_{n_i}(\Gamma_i)$, *where the division algebra* Γ_i *is the centralizer of a simple* CG-*module* S_i.

In particular, each simple module S_i is a complex vector space. In this context, the map $\mathcal{C} \to \mathcal{C}_l$, where $\mathcal{C}_l \leq End_Z(S_i)$ is the subring of left multiplications by elements of $\mathcal{C}$, embeds $\mathcal{C}$ as a subfield of the division ring Γ_i. Since $\dim_{\mathcal{C}} \mathcal{C}G = \mid G \mid = n$ is finite, so is the dimension of each Γ_i over $\mathcal{C}$. Thus, each $\gamma \in \Gamma_i$ must be algebraic over $\mathcal{C}$. But $\mathcal{C}$ is an algebraically closed field, so that $\Gamma_i = \mathcal{C}$ for each i. Thus, we can state the following more precise version:

Corollary 6.6.2 *Let G be a multiplicative group with $\mid G \mid = n$. Then $\mathcal{C}G \cong \prod_{i=1}^{k} M_{n_i}(\mathcal{C})$. Moreover, $n = \sum_{i=1}^{k} n_i^2$.*

The latter equality comes from computing the dimension of each side over $\mathcal{C}$.

6.7 Applications to finite groups

We close this chapter by illustrating how the results of the previous section can be applied to the study of finite groups. To make our exposition a little easier, we will apply the previous corollary to identify $\mathcal{C}G$ with the direct product $\prod_{i=1}^{k} M_{n_i}(\mathcal{C})$. Recall that e_j is the central idempotent of $\prod_{i=1}^{k} M_{n_i}(\mathcal{C})$ whose j-th component is the $n_j \times n_j$ identity matrix and which has zero component otherwise.

Definition 6.7.1 *For $1 \leq j \leq k$, define $\chi_j : G \to \mathcal{C}$ by the formula $\chi_j(g) = trace(e_j g)$.*

Note that $g = 1g \in \mathcal{C}G$ so that each $e_j g \in M_{n_i}(\mathcal{C})$. Thus, each $\chi_j : G \to \mathcal{C}$. The functions χ_j are called the **distinct irreducible characters** of the group G. The "distinct irreducible" part of the terminology comes from the fact that the functions χ_j were constructed using the matrix algebras $M_{n_i}(\mathcal{C})$. These matrix algebras, in turn, come from a complete set of representatives $\{S_1, ..., S_k\}$ of nonisomorphic irreducible (simple) $\mathcal{C}G$-modules (Corollary 6.5.1).

Let $Con(G) = \{[g_1], ..., [g_t]\}$ be the set of distinct conjugacy classes of G. The following lemma shows that each character χ_j induces a function $\chi_j' : Con(G) \to \mathcal{C}$.

Lemma 6.7.1 *If $x, y \in G$ are conjugate and $1 \leq j \leq k$, then $\chi_j(x) = \chi_j(y)$.*

Proof. See Problem 6.7.1 below. ∎

In Chapter 1 we proved that the number of conjugacy classes of a finite multiplicative group G is equal to the order of the factor group $G/C(G)$, where $C(G)$ is the center of G. Also note that, since $CG = \prod_{i=1}^{k} M_{n_i}(C)$, the center of the ring CG is the ring direct product of k copies of C. Thus, $\dim_C(centerCG) = k$.

We now compute $\dim_C(centerCG)$ in a different way, using a special C-basis for CG. For $g \in G$, let $C_g = \sum_{x \in [g]} 1x$. We call C_g the **class sum** of g. A simple computation shows that each element C_g commutes with anything in CG of the form $1h$, $h \in G$. It follows that each C_g is in the center of the ring CG. (See Problem 6.7.2.) Since the group elements $\{1h : h \in G\}$ are linearly independent over C, the elements C_g are also linearly independent over C.

We claim that the elements C_g span the C-subspace $centerCG$. Let $z = \sum_{i=1}^{n} \alpha_i g_i$ be an arbitrary element of $centerCG$. If $x \in G$, then $\sum_{i=1}^{n} \alpha_i g_i = z = xzx^{-1} = \sum_{i=1}^{n} \alpha_i(xg_ix^{-1})$. Since the group elements are linearly independent over C, we see that every conjugate of each g_i enters the sum that expresses z with the same complex coefficient as g_i itself. Thus, we can rewrite z as $z = \sum_{i=1}^{t} \alpha_i C_{g_i}$. We have shown that $\{C_{g_1}, ..., C_{g_t}\}$ is a C-basis for $centerCG$. Thus, t, which is equal to the number of distinct conjugacy classes of G, is equal to the C-dimension of $centerCG$. But the C-dimension of $centerCG$ is also equal to k, which equals the number of distinct irreducible characters of G. We have proved the following beautiful theorem. It's a fine way to end the chapter!

Theorem 6.7.1 *Let G be a finite multiplicative group. Then the number of distinct irreducible characters of G is equal to the number of conjugate classes of G.*

Forty-third Problem Set

Problem 6.7.1 *Prove Lemma 6.7.1.*

Problem 6.7.2 *Write out the details to show that each class sum C_g is in the center of CG.*

Further reading: Both [J2] and [Ro] contain many additional topics. For me [H] has the nicest and most compact introduction to this material. The second part of [Ka2] is also pleasant reading.

Chapter 7

Group extensions

7.1 Introduction

In this chapter, we will be concerned with the following kinds of problems:
(1) Given groups A, G, when is there a group C such that C contains a normal subgroup $A' \cong A$ and such that the factor group $C/A' \cong G$? Such a group C is (loosely) called an extension of A by G. A more precise definition of extension will follow. (2) If there is such a C, what can we say about its structure? (The group C is certainly not always completely determined; S_3 and $Z(6)$ are both extensions of $Z(3) \cong A_3$ by $Z(2)$.)

To make the problem more tractable, we require that the group A be abelian. The material here is inspired by Chapter 5 of Rotman's An Introduction to Homological Algebra, and I express my appreciation to have had such an excellent text as a guide.

7.2 Exact sequences and ZG-modules

Definition 7.2.1 *An **exact sequence** of groups is a diagram $E : 0 \to A \xrightarrow{\alpha} C \xrightarrow{\beta} G \to 1$ such that the image of each map coincides with the kernel of the preceding one. We say that such an exact sequence represents an extension of A by G.*

The group A will be an abelian group, written additively. The statement that E is exact at A just means that the image of the zero map is the kernel of α, that is the map α is an embedding. The group C need not be abelian; however it will be convenient to depart from standard practice and also

191

write C as an additive group. Exactness at C means that $\text{Im}\,\alpha = Ker\beta$. In particular $\alpha A \lhd C$. The group G will be written multiplicatively. Exactness at G means that $\text{Im}\,\beta = G$, that is β is an epimorphism.

We say E "represents" an extension of A by G rather than "is" an extension of A by G since our extensions, when precisely defined, actually will turn out to be equivalence classes of exact sequences.

To start, we will look at extensions represented by **standard** exact sequences, that is sequences of the form $E : 0 \to A \to C \to G \to 1$. Here A is a normal subgroup of C, the unlabeled monomorphism $A \to C$ is the inclusion map and the unlabeled epimorphism $C \to G$ is the natural factor map $\pi : C \to C/A$. Consequently, $G = C/A$. Later, we will consider more general exact sequences.

Before our first theorem, we need to introduce some notation:

1. Recall that $Aut(A)$ denotes the group of automorphisms of a group A. The binary operation is composition of automorphisms.

2. A **lifting** is a function $\lambda : G \to C$ that assigns to each coset $g \in G$ a representative $\lambda(g) \in C$ with $g = A + \lambda(g)$. Thus $\pi\lambda(g) = g$ for all $g \in G$. For convenience, we set $\lambda(1) = 0$.

3. Let $g \in G$ and let λ be a lifting. Define a map θ_g that acts on A by conjugation: $\theta_g(a) = \lambda(g) + a - \lambda(g)$ for all $a \in A$. (This looks odd, but remember that C is written additively.)

Theorem 7.2.1 *The map $\Theta : G \to Aut(A)$ defined by $\Theta(g) = \theta_g$ is a multiplicative group homomorphism (independent of the lifting λ).*

Proof. We first note that, since $A \lhd C$, conjugation by any fixed element $c \in C$ induces an automorphism of A. Thus Θ maps G into $Aut(A)$.

The next item of business is to check that the automorphism θ_g is independent of the lifting λ. If λ' is another lifting, then $\lambda'(g) = \alpha + \lambda(g)$ for some $\alpha \in A$. For $a \in A$, $\lambda'(g) + a - \lambda'(g) = (\alpha + \lambda(g)) + a + (-\lambda(g) - \alpha) = \alpha + [\lambda(g) + a - \lambda(g)] - \alpha = [\lambda(g) + a - \lambda(g)]$. For the last equality we used the fact that both α and the bracketed element are in the abelian group A.

Finally, we need to check that $\Theta(gh) = \Theta(g)\Theta(h)$ for all $g, h \in G$. This amounts to checking that $\theta_{gh} = \theta_g \circ \theta_h$. Let λ be a lifting, a be an arbitrary element of A, and g, h be fixed elements of G. Then $[\theta_g \circ \theta_h](a) = \theta_g[\lambda(h) + a - \lambda(h)] = [\lambda(g) + \lambda(h)] + a - [\lambda(g) - \lambda(h)]$. But, for any $x \in G$, we've already shown that the map θ_x is independent of the lifting. Hence, in computing the action of θ_{gh}, we can choose a lifting λ such that $\lambda(gh) = \lambda(g) + \lambda(h)$ for this pair g, h. It follows that $\theta_{gh} = \theta_g \circ \theta_h$ and the proof is complete. ∎

Definition 7.2.2 *Let G be a multiplicative group. The* **group ring** *ZG is the ring whose additive group is the free abelian group indexed by the elements of G; that is $< ZG, + > = \bigoplus_{g \in G} Z_g$. However, instead of writing a typical element as a finite formal sum $\bigoplus_{i=1}^{k} z_{g_i}$ we write it as $\sum_{g \in G} z_g g$ with the understanding that any such sum is finite (almost all of the integers $z_g = 0$). With this notation, we define a natural ring product in ZG, induced by using the product in G and requiring the left and right distributive laws.*

It is easy to check that ZG is a ring. We can regard Z as a subring of ZG by writing z for ze, e the identity of G.

The next example and exercise are essentially repeats from Chapter 6.

Example 7.2.1 *Let G be the cyclic group of order 3 written multiplicatively, $G = \{e, x, x^2 : x^3 = e\}$. Then in ZG $(z_0 e + z_1 x + z_2 x^2)(w_0 e + w_1 x + w_2 x^2) = (z_0 w_0 + z_1 w_2 + z_2 w_1)e + (z_0 w_1 + z_1 w_0 + z_2 w_2)x + (z_0 w_2 + z_2 w_0 + z_1 w_1)x^2$.*

Exercise 7.2.1 *Let $G = < e, a, b, c >$ be the Klein 4-group. Write the formula for the product of two typical elements of ZG.*

Let $E : 0 \to A \to C \to G \to 1$ be exact. For $a \in A, g \in G$ define $ga = \theta_g(a) \in A$. Then put $(\sum_{g \in G} z_g g)a = \sum_{g \in G} z_g(ga)$. Since A is abelian, each integral multiple $z_g(ga)$ has its natural meaning. In view of the fact that $\theta : G \to Aut(A)$ is a homomorphism, the following theorem is not hard to verify.

Theorem 7.2.2 *Each exact sequence $E : 0 \to A \to C \to G \to 1$ provides A with a left ZG-module structure.*

Proof. All the module axioms are easy to check except that $(rr')a = r(r'a)$ for all $r, r' \in ZG$ and $a \in A$. This needs to be written out (see Problem 7.2.1). ∎

Remark 7.2.1 *We can easily extend Theorem 7.2.2 to define the left ZG-module structure arising from an arbitrary exact sequence $E : 0 \to A \xrightarrow{\alpha} C \xrightarrow{\beta} G \to 1$. In this case we simply put $ga = \alpha^{-1}[\lambda(g) + a - \lambda(g)]$ for any lifting λ. Then A becomes a ZG-module as above.*

Definition 7.2.3 *If $_{ZG}A$ is a preassigned ZG-module, an exact sequence E* **realizes the operators** *if the given ZG action on A coincides with that inherited from E.*

Using standard terminology we say "G-module" instead of left ZG-module.

Definition 7.2.4 *Let* $E : 0 \to A \xrightarrow{\alpha} C \xrightarrow{\beta} G \to 1$ *and* $E_1 : 0 \to A \xrightarrow{\alpha_1} C_1 \xrightarrow{\beta_1}$ $G \to 1$ *be exact sequences representing extensions of* A *by* G. *We call* E, E_1 *equivalent, and write* $E \sim E_1$, *if there is a homomorphism* $\phi : C \to C_1$ *such that the following is a commutative diagram:*

(†)

$$
\begin{array}{ccccccccc}
0 & \to & A & \xrightarrow{\alpha} & C & \xrightarrow{\beta} & G & \to & 1 \\
 & & \downarrow 1 & & \downarrow \phi & & \downarrow 1 & & \\
0 & \to & A & \xrightarrow{\alpha_1} & C_1 & \xrightarrow{\beta_1} & G & \to & 1
\end{array}
$$

The commutativity conditions force ϕ to be an isomorphism. See Problem 7.2.3.

Exercise 7.2.2 *Write a short proof showing that our relation* $\sim$ *is an equivalence relation on the class of exact sequences representing extensions of* A *by* G.

Here is the key result of this section.

Theorem 7.2.3 *Let* $E : 0 \to A \xrightarrow{\alpha} C \xrightarrow{\beta} G \to 1$ *and* $E_1 : 0 \to A \xrightarrow{\alpha_1} C' \xrightarrow{\beta_1}$ $G \to 1$ *be exact sequences representing extensions of* A *by* G. *If* $E \sim E_1$, *then* E *and* E_1 *induce the same* ZG *module structure on* A.

Proof. Let $g \in G$ and $a \in A$ be arbitrary. Write ga for the G-action induced by E and $g \cdot a$ for the G-action induced by E_1. Since $E \sim E_1$, there is a commutative diagram (†) as in the definition above. Let λ, λ_1 be liftings for E, E_1. By definition, $ga = \alpha^{-1}[\lambda g + \alpha a - \lambda g]$ and $g \cdot a = \alpha_1^{-1}[\lambda_1 g + \alpha_1 a - \lambda_1 g]$.

We show that $ga = g \cdot a$ via a chain of equalities. Since ϕ is a homomorphism, $\phi \alpha(ga) = \phi[\lambda g + \alpha a - \lambda g] = \phi \lambda g + \phi \alpha a - \phi \lambda g$. By (†), we have $\phi \alpha(ga) = \alpha_1 ga$. Thus, $ga = \alpha_1^{-1}[\phi \lambda g + \phi \alpha a - \phi \lambda g]$. Since α_1 is monic and in view of the definition of $g \cdot a$ above, we can complete the proof by showing that $[\lambda_1 g + \alpha_1 a - \lambda_1 g] = [\phi \lambda g + \phi \alpha a - \phi \lambda g]$. But, by (†), we have $\alpha_1 a = \phi \alpha a$.

So what we really need to show is that, for all $g \in G$, conjugation by $\lambda_1 g$ and by $\phi \lambda g$ induce the same automorphism on $\alpha_1 A$. To show this, first note that, since λ_1 is a lifting for E', $\beta_1(\lambda_1 g) = g$. Then using (†), we have $\beta_1 \phi(\lambda g) = \beta(\lambda g) = g$, the latter equality since λ is a lifting for E. We have shown that $\lambda_1 g$ and $\phi \lambda g$ have the same image under β_1. By exactness of E_1, $\lambda_1 g = \phi \lambda g + \alpha_1 a'$ for some $a' \in A$. It follows that conjugation by $\lambda_1 g$ and by $\phi \lambda g$ coincide as automorphism of $\alpha_1 A$. The proof is now complete. ∎

We close this section with a brief discussion of the simplest kind of G-module.

Definition 7.2.5 *A G-module A is **trivial** if $ga = a$ for all $g \in G$ and $a \in A$.*

Theorem 7.2.4 *The following are equivalent for a G-module A. (a) A is trivial; (b) the map $\Theta : G \to Aut(A)$ is the trivial homomorphism ($\theta_g = 1$ for all g); (c) A is contained in the center of C.*

Proof. The equivalence of (a) and (b) follows immediately from the definitions. The implications $(c) \Rightarrow (a), (c) \Rightarrow (b)$ are also clear. The remaining implication, $(a) \Rightarrow (c)$, is Problem 7.2.5 below. ∎

Forty-fourth Problem Set

Problem 7.2.1 *Let $r = \sum_{i=1}^{k} z_i g_i, r' = \sum_{i=1}^{k} z_i' g_i'$ be elements of ZG and $a \in A$. Prove that $(rr')a = r(r'a)$. (If G is infinite, it still does no harm to write r, r' in this simpler form, since each element of ZG has zero coefficients off a finite subset of G.)*

Problem 7.2.2 *Prove that the map ϕ of Definition 7.2.4 is an isomorphism.*

Problem 7.2.3 *Suppose we had defined $\theta_g(a)$ to be $-\lambda(g) + a + \lambda(g)$. Show that this gives rise to a right, but not a left, ZG-module structure on A.*

Problem 7.2.4 *Recall the group D_4 of rotations of the square (Sect. 1.1). Let $V \lhd D_4$ be the subgroup $\{e, x^2, y, x^2 y\}$. Consider the exact sequence $0 \to A = Z(2) \oplus Z(2) \xrightarrow{\alpha} D_4 \xrightarrow{\pi} G \to 1$. The map $\alpha : Z(2) \oplus Z(2) \to D_4$ is given by $\alpha(1,0) = x^2, \alpha(0,1) = y$. The map π is the natural factor map from D_4 to the factor group $G = D_4/V = \{Ve, Vx\}$. Write down the induced G-module structure on A.*

Problem 7.2.5 *Prove the implication $(a) \Longrightarrow (c)$ of Theorem 7.2.4.*

7.3 Semidirect products

In this section we discuss a special kind of exact sequence.

Definition 7.3.1 *A standard exact sequence $0 \to A \to C \to G \to 1$ is **split** if there is a lifting $\lambda : G \to C$ that is a homomorphism. We call such a map λ a **splitting homomorphism**.*

Theorem 7.3.1 *The following are equivalent for a standard exact sequence* $E : 0 \to A \to C \xrightarrow{\pi} G \to 1$ *: (a) E is split; (b) there is a (not necessarily normal) subgroup $H \leq C$ with $H \cong G$, $C = A + H$ and $A \cap H = <0>$; (c) there is a subgroup $H \leq C$ with $H \cong G$ such that every $c \in C$ can be written uniquely as $c = (a + h) \in A + H$.*

Proof. $(a) \Rightarrow (b)$ If E is split, let $H = \lambda(G)$. Since λ is a homomorphism, $H \leq G$. Moreover, H contains a complete set of coset representatives of C/A so that $C = A + H$. If $c \in A \cap H$, then $c = \lambda(g)$ and $\pi(c) = 1$. Thus, $g = \pi\lambda(g) = \pi(c) = 1$, so that $c = \lambda(1) = 0$. By second group isomorphism theorem, $G = C/A = (A + H)/A \cong H/(A \cap H) \cong H$.

$(b) \Rightarrow (c)$ Suppose (b) holds and let $c \in C$. Then $c = (a + h) \in A + H$. If $c = (a' + h')$, then $h - h' = -a + a'$ is an element of $A \cap H$. It follows that $a = a', h = h'$.

$(c) \Rightarrow (a)$ If (c) holds, then each $c \in C$ can be uniquely written as $c = a + h$ for $a \in A, h \in H$. Thus, we can define $\lambda : G \to H \leq C$ by the stipulation $\lambda(c) = \lambda(a + h) = h$. Plainly, λ is a lifting. We need to check that λ is a homomorphism. Let $c = (a + h), c' = (a' + h')$ be elements of C. Since $A \lhd C$, we have $c + c' = a + (h + a') + h' = a + (h + a' - h) + (h + h') = (a + a_1) + (h + h')$. It follows directly that $\lambda(c + c') = \lambda(c) + \lambda(c')$ and the proof is complete. ∎

Definition 7.3.2 *If the situation of the above theorem holds, the group C is called a **semidirect product** of the subgroups A and H.*

More generally, we call an exact sequence $0 \to A \xrightarrow{\alpha} C \xrightarrow{\beta} G \to 0$ **split** if there exists a homomorphism $\lambda : G \to C$ such that $\beta\lambda = 1_G$. (Then λ is automatically a lifting.) In this case, we will say that C is a **semidirect product** of the groups A and G.

Exercise 7.3.1 *State Theorem 7.3.1 for this more general case.*

Example 7.3.1 *The exact sequence $E : 0 \to A_3 \to S_3 \to S_3/A_3 \to 1$ is split. A splitting homomorphism λ is given by $\lambda(A_3 e) = e$ and $\lambda(A_3(12)) = (12)$. Thus, S_3 is a semidirect product of the subgroups A_3 and $\{e, (12)\}$.*

Example 7.3.2 *Consider the sequence: $E' : 0 \to Z(3) \xrightarrow{\alpha} S_3 \xrightarrow{s} Z(2) \to 1$ Here $\alpha : Z_3 \to S_3$ is defined by $\alpha(1) = (123)$. The group $Z(2)$ is written multiplicatively, say $Z(2) = \{\pm 1\}$, and $s(p) = sgn(p)$ for $p \in S_3$. In this case, a splitting homomorphism λ is given by $\lambda(1) = e$ and $\lambda(-1) = (12)$. Thus, we can equally well call S_3 a semidirect product of $Z(3)$ by $Z(2)$.*

Example 7.3.3 *Check that the abelian group $Z(6)$ is also a semidirect product of $Z(3)$ by $Z(2)$.*

The above two examples show that a semidirect product, as we have defined it, need not be unique. See the three problems at the end of this section for an amplification and clarification of this fact.

Exercise 7.3.2 *Give an example of an exact sequence* $0 \to A \overset{\alpha}{\to} C \overset{\beta}{\to} G \to 0$ *such that* C *is not a semidirect product of* A *by* G.

We close this section with another result on splitting, to be used later. We stick to standard exact sequences since that's all we will need.

Theorem 7.3.2 *A standard exact sequence* $E : 0 \to A \to C \to G \to 1$ *is split if and only if the following two conditions hold: (i) there is a group* C' *on the Cartesian product* $A \times G$ *with binary operation* $(a, x) + (a', x') = (a + xa', xx')$; *(ii) there is a lifting* λ *such that the function* $\phi : C \to C'$ *defined by* $\phi(a + \lambda x) = (a, x)$ *is an isomorphism from* C *to* C'. *(In (i) the term* xa' *comes from the* G-*module action on* A *induced by* E.)

Proof. Suppose that E is split with splitting homomorphism λ. In Theorem 7.2.4 we proved that each $c \in C$ can be expressed uniquely in the form $c = a + h = a + \lambda x$. Thus the map ϕ, as defined in (ii) above, is a bijection from the set C to the set $A \times G$. If $c' = (a' + \lambda x') \in C$, then $c + c' = (a + \lambda x) + (a' + \lambda x') = a + \lambda x + a' - \lambda x + \lambda x + \lambda x' = a + (\lambda x + a' - \lambda x) + \lambda x x'$. The final term in our string of equalities is equal to $(a + xa') + \lambda x x'$ so, by definition of ϕ, $\phi(c + c') = (a + xa', xx')$.

Now $\phi(c) = (a, x)$ and $\phi(c') = (a', x')$. Define an addition on $A \times G$ by the formula $(a, x) + (a', x') = (a + xa', xx')$ above and let C' be $A \times G$ with this addition. Since C is a group, it is now not hard to check both that C' is a group and that $\phi : C \cong C'$.

Conversely, suppose that (i) and (ii) hold. Let C', ϕ, λ be as in conditions (i) and (ii). Define a group embedding $i_2 : G \to C'$ by $i_2(g) = (0, g)$. Then i_2 is a splitting homomorphism for the sequence $E' : 0 \to A \overset{i_1}{\to} A \times G \overset{\pi_2}{\to} G \to 1$, where $i_1(a) = (a, 1)$ and π_2 is projection onto the second coordinate. (Check this.) It easily follows that $\phi^{-1} i_2$ is a splitting homomorphism for E. ■

Exercise 7.3.3 *Verify that* $\phi^{-1} i_2$ *is a splitting homomorphism for* E.

Forty-fifth Problem Set

Problem 7.3.1 *Prove that, if* E, E' *are equivalent sequences representing extensions of* A *by* G, *and* E *is split, then so is* E'.

Problem 7.3.2 *Let A be a G-module and let E and E' be split sequences of A by G, **each of which realizes the operators**. Prove that $E \sim E'$.*

Problem 7.3.3 *Why aren't Examples 7.3.2 and 7.3.3 contradicting Problem 7.3.2? Justify your claim.*

7.4   Extensions and factor Sets

Definition 7.4.1 *Let A be a G-module. An **extension** of A by G is a $\sim$ equivalence class of an exact sequence $0 \to A \xrightarrow{\alpha} C \xrightarrow{\beta} G \to 0$ which realizes the operators.*

Here $\sim$ is our equivalence relation of Definition 7.2.4. We noted that, if one sequence realizes the operators for a given module $_{ZG}A$, so does any equivalent sequence. Thus, our definition of extension makes sense.

For a given multiplicative group G and G-module A, let $Ext(G, A)$ be the set of all extensions of A by G, that is the set of equivalence classes of exact sequences that realize the operators. Our **extension problem** is this: Given a left module $_{ZG}A$, characterize the elements of $Ext(G, A)$.

Remark 7.4.1 *In case you're alert and wondering, Theorem 7.5.2 below will allow us to call $Ext(G, A)$ a set, rather than a class. This should not really be a surprise, since A and G are fixed. Thus, the underlying sets of the groups A, G, C have bounded cardinality. Modulo equivalence, we can work in the power set of a sufficiently large fixed set X.*

(If you weren't wondering, feel free to ignore the remark.)

We are now ready to take the first steps in the solution of the extension problem. Let's consider a standard exact sequence $E : 0 \to A \to C \to G \to 1$. Let $\lambda : G \to C$ be a lifting (not necessarily a homomorphism since E need not split). We display for future reference the formula relating the ZG-structure on A with a particular choice of lifting λ :

$$(*) : xa = \lambda x + a - \lambda x.$$

Here is a simple lemma that we could have mentioned earlier.

Lemma 7.4.1 *Let $E : 0 \to A \xrightarrow{\alpha} C \xrightarrow{\beta} G \to 1$ be exact and let λ be an arbitrary lifting. Then each $c \in C$ determines a unique pair of elements $a \in A, g \in G$ such that $c = \alpha a + \lambda g$.*

Proof. Given $c \in C$ there is a unique coset $\alpha A + \lambda g$ of which c is a member. Then, there is a unique $a \in A$ such that $c = \alpha a + \lambda g$. ∎

Let E be an exact sequence representing an extension of A by G and let λ be a lifting. We remark that, for all $x, y \in G$, since $\beta\lambda(xy) = xy = (\beta\lambda x)(\beta\lambda y) = \beta(\lambda x + \lambda y)$, both $\lambda(xy)$ and $\lambda x + \lambda y$ represent the same coset mod αA. Thus, $[\lambda x + \lambda y - \lambda(xy)] \in \alpha A$. This fact motivates the following.

Definition 7.4.2 *Let A be a G-module and $E : 0 \to A \xrightarrow{\alpha} C \xrightarrow{\beta} G \to 1$ be an exact sequence which realizes the operators. Let $\lambda : G \to C$ be a lifting. The function $[\_,\_] : G \times G \to A$ defined by the equation $[x, y] = \alpha^{-1}[\lambda x + \lambda y - \lambda(xy)]$ is called a **factor set**.*

A factor set is determined both by E and the lifting λ. We emphasize that a factor set, by definition, must come from a sequence E that realizes the operators on a preassigned ZG-module A. The next result distinguishes the factor sets among the functions from $G \times G$ to A.

Theorem 7.4.1 *Let A be a G-module. A function $[\_,\_] : G \times G \to A$ is a factor set if and only if (i) $[x, 1] = [1, y] = 0$ for all $x, y \in G$ and (ii) $x[y, z] - [xy, z] + [x, yz] - [x, y] = 0$ for all $x, y, z \in G$.*

Proof. Suppose $[\_,\_]$ is a factor set obtained from a sequence E and lifting λ as above. We have $[x, 1] = \alpha^{-1}[\lambda x + \lambda 1 - \lambda x] = \alpha^{-1}[\lambda x + 0 - \lambda x] = \alpha^{-1}[0] = 0$. Similarly, $[1, y] = 0$. A straightforward but slightly involved calculation, reveals that equation (ii) follows from the associative law in the middle group C. (See Problem 7.4.3.)

Now suppose A is a G-module and we have a function $[\_,\_]$ satisfying (i) and (ii). We will construct an exact sequence E representing an extension of A by G. Then we will choose a lifting λ satisfying equation (∗). Thus, E will realize the operators.

Define an addition on the set $A \times G$ by the formula: $(a, x) + (a', y) = (a + xa' + [x, y], xy)$. The term xa' comes from the G-module structure on A. Let C be $A \times G$ with this addition. Formula (ii) shows that this addition is associative. (See the proof of Problem 7.4.3.) Furthermore, a simple calculation reveals that $(0, 1)$ is a two-sided additive identity for C. We'll prove that the additive inverse for an element $(a, x) \in C$ is given by the formula $-(a, x) = (-x^{-1}a - x^{-1}[x, x^{-1}], x^{-1})$. Actually, in view of Problem 1.3.4, since $(C, +)$ has an identity and $+$ is associative, we need only check that the element in question is a left inverse for (a, x). We verify the equality $(-x^{-1}a - x^{-1}[x, x^{-1}], x^{-1}) + (a, x) = (0, 1)$. The equality of the first components is all that's in doubt.

By definition of addition in C, the first component of the sum $(-x^{-1}a - x^{-1}[x, x^{-1}], x^{-1}) + (a, x)$ is the expression $(\dagger)$: $(-x^{-1}a) - x^{-1}[x, x^{-1}] + x^{-1}a + [x^{-1}, x]$. Using (ii) and (i), we have $-x^{-1}[x, x^{-1}] = -([1, x^{-1}] + [x^{-1}, x]) = -[x^{-1}, x]$. Since A is abelian, substitution yields directly that $(\dagger)$ is zero. We have shown that C is a group.

The following sequence is easily seen to be exact: $E : 0 \to A \xrightarrow{i_1} C \xrightarrow{\pi_2} G \to 1$. Here $i_1(a) = (a, 1)$ and $\pi_2(a, g) = g$. Also, it is not hard to check that $i_1 A = \{(a, 1) : a \in A\}$ is a normal subgroup of C. Thus, E can serve as a representative of an extension of A by G. Define $\lambda g = (0, g)$ for $g \in G$. The map λ is certainly a lifting.

Let $\cdot$ denote the G-action on $i_1 A$ induced by E. This action is given by $g \cdot (a, 1) = \lambda g + (a, 1) - \lambda g = (0, g) + (a, 1) - (0, g)$. We have $\{(0, g) + (a, 1)\} - (0, g) = (ga + [g, 1], g) - (0, g) = (ga, g) + (-g^{-1}0 - g^{-1}[g, g^{-1}], g^{-1})$. Adding again, we obtain, $(ga, g) + (-g^{-1}[g, g^{-1}], g^{-1}) = (ga + g[(-g^{-1}[g, g^{-1}] + [g, g^{-1}]), 1) = (ga, 1)$. Thus, $g \cdot (a, 1) = (ga, 1)$.

Finally, applying i_1^{-1}, we see that the natural G-action on A induced by E is $ga = i_1^{-1}[g \cdot (a, 1)] = i_1^{-1}(ga, 1)$. Thus, the induced E-action coincides with the original G-action on A. The proof is complete. ∎

Forty-sixth Problem Set

Problem 7.4.1 *Let* $[\_, \_]_0$ *be the map from* $G \times G$ *into* A *sending all pairs* (x, y) *to* 0. *(i) Prove that* $[\_, \_]_0$ *is a factor set. (ii) Prove that the exact sequence corresponding to* $[\_, \_]_0$, *constructed as in Theorem 7.4.1, is split.*

Problem 7.4.2 *Prove that, for every exact sequence:* $0 \to A \xrightarrow{\alpha} C \xrightarrow{\beta} G \to 1$, *there is a standard exact sequence:* $0 \to A_1 \to C_1 \to G_1 \to 1$ *and isomorphisms* $\theta : A \cong A_1, \sigma : C \cong C_1, \phi : G \cong C_1/A_1 = G_1$ *such that the following diagram is commutative:*

$$
\begin{array}{ccccccccc}
0 & \to & A & \xrightarrow{\alpha} & C & \xrightarrow{\beta} & G & \to & 1 \\
& & \downarrow \theta & & \downarrow \sigma & & \downarrow \phi & & \\
0 & \to & A_1 & \to & C_1 & \to & G_1 & \to & 1
\end{array}
$$

Problem 7.4.3 *Let* $[\_, \_]$ *be a factor set obtained from an standard exact sequence* $E : 0 \to A \to C \to G \to 1$ *and a lifting* λ. *Prove that condition (ii) in Theorem 7.4.1 follows from the associative law in the group* C. *(We made the assumption that* E *is standard solely for your notational convenience in writing out the proof.)*

7.5 Solution of the extension problem

Let A be an additive abelian group and G a multiplicative group. Consider the set $Z^2(G, A)$ of all factor sets arising from extensions of A by G (equivalently, maps from $G \times G$ to A satisfying the conditions (i), (ii) of Theorem 7.4.1). Define the sum of two factor sets pointwise; $([\_, \_] + [\_, \_]')(x, y) = [x, y] + [x, y]'$. It is immediate that, since $[\_, \_]$ and $[\_, \_]'$ each satisfy (i), (ii) of Theorem 7.4.1, then so does their pointwise sum. Thus, $Z^2(G, A)$ is closed under this addition. Since the addition in A is both associative and commutative, then so is our addition in $Z^2(G, A)$. The zero element of $Z^2(G, A)$ is the factor set $[\_, \_]_0$ defined by the stipulation $[x, y]_0 = 0$ for all $(x, y) \in G \times G$. The additive inverse of a factor set $[\_, \_]$ is the factor set which maps each $(x, y) \in G \times G$ to $-[x, y] \in A$. We have outlined the proof that $Z^2(G, A)$ with pointwise addition is an additive abelian group.

Definition 7.5.1 *The elements of $Z^2(G, A)$ are called **cocyles**.*

Our definition of a factor set depended both on the given extension E and the lifting λ. The next project is to construct an object independent of the choice of λ. The notation $< x >$, which will be used exclusively in this section, should not be confused with our previous notation for the cyclic subgroup generated by an element x. Here $<>$ will be a function from G into A. Thus, $< x >$ will represent the element of A determined by some $x \in G$. We start the project by proving the following theorem.

Theorem 7.5.1 *Let $E : 0 \to A \xrightarrow{\alpha} C \xrightarrow{\beta} G \to 1$ be an extension with associated G-module structure on A. Let $\lambda, \lambda' : G \to C$ be liftings with corresponding factor sets $[\_, \_]$ and $[\_, \_]'$. Then there is a function $<>$: $G \to A$ satisfying (i) $< 1 >= 0$ and (ii) $[x, y]' - [x, y] = x < y > - < xy > + < x >$ for all $x, y \in G$.*

Proof. For simplicity, we will restrict our proof to the case of a standard exact sequence $E : 0 \to A \to C \to G \to 1$. The proof for the general case is similar, but more cumbersome. For $x \in G$, the elements λx and $\lambda' x$ are in the same coset mod A; that is, $\lambda' x - \lambda x =< x >$ for some $x \in A$. This equation provides the definition of our function $<>$: $G \to A$. Note that $< 1 >= \lambda' 1 - \lambda 1 = 0 - 0 = 0$, so condition (i) holds.

To prove (ii) we need to work with a series of equalities:

$$\lambda' x + \lambda' y = (< x > + \lambda x) + (< y > + \lambda y)$$

$$=< x > + x < y > + (\lambda x + \lambda y)$$

$$=< x > + x < y > + ([x, y] + \lambda xy)$$

$$= < x > +x < y > +[x,y] + (- < xy > +\lambda'xy).$$

Equate the first and last terms in this series of equalities to obtain the equation:

$$\lambda'x + \lambda'y = \{< x > +x < y > +[x,y]- < xy >\} + \lambda'xy.$$

Subtract $\lambda'xy$ from the right of both sides directly above to obtain:

$$[x,y]' = < x > +x < y > +[x,y]- < xy > .$$

Since all of the terms in the equation directly above are in the abelian group A, equation (ii) follows. The proof is complete. ∎

Let A be a fixed G-module. Denote by $B^2(G,A)$ the subset of the abelian group $Z^2(G,A)$ consisting of all factor sets $[\_,\_]$ such that there exists a function $<>: G \to A$ satisfying the equation $[x,y] =< x > +x < y > - < xy >$ for all $(x,y) \in G \times G$. The elements of $B^2(G,A)$ are called **coboundaries**. In Problem 7.5.1 we ask you to show that $B^2(G,A)$ is a (normal) subgroup of $Z^2(G,A)$.

Definition 7.5.2 *For a G-module A, define the group $e(G,A)$ as the factor group $e(G,A) = Z^2(G,A)/B^2(G,A)$.*

Remark 7.5.1 *The readers who have seen some algebraic topology will note that $e(G,A)$ looks like a homology or cohomology group. See Problem 7.5.2.*

To summarize our results so far: Let A be a G-module, E an exact sequence that realizes the operators and λ, λ' liftings that determine factor sets $[x,y], [x,y]'$. Then $([x,y]'-[x,y]) \in B^2(G,A)$. Thus, $[x,y]$ and $[x,y]'$ determine the same coset in the factor group $Z^2(G,A)/B^2(G,A) = e(G,A)$.

Here is our main and final theorem, one that characterizes the extensions of a given G-module A by G. The computations are quite intricate, but having come this far you won't fail now.

Theorem 7.5.2 *Let G be a multiplicative group and A a G-module. There is a bijection between the underlying set of the abelian group $e(G,A)$ and the set $Ext(G,A)$ of equivalence classes of extensions of A by G that realize the operators.*

Proof. Let E be an exact sequence representing an extension of A by G that realizes the operators and let $[E]$ be the equivalence class containing E. If we choose a lifting λ, the pair (E, λ) determines a factor set $[\_,\_]_{(E,\lambda)}$

We want to define a function Φ: $Ext(G, A) \to e(G, A)$ by the formula $\Phi([E]) = B^2(G, A) + [\_, \_]_E$. By Theorem 7.5.1, when we take cosets mod $B^2(G, A)$, we can omit the subscript indicating dependence on λ. By definition, every factor set in $Z^2(G, A)$ arises from a sequence E that realizes the operators. Thus, if Φ is well defined, Φ is epic. We prove that Φ is well defined.

Suppose $E_1 : 0 \to A \overset{\alpha}{\to} C_1 \overset{\beta}{\to} G \to 1$ is equivalent to $E_2 : 0 \to A \overset{\gamma}{\to} C_2 \overset{\delta}{\to} G \to 1$. There is an isomorphism $\theta : C_1 \to C_2$ such that the following diagram (Δ) commutes:

$$
\begin{array}{ccccccccc}
(E_1) : 0 & \to & A & \overset{\alpha}{\to} & C_1 & \overset{\beta}{\to} & G & \to & 1 \\
(\Delta) & & \downarrow 1_A & & \downarrow \theta & & \downarrow 1_G & & \\
(E_2) : 0 & \to & A & \overset{\gamma}{\to} & C_2 & \overset{\delta}{\to} & G & \to & 1
\end{array}
$$

Say that $\lambda_i : G \to C_i$ is a lifting, $i = 1, 2$, and let $x, y \in G$. Let $[\_, \_]'_1$ (resp.$[\_, \_]'_2$) be the maps from $G \times G$ to αA (resp. γA) determined by the factor set equation applied to the liftings λ_1, λ_2. Then the functions $[x, y]_1 = \alpha^{-1}[\_, \_]'_1$ and $[x, y]_2 = \gamma^{-1}[\_, \_]'_2$ are the factor sets determined by (E_1, λ_1) and (E_2, λ_2).

By commutativity of the right hand square, $(\theta\lambda_1)$ is a lifting for E_2. Thus, being the difference of factor sets for E_2 that were constructed via different liftings,

$$(*) : [x, y]_2 - \gamma^{-1}\{(\theta\lambda_1)(x) + (\theta\lambda_1)(y) - (\theta\lambda_1)(xy)\} \in B^2(G, A).$$

Apply $\theta\alpha$ to the defining equation $[x, y]_1 = \alpha^{-1}[\lambda_1 x + \lambda_1 y - \lambda_1(xy)]$ to obtain $\theta\alpha[x, y]_1 = (\theta\lambda_1)x + (\theta\lambda_1)y - (\theta\lambda_1)(xy)$. Now apply γ^{-1} to this latter equation to conclude that

$$(**) : \gamma^{-1}\theta\alpha[x, y]_1 = \gamma^{-1}\{(\theta\lambda_1)(x) + (\theta\lambda_1)(y) - (\theta\lambda_1)(xy)\}.$$

It follows from $(*)$ and $(**)$ that $[x, y]_2 - \gamma^{-1}\theta\alpha[x, y]_1$ is an element of $B^2(G, A)$. But, by commutativity of the diagram, $\gamma^{-1}\theta\alpha = 1_A$. We have shown that the map Φ is well defined.

Finally, we show that Φ is monic. Let $[\_, \_]_2$ and $[\_, \_]_1$ be factor sets associated with sequences E_1, E_2 such that $[\_, \_]_2 - [\_, \_]_1$ is an element of $B^2(G, A)$. We prove that $E_1 \sim E_2$.

Imagine a new diagram (Δ), as above, with E_1, E_2 this new pair of exact sequences and without the homomorphism θ. To show $E_1 \sim E_2$, we construct a θ that will make our new (Δ) commute. Let $c = \alpha a + \lambda_1 g, c' = \alpha a' + \lambda_1 g'$ (Lemma 7.4.1) be arbitrary elements of C_1.

We have:

$$c + c' = \alpha a + \lambda_1 g + \alpha a' + \lambda_1 g' = \alpha a + (\lambda_1 g + \alpha a' - \lambda_1 g) + \lambda_1 g + \lambda_1 g'.$$

Thus, $c + c' = (\alpha a + g\alpha a') + \lambda_1 g + \lambda_1 g'$. This follows since E_1 realizes the operators. Since $\lambda_1 g + \lambda_1 g' = \alpha[g, g']_1 + \lambda_1 gg'$, it follows that:

$$(a) \; c + c' = \alpha a + g\alpha a' + \alpha[g, g']_1 + \lambda_1 gg'.$$

Similarly, for arbitrary elements $d = \gamma a + \lambda_2 g, d' = \gamma a' + \lambda_2 g' \in C_2$, we have:

$$(b) \; d + d' = \gamma a + g\gamma a' + \gamma[g, g']_2 + \lambda_2 gg'.$$

We now bring in our hypothesis. There exists a function $<>: G \to A$ such that:

$$(c) \; [x, y]_2 = [x, y]_1 + x < y > - < xy > + < x >.$$

Define a function $\theta : C_1 \to C_2$ by $\theta(\alpha a + \lambda_1 g) = \gamma(a + < g >) + \lambda_2 g$. It is easy to prove that, with this θ, the new diagram (Δ) commutes. Thus, to finish the proof that $E_1 \sim E_2$, it only remains to check that θ is a homomorphism. (This is the only place where we need displayed equations (a),(b),(c).) We leave this final computation to you as Problem 7.5.4. ∎

Forty-seventh Problem Set

Problem 7.5.1 *Prove that $B^2(G, A)$ is a subgroup of $Z^2(G, A)$.*

Problem 7.5.2 *Let $M(G, A)$ be the additive abelian group consisting of all functions f from a multiplicative G into an additive abelian group A with $f(1) = 0$. Then $M(G, A)$ with binary operation pointwise addition of functions is an additive abelian group. Find homomorphisms θ_1, θ_2 which map as follows: $M(G, A) \xrightarrow{\theta_1} M(G \times G, A) \xrightarrow{\theta_2} M(G \times G \times G, A)$ such that $B^2(G, A) = \text{Im } \theta_1$ and $Z^2(G, A) = \ker \theta_2$. (First you have to give sensible definitions for the additive abelian groups $M(G \times G, A)$ and $M(G \times G \times G, A)$.)*

Problem 7.5.3 *Use the flavor of our theory of extensions (not necessarily all the technical detail) to prove that the only two nonabelian groups of order 8 are D_4 (the group of rotations of the square) and $\{\pm 1, \pm i, \pm j, \pm k\}$ with quaternion multiplication. Start with the fact that every group of order $8 = 2^3$ has a non-trivial center (Corollary 1.9.2).*

Problem 7.5.4 *Using equations (a),(b),(c), prove that the map θ of Theorem 7.5.2 is a homomorphism.*

Problem 7.5.5 *Let A be a G-module. Then $Ext(G, A)$ can be made into an additive abelian group.*

Further reading: [R] is a great text! I often look at it.

Chapter 8

Topics in abelian groups

In this final chapter we consider four nice classes of abelian groups, all of which are determined by cardinal invariants. Throughout, G will be an abelian group written additively and the word "group" will mean abelian group. Our investigations just scratch the surface of modern day abelian group theory. For an excellent comprehensive introduction, see [F].

8.1 Direct sums and products

For T a torsion group and p a prime, let T_p be the subgroup of all elements of p-power order. We start with a generalization of Theorem 1.12.1. A subscript p following a direct product or sum means that the product or sum is to be taken over all positive primes $p \in Z$.

Theorem 8.1.1 *Let T be torsion. Then $T = \bigoplus_p T_p$.*

Proof. Since each $T_p \leq T$, then $\sum T_p$, the set of all finite sums $t_{p_1} + \cdots + t_{p_k}$, will be the subgroup of T generated by all the individual subgroups T_p. We show that, for any fixed prime q, $T_q \cap \sum_{p \neq q} T_p = <0>$ and that $\sum T_p = T$. So, actually, what we're doing is proving that T is the internal direct sum of its p-torsion subgroups. (Refer back to Definition 1.10.2.) The proof is simple and follows that of Lemma 3.4.1.

First, if $x \in T_q \cap \sum_{p \neq q} T_p$, then $q^s x = 0$ and $x = t_{p_1} + \cdots + t_{p_k}$ with $t_{p_i} \in T_{p_i}, p_i \neq q, 1 \leq i \leq k$. We have $p_i^{s_i} t_{p_i} = 0$ for each i. Thus, since T is abelian, $(p_1^{s_1} \cdots p_k^{s_k})x = 0$. Hence, the order of x is a simultaneous divisor of q^s and of $p_1^{s_1} \cdots p_k^{s_k}$. It follows that $|x| = 1$ and $x = 0$. Second, if $t \in T$, then $nt = 0$ for some positive integer $n = p_1^{e_1} \cdots p_j^{e_j}$. Let $n_i = n/p_i^{e_i}$. Since gcd $\{n_1, ..., n_j\} = 1$, we can write $\sum_{i=1}^{j} a_i n_i = 1$ for suitable integers a_i. We have $x = 1x = \sum_{i=1}^{j} a_i(n_i x) \in \sum_{i=1}^{j} T_{p_i} \subset \sum_p T_p$. ∎

Forty-eighth Problem Set

Problem 8.1.1 *For G an abelian group, the **torsion subgroup** $T(G) = \{g \in G :| g |< \infty\}$. Prove $T(G) \leq G$ and that $G/T(G)$ is torsion-free.*

Problem 8.1.2 *Show that $\bigoplus_p Z(p)$ is the torsion subgroup of $\prod_p Z(p)$.*

Problem 8.1.3 *(a) If A is an abelian group check that its **endomorphism ring** $E(A) = Hom(A, A)$ is a ring under the operations of pointwise addition and function composition. (b) Now let T be a torsion group and write $T = \bigoplus_p T_p$. Show that, as rings, $E(T) \cong \prod_p E(T_p)$. Recall that the direct product of the rings $E(T_p)$ is a ring via componentwise addition and componentwise composition of functions.*

8.2   Structure theorem for divisible groups

Definition 8.2.1 *A group G is called **divisible** if $nG = G$ for all positive integers n. Here $nG = \{ng : g \in G\}$.*

One way to think about divisible groups is that, given any $g \in G$ and positive integer n, there exists an $x \in G$ such that $nx = g$. Given g and n the element x need not be unique. Any $x' \in G$ with $n(x' - x) = 0$ will also work. However if G is torsion-free and divisible then there is a unique solution in G to any equation of the form $nx = g$. (Check this.) Thus, in this case, $(1/n)g$ has a unique meaning for each natural number n and group element g. It follows that a torsion-free divisible group can be made into a Q-vector space in the natural way.

Exercise 8.2.1 *Show that the class of divisible groups is closed under direct sums, summands and homomorphic images.*

Some examples of divisible groups are $Q, Q/Z$ and $(Q/Z)_p$. The group $(Q/Z)_p$ is usually denoted $Z(p^\infty)$.

Exercise 8.2.2 *Show that $Z(p^\infty)$ is isomorphic to the group generated by a countable set of elements $\{a_1, a_2, ..., a_n, ...\}$ with relations $pa_1 = 0$, $pa_2 = a_1, ..., pa_n = a_{n-1}, ...$*

The proper subgroups of $Z(p^\infty)$ can be arranged in an ascending chain of cyclic groups $< a_1 > \subset < a_2 > \subset \cdots \subset < a_n > \subset \cdots$ (see Problem 8.2.1).

Theorem 8.2.1 *Any group can be embedded as a subgroup of a divisible group.*

Proof. The class of abelian groups coincides with the class of Z-modules. Thus, by Theorem 3.2.1, any group G is a epimorphic image of a free group F. Say $G \cong F/N$ with $F = \bigoplus_{i \in I} Zx_i$. If $D = \bigoplus_{i \in I} Qx_i$, then $G \cong F/N \leq D/N$ with D/N divisible. ■

We next introduce the concept of injectivity, dual to the notion of projectivity which was discussed in Section 3.2. For the class of Z-modules (but not for R-modules in general), injectivity is equivalent to divisibility.

Definition 8.2.2 *An R-module M is called **injective** if, whenever $f : M \to N$ is an R-module embedding, then there is a direct sum decomposition of R-modules $N = f(M) \oplus N'$.*

For an alternate definition of injectivity, see Problem 8.2.2.

Theorem 8.2.2 *An abelian group is injective if and only if it is divisible.*

Proof. Suppose G is an injective Z-module and let $f : G \to D$ be an embedding of G into a divisible group. Then $D = f(G) \oplus D'$ so that $G \cong f(G)$ is divisible.

Conversely, let G be divisible and suppose that $G \leq H$. (For convenience we have suppressed the embedding.) We need to show that G is a direct summand of H. Let $\mathcal{S} = \{X \leq H : X \cap G = <0>\}$. Then $\mathcal{S}$ with inclusion is a partially ordered set. If $\{X_i : i \in I\}$ is a chain in $\mathcal{S}$, then $\cup_{i \in I} X_i$ is an upper bound. (Check this.) Hence, by Zorn's Lemma, $\mathcal{S}$ contains a maximal element X'.

We claim $G \oplus X' = H$. We suppose not and contradict the maximality of X'. Let $h \in H, h \notin (G \oplus X')$. Then $< X', h >= \{x + nh : x \in X', n \in Z\}$ properly contains X', so $< X', h > \cap G \neq <0>$. Let $0 \neq g \in G$ be such that $g = x' + nh$. Then $nh = (g - x') \in G \oplus X'$. At this point we have shown that $H/(G \oplus X')$ is a torsion group. Let m be the order of $(G \oplus X') + h$ in $H/(G \oplus X')$. If $m = pm_1$ replace h with $m_1 h$ to get an element such that $h \notin (G \oplus X')$ but $ph \in (G \oplus X')$. Put $ph = g + x'$ and let $g_1 \in G$ be such that $pg_1 = g$. Since $h \notin (G \oplus X')$, we have $h - g_1 \notin X'$. But $p(h - g_1) = (g + x') - g = x' \in X'$. Thus, $< X', h - g_1 >= \{x + th : x \in X', 0 \leq t < p\}\}$ properly contains X'. But $< X', h - g_1 > \cap G = <0>$. Our final claim holds since, if $0 < t < p$, then $t(h - g_1) \notin X'$. (Why not?) We have contradicted the maximality of X'. The proof is complete. ■

Exercise 8.2.3 *Let $\{D_i : i \in I\}$ be a collection of divisible subgroups of a group G. Show that $\sum_{i \in I} D_i = \{d_{i_1} + \cdots + d_{i_k} : d_{i_j} \in D_{i_j}, \{i_1, ..., i_k\} \subset I\}$ is a divisible subgroup of G.*

By the above exercise, there is a unique maximal divisible subgroup D_G of a group G, namely the subgroup generated by all divisible subgroups of G. By Theorem 8.1.3, we have $G = D_G \oplus R$ for some complementary summand R. The group R will be **reduced**, meaning that R can have no nonzero divisible subgroups. (Why not?) We close this section with a structure theorem for divisible groups.

Theorem 8.2.3 *(structure theorem for divisible groups) Any divisible group D is the direct sum of copies of Q and copies of $Z(p^\infty)$ for various primes p. Let $r_0(D)$ be the number of copies of Q and, for each prime p, let $r_p(D)$ be the number of copies of $Z(p^\infty)$ in **any** such direct sum decomposition of D. Then the cardinals $\{r_0(D), r_p(D) : p$ a prime$\}$ form a complete set of isomorphism invariants for D.*

Proof. Let T be the torsion subgroup of D. If $t \in T$ has order m and $t = nd$ for $d \in D$, then $nmd = 0$ so that $d \in T$. Thus, T is itself a divisible subgroup of D and we can write $D = T \oplus X$, where X is torsion-free and divisible. By our earlier remarks, X has the natural structure of a vector space over Q. Hence, as an additive group, X will be isomorphic to a direct sum of copies of Q. The number of these copies is simply the vector space dimension, $\dim_Q X = \dim_Q(D/T)$, which is an invariant of D.

Write T as the direct sum of its p-torsion subgroups, $T = \bigoplus_p T_p$. Each T_p is divisible, being a summand of D, and is uniquely determined by D. It remains to show that each T_p is the direct sum of copies of $Z(p^\infty)$ and that the number of copies in any such direct sum decomposition is an invariant of T_p.

For a fixed prime p, consider $T_p[p] = \{t \in T_p : pt = 0\}$. Let $\{t_1^i : i \in I\}$ be a basis for $T_p[p]$ as a (Z/pZ)-vector space. For each i, choose elements t_j^i, $j \geq 2$, as follows: $pt_2^i = t_1^i, pt_3^i = t_2^i, .., pt_{n+1}^i = t_n^i, ...$ Then, for each i, the subgroup C_i generated by $\{t_j^i : j \geq 1\}$ is isomorphic to $Z(p^\infty)$.

We show that, as subgroups of T_p, $\sum_{i \in I} C_i = \bigoplus_{i \in I} C_i$. To show this, suppose that $(s) : y_{i_1} + \cdots + y_{i_k} = 0$ with $0 \neq y_{i_j} \in C_{i_j}$. We'll derive a contradiction. Without loss, assume that the y_{i_j} are arranged so that their order is non-decreasing. Say that $y_{i_s}, ..., y_{i_k}$ have maximal order p^n. Then, for $s \leq j \leq k$, each y_{i_j} must be of the form $y_{i_j} = m_{i_j} t_n^{i_j}$ with the integer m_{i_j} prime to p. (See Problem 8.2.1.) Thus, multiplying our sum (s) by p^{n-1} gives the equation $m_{i_s} t_1^{i_s} + \cdots + m_{i_k} t_1^{i_k} = 0$. This latter equation can be regarded as a linear dependence in the (Z/pZ)-vector space $T_p[p]$. Since the set $\{t_1^i : i \in I\}$ is independent over Z/pZ, we have $p \mid m_{i_j}$, $s \leq j \leq k$, a contradiction.

Next we show that $T_p = \sum_{i \in I} C_i$. We show $t \in T \to t \in \sum_{i \in I} C_i$ by induction on n with $\mid t \mid = p^n$. Since $T_p[p]$ is spanned by $\{t_1^i : i \in I\}$, certainly $T_p[p] \leq \sum_{i \in I} C_i$. Thus, our implication holds for $n = 1$. Let $t \in T$ with $|t| = p^n, n > 1$. Then $p^{n-1}t = \sum c_j t_1^j$. Hence, $p^{n-1}(t - \sum c_j t_n^j) = 0$. By induction, $t - \sum c_j t_n^j \in \sum_{i \in I} C_i$. Hence, $t \in \sum_{i \in I} C_i$.

Next, note that if $T_p = \bigoplus_{j \in J} Z(p^\infty)_j$ is any direct sum decomposition, then $|J| = dim_{Z/pZ} T_p[p]$. Thus, each r_p is an invariant of D.

Finally, if D, D' are divisible groups with $r_0(D) = r_0(D')$ and $r_p(D) = r_p(D')$ for each prime p, it takes very little ingenuity to construct an isomorphism from D to D'. Our proof is complete. ∎

Forty-ninth Problem Set

Problem 8.2.1 *Show that the proper subgroups of $Z(p^\infty)$ form a chain: $< a_1 > \subset < a_2 > \subset \cdots \subset < a_n > \subset \cdots$ whose union is $Z(p^\infty)$. Show that $z \in Z(p^\infty)$ has order p^n if and only if $z = ca_n$ with $(c, p) = 1$.*

Problem 8.2.2 *Let $i : A \rightarrow B$ be an embedding of abelian groups and let $f : A \rightarrow D$ with D divisible. Show that there is a map $f' : B \rightarrow D$ such that $f'i = f$. Hint: Start with the set $S = \{(C, g) : i(A) \leq C \leq B, g : C \rightarrow D, gi |_A = f\}$. Let $(C, g) \leq (C', g')$ if $C \leq C'$, $g' |_C = g$. Show that S has a maximal element (Y, f'). If $Y \neq B$, contradict the maximality. (Remember, you'd better use the fact that D is divisible somewhere in the proof.)*

Problem 8.2.3 *What is the structure of $[\prod_p Z(p)]/[\bigoplus_p Z(p)]$?*

Problem 8.2.4 *Give an example of a group $G = D \oplus R = D \oplus R'$ with D divisible and $R \neq R'$ reduced groups. But prove that, in any event, any pair of reduced complements R, R' must be isomorphic.*

8.3 Rank one torsion-free groups

Definition 8.3.1 *A subset $\{g_i : i \in I\}$ of a torsion-free group G is called **independent** if, whenever a finite integral combination $\sum t_i g_i = 0$, then all the $t_i = 0$. Define the **rank** of a torsion-free group G to be the cardinality of a maximal independent subset of G.*

If G is torsion-free then $rankG$ can be defined as the dimension of the rational vector space $Q \otimes_Z G$. Thus, the rank is an invariant of a torsion-free group. See Problem 8.3.1 below.

In this section we consider the simplest kind of torsion-free groups, those of rank one. Suppose rank $A = 1$ and let $0 \neq a \in A$. For any $a_1 \in A$ there is an integral dependence relation $na_1 - ma = 0$ with $n \neq 0$. Since the equation

$nx = ma$ has a unique solution in A, we can write a_1 unambiguously as $a_1 = (m/n)a$. Then the map induced by the stipulation $a_1 \to m/n$ is an embedding of A as a subgroup of Q. Conversely, any nonzero subgroup of Q has rank one. Our first task is to produce an invariant that classifies rank one groups, equivalently subgroups of Q.

Definition 8.3.2 *Let A be torsion-free and $a \in A$. For p a prime, the p-**height of** a in A (denoted $h_p^A(a)$) is the largest nonnegative integer k such that the equation $p^k x = a$ has a solution in A. If $p^k x = a$ is solvable in A for all k we set $h_p^A(a) = \infty$. The **height sequence of** a in A, denoted $H^A(a)$, is the sequence $< h_p^A(a) >_p$, indexed by the positive integral primes. An **abstract height sequence** is simply a sequence $< h_p >_p$, where each h_p is either a nonnegative integer or the symbol ∞ .*

Exercise 8.3.1 *For height sequences and $< h_p >$ and $< k_p >$ define $< h_p > \sim < k_p >$ if $h_p = k_p$ for all but a finite number of primes p and when $h_p \neq k_p$ neither h_p nor k_p equals ∞. Show that $\sim$ is an equivalence relation on the set of all height sequences.*

Definition 8.3.3 *Let A be torsion-free of rank one. Define the **type** of A (written $t(A)$) to be the equivalence class of $H^A(a)$, where a is any nonzero element of A.*

We first will show that $t(A)$ is an invariant of a rank one group A. Let a, b be nonzero elements of A with height sequences $H^A(a) = < h_p >$, $H^A(b) = < k_p >$, respectively. There are relatively prime positive integers m, n such that $ma = nb$. Write m, n as products of powers of distinct primes, $m = p_1^{e_1} \cdots p_t^{e_t}$, $n = q_1^{f_1} \cdots q_j^{f_j}$. Now check out the following observations.

Exercise 8.3.2 *Show that: (1) If $p \nmid mn$ then $h_p = k_p$. (2) If $p = p_i$ then $k_p = h_p + e_i$. (And, by symmetry, if $p = q_i$ then $h_p = k_p + f_p$.*

The previous exercise shows that $< h_p > \sim < k_p >$. It follows that, for rank one groups A, the type $t(A)$ is well defined.

Furthermore, let $\theta : A \cong B$, with $< 0 > \neq A, B \leq Q$, and let $0 \neq a \in A$. It is easy to check that $H^A(a) = H^B(\theta(a))$. Thus, $t(A)$ is an invariant for rank one torsion-free groups.

Example 8.3.1 *The type of Z is the equivalence class of the sequence of all zeros. This is also the type of $(1/n)Z$ for any positive integer n. The type of Q is the equivalence class of the sequence of all infinities. Let $A = \{r/s \in Q : (r, s) = 1, s \text{ involves no } p^2 \text{ in its factorization}\}$.*

The type of A is the equivalence class of the sequence of all ones. Let $B = \{r/5^k : k \geq 0\}$. The type of B is the equivalence class of the sequence $< 0, 0, \infty, 0..., 0, ... >$.

Theorem 8.3.1 *Let A, B be torsion-free rank one. Then $A \cong B$ if and only if $t(A) = t(B)$.*

Proof. The "only" if direction of the proof having been shown, suppose that A, B are rank one with $t(A) = t(B)$. Say that $t(A)$ is the equivalence class of the height sequence $H^A(a) = < h_p >$, $t(B)$ is the equivalence class of the height sequence $H^B(b) = < k_p >$. For the finitely many primes p where $h_p > k_p$, let $h_p = k_p + e_p$. For the finitely many primes p where $k_p > h_p$, let $k_p = h_p + f_p$. Let $m = \prod p^{e_p}$ and $n = \prod p^{f_p}$. Choose $a' \in A, b' \in B$ such that $ma' = a, nb' = b$. Then $H^A(a') = H^B(b')$. As previously noted, any $x \in A$ can be uniquely written in the form $x = (m/n)a'$. Similarly, any $y \in B$ is of the form $y = (m/n)b'$. By construction, $(m/n)a'$ represents an element of A if and only if $(m/n)b'$ is an element in B. Hence, the map $(m/n)a' \to (m/n)b'$ provides an isomorphism from A to B. ∎

Fiftieth Problem Set

Problem 8.3.1 *Prove that the rank of a torsion-free group G is equal to the dimension of the rational vector space $Q \otimes_Z G$.*

Problem 8.3.2 *Prove that the set of types can be made into a partially ordered set by defining $\tau \leq \sigma$ if and only if there are representative height sequences $(h_p) \in \tau, (k_p) \in \sigma$ such that $h_p \leq k_p$ for all p.*

Problem 8.3.3 *(a) Prove that the set of types can be made into a lattice as follows. If $\tau = [(h_p)]$ and $\sigma = [(k_p)]$, put $\tau \vee \sigma = [(h_p \vee k_p)]$ and $\tau \wedge \sigma = [(h_p \wedge k_p)]$. You'll need to check that the operations of sup and inf are well defined. (b) Then check that the sup operation has the following desired property: For all τ, σ the type $\tau \vee \sigma$ is greater than or equal to both τ and σ and, if a type δ is greater than or equal to both τ and σ, then $\delta \geq \tau \vee \sigma$. (c) State and prove an analogous property for the inf.*

Problem 8.3.4 *Prove that there are uncountably many nonisomorphic subgroups of Q. Hints: (1) Each real number in the interval $[0, 1)$ determines a unique dyadic expansion. (2) Use these expansions to create a set of types $\{\tau(r) : r \in [0, 1)\}$. Choose types with representative height sequences consisting of zeros and infinities. (3) Show that the subgroups of Q with these types are pairwise nonisomorphic.*

8.4 Structure of completely decomposable groups

In this section we provide a reasonably straightforward structure theorem for the most basic class of torsion-free groups.

Definition 8.4.1 *A group G is called **completely decomposable** if $G = \bigoplus_{i \in I} G_i$ where each G_i is torsion-free of rank one.*

Definition 8.4.2 *Let $G = \bigoplus_{i \in I} G_i$ be completely decomposable. For each type (equivalence class of an abstract height sequence) τ define $r_\tau(G)$ to be the cardinality of the set $\{i \in I : t(G_i) = \tau\}$.*

Rank one groups are isomorphic if and only if they have the same type. It is plain then that, if we can show that the numbers $r_\tau(G)$ are well defined (independent of the direct sum decomposition chosen), then these will be a complete set of isomorphism invariants for completely decomposable groups. We need one additional definition before presenting our theorem.

Definition 8.4.3 *Let G be torsion-free and $0 \neq x \in G$. Let $< x >_*$ denote the rank one subgroup $\{(r/s)x : (r/s)x \in G\}$. The group $< x >_*$ is called the **pure rank one subgroup generated by** x. The **type in** G **of** x (written $t^G(x)$) is the type of the rank one group $< x >_*$.*

Exercise 8.4.1 *Show that $t^G(x)$ is the type of the height sequence $H^G(x)$.*

Theorem 8.4.1 *The cardinal numbers $\{\ r_\tau(G) : \tau$ a type$\}$ are a complete set of isomorphism invariants for completely decomposable groups.*

Proof. Let G be completely decomposable and let τ be a type. Write $G = \bigoplus_{i \in I} G_i$ for a collection of rank one subgroups G_i. The subgroup $G(\tau) = \{g \in G : t^G(g) \geq \tau\}$ coincides with the direct sum of all G_i such that $t(G_i) \geq \tau$. This is because for an element $g = (g_{i_1} \oplus \cdots \oplus g_{i_k}) \in \bigoplus_{i \in I} G_i$ to be divisible by p^r in G, it is necessary and sufficient that each component g_{i_j} be divisible by p^r in G_{i_j}. It follows that $H^G(g) = \wedge_{j=1}^k H^{G_{i_j}}(g_{i_j})$ and, hence, that $t^G(g) = \wedge_{j=1}^k t(G_{i_j})$. Similarly, the subgroup $G^*(\tau) = < g \in G : t^G(g) > \tau >$ (the subgroup of G generated by all elements of type greater than τ) coincides with the direct sum of all G_i such that $t(G_i) > \tau$. Hence, $G(\tau)/G^*(\tau) \cong \bigoplus G_i$ such that $t(G_i) = \tau$. Therefore, for any type τ, $r_\tau(G) = rank[G(\tau)/G^*(\tau)]$. Since both $G(\tau)$ and $G^*(\tau)$ are invariants of G, the proof is complete. ∎

Fifty-first Problem Set

Problem 8.4.1 *Let A (resp. B) be the subrings of Q generated by Z and the elements $1/2$ (resp. $1/3$). If $G = A \oplus B$, list the possible types of nonzero elements of G.*

Problem 8.4.2 *Let $A = \bigoplus_{i=1}^{n} A_i, B = \bigoplus_{i=1}^{n} B_i$ be completely decomposable groups. Prove that if $A \bigoplus A \cong B \bigoplus B$ then $A \cong B$.*

8.5   Algebraically compact groups

In this section we follow the excellent treatment in [F]. Recall that, for an abelian group A and positive integer n, nA is the subgroup $\{na : a \in A\}$.

Definition 8.5.1 *A subgroup H of a group G is called **pure in** G if, for all positive integers n, $nG \cap H = nH$.*

Plainly $nH \subset nG \cap H$. What we are saying is that, if an element $h \in H$ can be written in the form $h = ng$ for some $g \in G$, then $h = nh'$ for some $h' \in H$.

Exercise 8.5.1 *Show that if $G = H \bigoplus K$ then H is pure in G.*

Exercise 8.5.2 *Show that if G/H is torsion-free then H is pure in G.*

Exercise 8.5.3 *Show if G is torsion-free and H is pure in G, then G/H is torsion-free.*

Definition 8.5.2 *A group A is **algebraically compact** if it is a direct summand in any group in which it is isomorphically embedded as a pure subgroup.*

By Theorem 8.2.2, divisible groups are algebraically compact. Also, we have already noted in Problem 8.1.3 that, if D is a divisible group, $i : U \to V$ is an embedding of U as a subgroup of V and $f : U \to D$ is a homomorphism, then there exists a homomorphism $f' : V \to D$ such that $f'i = f$. To describe this situation we say that divisible groups are **injective with respect to group embeddings**. The following diagram illustrates our situation:

$$U \xrightarrow{i} V$$
$$\downarrow f \quad \diagup f'$$
$$D$$

Definition 8.5.3 *An embedding* $i : U \rightarrow V$ *is called a* **pure embedding** *if* $i(U)$ *is a pure subgroup of* V.

Definition 8.5.4 *A group* A *is called* **pure injective** *if it is injective with respect to all pure embeddings.*

We are almost ready to present a theorem which gives a number of interesting alternate descriptions of the class of algebraically compact groups. We need one preparatory lemma.

Lemma 8.5.1 *Let* A *be a pure subgroup of* B *such that* B/A *is finitely generated. Then* A *is a direct summand of* B.

Proof. The group B/A is a direct sum of cyclic groups of infinite or prime power order, say $B/A = \bigoplus_{i=1}^{n} < A + b_i >$. If $| A + b_i | = p^k$, then $p^k b_i = a_i \in A$. By purity, there exists $a_i' \in A$ with $p^k a_i' = a_i$. Let $b_i' = b_i - a_i'$. Then $| b_i' | = p^k$ and $A + b_i' = A + b_i$. If $| A + b_i | = \infty$, put $b_i' = b_i$. We claim that (1) $< b_i' : 1 \leq i \leq n >= \bigoplus_{i=1}^{n} < b_i' >$ and that (2) $(\bigoplus_{i=1}^{n} < b_i' >) \oplus A = B$.

For (1), suppose $\sum t_i b_i' = 0$ for integers $\{t_i\}$. Then $\sum t_i (A + b_i) = A + 0$ so that, if $| A + b_i | = p^k$, then p^k divides t_i and, if $| A + b_i | = \infty$, then $t_i = 0$. Hence, $t_i b_i' = 0$ for all i. For (2), first note that, plainly, the subgroup generated by $\bigoplus_{i=1}^{n} < b_i' >$ and A is all of B. Furthermore, if $\sum t_i b_i' = a \in A$, we can argue as in (1) to conclude that $t_i b_i' = 0$ for all i. ∎

Theorem 8.5.1 *For an abelian group* A *the following are equivalent:*

1. A *is pure injective.*

2. A *is algebraically compact.*

3. A *is a direct summand of a direct product of groups* $\prod A_p$, *the product taken over a subset of the primes, where each* A_p *is a direct product of groups, each isomorphic to a group of the form* $Z(p^\infty)$ *or* $Z(p^k)$.

4. A *is a direct summand of a group which admits a compact topology (hence the genesis of the name).*

5. *If a system of linear equations* $(i \in I)$ $(j \in J)$ $\sum_{finite} n_{ij} x_j = a_i$, *where the* n_{ij} *are integers and the* a_i *are in* A, *has the property that every finite subsystem is solvable, then the system is solvable.*

Proof. 1.→2. Assume 1. and let A be pure in G. Since the inclusion map $i : A \rightarrow G$ is a pure embedding, there exists a map $f : G \rightarrow A$ such that f restricted to A is the identity map $1_A : A \to A$. It is not hard to check (see Problem 8.5.1 below) that $G = i(A) \oplus Ker\ f$.

2.$\to$3. We will show that any group can be embedded as a pure subgroup of a direct product as described in 3. Hence, assuming 2., A will be a direct summand of such a product.

Let G be a group. For $0 \neq g \in G$, let S_g be the collection of all subgroups $H \leq G$ such that $g \notin H$. The set S_g is nonempty since $< 0 >\in S_g$. If we make S_g a partially ordered set via inclusion, it is plain that every chain in S_g has an upper bound, namely the union of all its elements. Invoking Zorn's lemma, S_g has a maximal element H_g. Consider the map $i : G \to \prod_{0 \neq g \in G} G/H_g$ given by $i(x) = (H_g + x)_{g \in G}$. Plainly, i is a homomorphism. Since, for $0 \neq x \in G$, the coset $H_x + x$ is nonzero, it follows that i is an embedding of abelian groups.

We next focus our attention on the structure of the groups G/H_g. Let $g \neq 0$ be arbitrary but fixed and set $\bar{G} = G/H_g$. If $\bar{G}$ decomposes as a nontrivial direct sum $\bar{G} = U/H_g \bigoplus V/H_g$, we have $g \in U, g \in V$. Thus, $g \in U \cap V = H_g$, a contradiction. Hence, each G/H_g is an **indecomposable** abelian group. Suppose that $\bar{G}$ has a nonzero torsion subgroup T/H_g. Then $g \in T$ and, since $< H_g, g >= G$, we have that $\bar{G} = G/H_g$ is a torsion group. In view of Theorem 8.1.1, $\bar{G}$ must be a p-group for some prime p. If $\bar{G}$ is divisible, then, by Theorem 8.2.3, $\bar{G} \cong Z(p^\infty)$. If $\bar{G}$ is not divisible (which in this case just means not p-divisible), we claim that there is an element in $y \in \bar{G}[p]$ such that $h_p^{\bar{G}}(y) = k < \infty$. (See Problem 8.5.2 below.) Let $z \in \bar{G}$ be such that $p^k z = y$. We claim that $< z >$ is a direct summand of $\bar{G}$. To see this, we refer back to Section 1.11. Using the notation of that section, let H be a $< z >$-high subgroup of $\bar{G}$ and apply Lemmas 1.11.1 and 1.11.2. Since $\bar{G}$ is indecomposable we must have $\bar{G} = < z >\cong Z(p^{k+1})$.

We next enlarge our direct product to include all factors G/H such that, for some prime p, $G/H \cong Z(p^k), 1 \leq k \leq \infty$. Let P be this enlarged direct product and let $i : G \to P$ be the natural map, which remains an embedding. Now, for each fixed prime p, let A_p be the direct product of all those groups G/H isomorphic to some $Z(p^k)$, $1 \leq k \leq \infty$. We have G embedded as a subgroup of $P = \prod_p A_p$.

We finally show that $i(G)$ is a pure subgroup of P. We use Problem 8.5.3 below to reduce our job to proving that $p^k P \cap i(G) = p^k i(G)$ for all primes p and natural numbers k. Equivalently, we show that if $i(x) \notin p^k i(G)$, then $i(x) \notin p^k P$. If $i(x) \notin p^k i(G)$ then, since i is a homomorphism, $x \notin p^k G$. Let H be a subgroup of G maximal with respect to the properties: $p^k G \subset H, x \notin H$. We have $< H, x >= G$. It follows that G/H is a cyclic p-group of order $p^t, t \leq k$. Hence, $H + x$ cannot be divisible by p^k in G/H. Since an element of a direct product is divisible by p^k if and only if all its factors are divisible by p^k, $i(x) \notin p^k P$.

Having shown that $i(G)$ is a pure subgroup of P, the proof of 2.$\to$ 3. is complete.

3.$\to$ 4. Each finite cyclic group is compact in the discrete topology. So it's enough to show that $Z(p^\infty)$ is a summand of a group which admits a compact topology. But $Z(p^\infty)$ is a summand of Q/Z, which in turn is a summand of $\mathcal{R}/Z$. The additive group $\mathcal{R}/Z$ is isomorphic to the multiplicative group C of all complex numbers of absolute value one via the map $Z + r \to e^{2\pi r i}$. The points of C can be identified with the points of the unit circle. The group C is compact in the topology induced from the plane.

4.$\to$ 5. Suppose $A \oplus B = C$, where C is a group with a compact topology. Since we have not discussed topological groups before, we take a second to explain how to construct one. We start with a collection of subgroups $\{N_j \leq C : j \in J\}$, chosen such that $\cap_{j \in J} N_j = <0>$. This collection is defined to be a neighborhood base of the element 0. The neighborhood base of any other $c \in C$ is defined to be the collection of cosets $\{N_j + c : j \in J\}$. The topology τ thus defined is Hausdorff. (Check this.) Moreover, the maps $c \to -c$ and $(c_1, c_2) \to c_1 + c_2$ will be continuous maps from C, resp. $C \times C$, to C. For the second map, the Cartesian product $C \times C$ is endowed with the product topology $\tau \times \tau$.

To say that C is a compact topological group means that there exists a topology τ of this type on C such that (C, τ) is compact.

To prove 4.$\to$5., consider

$$(i \in I)(j \in J) \sum_{finite} n_{ij} x_j = a_i$$

a system of linear equations with integer coefficients such that, for any finite subset $I_0 \subset I$, the system

$$(i \in I_0)(j \in J) \sum_{finite} n_{ij} x_j = a_i$$

has a solution in A. Regard the system as a system in the containing group C. For each $i \in I$, the set of all solutions to the i-th equation is a subset, call it P_i, of the direct product $P = \prod_{j \in J} C_j$ of copies of the group C, indexed by the set J. If P is endowed with the product topology, each solution set P_i will be closed, since addition in C is continuous. Our assumption on the solutions is saying that the collection $\{P_i : i \in I\}$ has the finite intersection property. By 4. and the compactness theorem for products, P is compact.

Hence, $\cap_{i \in I} P_i$ is nonempty. Thus, our system has a solution $(c_j)_{j \in J}$. If $c_j = (a_j \oplus b_j) \in A \oplus B$, then $(a_j)_{j \in J}$ is a solution in $\prod_{j \in J} A_j$, as desired.

5.$\rightarrow$ 1. Let A satisfy 5. and suppose that we have:

$$U \overset{\theta}{\rightarrow} V$$
$$\downarrow f$$
$$A$$

Here θ is a pure embedding and f is a homomorphism. We need to construct $f' : V \rightarrow A$ to make the diagram commute.

Let $\{v_j : j \in J\}$ be such that $< \theta(U), v_j, j \in J >= V$. Let $(*)$: $(i \in I)$ $(j \in J)$ $\sum_{finite} n_{ij} v_j = \theta(u_i)$ be all the relations in V between the generators v_j and elements of $\theta(U)$. To define $f' : V \rightarrow A$ such that $f'\theta = f$, it suffices to define the elements $f'(v_j)$. The difficulty here is that, if $f'(v_j) = a_j \in A$, then all existing relations must be preserved. That is $(a_j)_{j \in J}$ must be a solution to the system: $(f'*) : (i \in I)$ $(j \in J)$ $\sum_{finite} n_{ij} a_j = f(u_i)$.

Any finite subset of equations from $(f'*)$ is of the form $f'S$, for S a finite subset of equations from $(*)$. The set S involves only involves finitely many v_j, say $\{v_{j_1}, \cdots v_{j_k}\}$. Put $V' =< \theta(U), v_{j_1}, \cdots v_{j_k} >$. Since $\theta(U)$ is pure in V', we can invoke Lemma 8.3.1 to conclude that $V' = \theta(U) \oplus V''$. Note that the k-tuple $\{\theta u_{j_1}, \cdots, \theta u_{j_k}\}$ of the $\theta(U)$ components of the v_{j_t} with respect to this direct sum decomposition will be a solution to the finite set of equations S. It follows that the set $\{a_{j_1} = f u_{j_1}, \cdots, a_{j_t} = f u_{j_k}\}$ will be a solution to the system $f'S$. We have shown that any finite subset of $(f'*)$ has a solution.

By 5., the complete system $(f'*)$ has a solution $(a_j)_{j \in J}$. The map f' : $V \rightarrow A$ given by $f'(\theta(u) + \sum t_j v_j) = f(u) + \sum t_j a_j$ is our desired map. The proof is complete. ∎

Corollary 8.5.1 *A reduced algebraically compact group is a direct summand of a direct product of cyclic p-groups.*

Fifty-second Problem Set

Problem 8.5.1 *Suppose $A \leq G$ and there exists a homomorphism f : $G \rightarrow A$ such that $f \mid_A = 1_A$. Prove that $G = A \oplus \ker f$.*

Problem 8.5.2 *Suppose that G is a p-group and that $h_p(g) = \infty$ for all $g \in G[p]$. Prove that G is divisible. (Note that here divisibility is equivalent to p-divisibility.)*

Problem 8.5.3 *Prove that if $A \leq G$ is such that $p^k A = p^k G \cap A$ for all primes p and natural numbers k, then A is pure in G.*

Problem 8.5.4 *Prove that $\bigoplus_p Z(p)$ is not algebraically compact.*

Problem 8.5.5 *Give an example of an algebraically compact group A with a pure subgroup B which is not algebraically compact.*

8.6   Structure of algebraically compact groups

Definition 8.6.1 *The Z-adic topology on an abelian group G is the one induced by setting $\{nG : n > 0\}$ as a neighborhood base of $0 \in G$.*

An abelian group G is Hausdorff in the Z-adic topology if and only if $G^1 = \cap_{n>0} nG = < 0 >$. Any finite abelian group is discrete, thus both Hausdorff and complete, in its Z-adic topology. We use "complete" to mean both Hausdorff and complete. Any direct product of finite groups $G = \prod_\alpha G_\alpha$ is also complete in its Z-adic topology. Note that, if $(x_n) \in G$ is Cauchy, then, for each k and sufficiently large i, j, $(x_i - x_j) \in k!G = \prod_\alpha k!G_\alpha$.

Exercise 8.6.1 *Prove that $G = \prod_\alpha G_\alpha$ as above is Hausdorff in its Z-adic topology.*

Exercise 8.6.2 *With notation as above, prove that there exists $x \in \prod_\alpha G_\alpha$ such that $(x_n) \to x$.*

Exercise 8.6.3 *Prove that a reduced algebraically compact group is complete in its Z-adic topology.*

At this point we can give an improvement on Corollary 8.5.1.

Theorem 8.6.1 *Let A be reduced and algebraically compact. Then $A = \prod_p A_p$, the product taken over all primes p, where each A_p is p-**local** (divisible by all primes $q \neq p$). The subgroups A_p are uniquely determined by A.*

Proof. First, by Corollary 8.5.1, $A \bigoplus B = \prod_p C_p$, where each C_p is a direct product of cyclic groups of p-power order. Let $A_p = C_p \cap A$, $B_p = C_p \cap B$. Then $(A_p \oplus B_p) \leq C_p$. Let π_A, π_B be projections of $A \bigoplus B$ onto A, B. By inspection, each C_p is the unique maximal p-local subgroup of $A \bigoplus B$. Thus, each C_p is mapped into itself by every endomorphism of $A \bigoplus B$. In particular, $\pi_A(C_p) \leq C_p, \pi_B(C_p) \leq C_p$. So if $c \in C_p$, then

$c = (\pi_A + \pi_B)(c) = [\pi_A(c) + (\pi_B)(c)] \in A_p \oplus B_p$. Hence, for each p, $C_p = A_p \oplus B_p$.

We have $\bigoplus_p A_p \leq A$. By Exercise 8.6.3, A is complete in its Z-adic topology. Since $\prod_p A_p / \bigoplus_p A_p$ is divisible (check this), every element in $\prod_p A_p$ is a limit of a Cauchy sequence in $\bigoplus_p A_p$. It follows that $\prod_p A_p \leq A$. Similarly, $\prod_p B_p \leq B$. Hence, $A = \prod_p A_p$. Each A_p is the unique maximal p-local subgroup of A, thus is determined by A. ∎

Definition 8.6.2 *For a prime p, the **ring of p-adic integers** $\hat{Z}_p$ is the set of all formal power series $\{\sum_{i=0}^{\infty} c_i p^i : 0 \leq c_i < p\}$. Addition and multiplication in $\hat{Z}_p$ are the natural extensions of addition and multiplication of natural numbers written in base p.*

The set of nonnegative integers written in base p coincide with the set of finite power series in $\hat{Z}_p$. It is reasonably clear that addition and multiplication in $\hat{Z}_p$ are both commutative and associative and that multiplication distributes over addition. We show that each $c = \sum_{i=0}^{\infty} c_i p^i$ has an additive inverse $d = \sum_{i=0}^{\infty} d_i p^i$ in $\hat{Z}_p$. If $c = 0$, then $d = 0$. Otherwise, say that c_k is the first nonzero coefficient of c. Put $d_i = 0$ for $i < k$. Let $d_k = p - c_k$. Then let $0 \leq d_{k+1} < p$ be the unique integer such that $c_{k+1} + 1 + d_{k+1} = p$. Let $0 \leq d_{k+2} < p$ be the unique integer such that $c_{k+2} + 1 + d_{k+2} = p$. Continue inductively to compute the coefficients of a p-adic integer d such that $d + c = 0$.

Exercise 8.6.4 *Compute -2 in the ring $\hat{Z}_5$.*

In Problem 8.6.1 below you're asked to show that $\sum_{i=0}^{\infty} c_i p^i$ is a unit in the ring $\hat{Z}_p$ if and only if its constant term $c_0 \neq 0$.

Let's now return to the groups A_p of Theorem 8.6.1. Since A_p is p-local, the Z-adic neighborhood base of 0 in A_p coincides with the set $\{p^k A_p : k > 0\}$. This neighborhood base defines what's called the p-**adic topology** for any abelian group. Since A_p is complete in its p-adic ($= Z$-adic) topology, we can make A_p into a module over the ring $\hat{Z}_p$ in the natural way. For $a \in A_p$ and $c = \sum_{i=0}^{\infty} c_i p^i \in \hat{Z}_p$, define ca to be the (unique) p-adic limit in A of the Cauchy sequence $C_k a$, where $C_k \in Z$ is the k-th partial sum of c, $C_k = \sum_{i=0}^{k-1} c_i p^i$.

Definition 8.6.3 *Let M be a reduced $\hat{Z}_p$-module. A **basic submodule** $B \leq M$ is a submodule such that (1) B is a direct sum of copies of $Z(p^k)$ and $\hat{Z}_p$, (2) B is pure in M and (3) M/B is divisible.*

Theorem 8.6.2 *Any reduced $\hat{Z}_p$-module M has a basic submodule. Any two basic submodules for M are isomorphic.*

Proof. Let $S = \{m_i : i \in I\}$ be a subset of M maximal with respect to the **pure-independence** (pi) property: $\sum_{finite} c_i m_i \in p^k M$, $c_i \in \hat{Z_p}$, implies $c_i = p^k c_i'$ for all i. The existence of such a set is provided by an application of Zorn's lemma. If $\sum_{finite} c_i m_i = 0$, then $\sum_{finite} c_i m_i \in p^k M$ for all positive integers k. Hence $p^k \mid c_i$ for all k, i. It follows that each $c_i = 0$ so that the submodule B generated by S is the direct sum of cyclic submodules $B = \bigoplus_{i \in I} \hat{Z_p} m_i$. In Problem 8.6.2 below, you show that $\hat{Z_p}$ is a pid with ideals $\{p^k \hat{Z_p} : k \geq 0\}$. Referring back to Section 3.4, we have that the order of each m_i is either infinite or some p^k. If $\mid m_i \mid = \infty$, then $\hat{Z_p} m_i \cong \hat{Z_p}$. If $\mid m_i \mid = p^k$, then $\hat{Z_p} m_i \cong \hat{Z_p}/p^k \hat{Z_p} \cong Z(p^k)$. Hence, B satisfies (1).

Since both B and M are p-local, purity amounts to the assertion that $B \cap p^k M = p^k B$ for all $k > 0$. But this is precisely the property (pi). Thus, B satisfies (2).

We show that M/B is divisible. Since M/B is p-local, we need only show that $p(M/B) = M/B$. The preceding equality is equivalent to the fact that $M = pM + B$. Suppose that $m \in M, m \notin pM + B$. We claim that $S \cup \{m\}$ has property (pi), contradicting the maximality of S. Say $(\sum_{finite} c_i m_i + cm) \in p^k M$. Then $cm \in p^k M + B \subset pM + B$. If $p \nmid c$, then c has a multiplicative inverse in $\hat{Z_p}$ so that $m \in pM + B$, a contradiction. Hence, $c = pc'$. We then have $\sum_{finite} c_i m_i \in pM$. Thus, each $c_i = pc_i'$. Division by p yields the equation $(\sum_{finite} c_i' m_i + c'm) \in p^{k-1} M$. Continue the argument to eventually conclude that all the c_i and c are divisible by p^k. We have shown that $S \cup \{m\}$ has property (pi), a contradiction. Thus, M/B is divisible. We have proved that any submodule B constructed via a maximal pi set S is a basic submodule.

Note that, for all $k > 0$, the number of $Z(p^k)$ summands of B is equal to the number of $Z(p^k)$ summands of $B/p^n B$ for any $n > k$. But $B/p^n B = B/(p^n M \cap B) \cong (B + p^n M)/p^n M = M/p^n M$. Hence, $r_p^k(M)$, the number of $Z(p^k)$ summands of **any** p-basic submodule B, is an invariant of M. Let B_0 be the direct sum of the $\hat{Z_p}$ summands of B. If T is the torsion submodule of M, then $\bar{B_0} = (B_0 + T)/T$ is a pure free submodule of $\bar{M} = M/T$ such that $\bar{M}/\bar{B_0}$ is divisible. Thus,

$(\bar{B_0}/p\,\bar{B_0}) = \bar{B_0}/(p\,\bar{M} \cap \bar{B_0}) \cong [(\bar{B_0} + p\,\bar{M})/p\,\bar{M}] = (\bar{M}/p\,\bar{M})$. Since $\bar{B_0} \cong B_0$, the number of $\hat{Z_p}$ summands of B, equivalently of B_0, is equal to the Z/pZ dimension of $(\bar{B_0}/p\,\bar{B_0})$. This dimension in turn is equal to the Z/pZ dimension of $\bar{M}/p\,M$. Hence, $r_p^0(M)$, the number of $\hat{Z_p}$ summands of any p-basic submodule B, is an invariant of M.

It follows that any two basic submodules for M are isomorphic. ∎

We are now ready for our structure theorem for reduced algebraically compact groups.

Theorem 8.6.3 *(structure theorem for algebraically compact groups)*
Let $A = \prod A_p$ be a reduced algebraically compact group. For each prime p, let B_p be a p-basic submodule of A_p. For $k > 0$, let $r_p^k(A)$ be the number of $Z(p^k)$ summands of B_p. Let $r_p^0(A)$ be the number of $\hat{Z_p}$ summands of B_p. Then the set of cardinal numbers $\{r_p^k(A) : p \text{ a prime}, k \geq 0\}$ is a complete set of isomorphism invariants of A.

Proof. We have already shown that, for each p, the numbers $r_p^k(A)$ are invariants of A_p, hence of A. Conversely, for each prime p, the numbers $r_p^k(A)$ determine the structure of a p-basic submodule B_p for A_p. Let $A' = \prod A_p'$ be another reduced algebraically compact group with p-basic submodules B_p' for A_p'. If $r_p^k(A) = r_p^k(A')$ for all p, k, then $B_p \cong B_p'$ for all p. These isomorphisms extend to isomorphisms from A_p, the p-adic completion of B_p, to A_p', the p-adic completion of B_p'. It follows that $A \cong A'$ and the proof is complete. ∎

Fifty-third Problem Set

Problem 8.6.1 *(a) Prove that $\sum_{i=0}^{\infty} c_i p^i$ is a unit in the ring $\hat{Z_p}$ if and only if $c_0 \neq 0$. (b) Recall that Z_p is the ring $\{r/s \in Q : (r, s) = 1, (p, s) = 1\}$. Prove that Z_p can be regarded as a subring of $\hat{Z_p}$.*

Problem 8.6.2 *Show that the set of ideals of $\hat{Z_p}$ is $\{p^k \hat{Z_p} : k \geq 0\}$.*

Problem 8.6.3 *Prove that if a reduced algebraically compact group is torsion, then it is bounded.*

Problem 8.6.4 *Prove that, if A is reduced algebraically compact and $A \oplus A \cong A' \oplus A'$, then $A \cong A'$.*

Problem 8.6.5 *Prove that $E(Z(p^\infty)) \cong \hat{Z_p}$. Hint: Let $\{a_i : 1 \leq i < \infty\}$ be the standard generators for $Z(p^\infty)$ and f be an endomorphism. Consider $f(a_1), f(a_2), \dots$ and use them to construct a p-adic integer.*

8.7 Structure of countable torsion groups

Let T be a countable torsion group. We will find a complete countable set of cardinal isomorphism invariants for T. Since T is the direct sum of its p-torsion subgroups, $T = \bigoplus T_p$, it suffices to find invariants for each T_p.

Hence, assume that T is a p-group. Since the number of $Z(p^\infty)$ summands is an invariant for the maximal divisible subgroup of T, it does no harm to assume that T is reduced. To construct our invariants, we need the notion of a countable ordinal number.

Remark 8.7.1 *Countable ordinals. Let the ordinal 0 be the empty set ϕ, the ordinal 1 be $\{0\}$, the ordinal 2 be $\{0,1\}$. Suppose that the ordinal n has been defined for a natural number n. Let $n+1$ be the set of all previously defined smaller ordinals, $n+1 = \{0,1,...,n\}$. Each finite ordinal is an ordered set with the natural ordering. Let ω, the first infinite ordinal, be the ordered set of all its predecessors, $\omega = \{0,1,2,....\}$. Up to this point there has been really no difference between ordinals and cardinals. But now we define $\omega+1$ to be the ordered set $\{0,1,2,...,\omega\}$. Note that ω and $\omega+1$ both are countable sets, but $\omega < \omega+1$ in the sense that there is an order preserving embedding from ω to $\omega+1$ but none from $\omega+1$ to ω. We continue this procedure for every natural number $n > 1$ to define $\omega+n = \{0,1,2,...,\omega,\omega+1,...,\omega+n-1\}$. Then $2\omega = \{0,1,2,...,\omega,\omega+1,...,\omega+n,...\}$ will be the ordered set consisting essentially of two copies of ω, everything in the left hand copy being less than anything in the right hand copy. In this fashion we can generate the set of all countable ordinals.*

A well ordered set is one in which each nonempty subset has a least element. Each countable ordinal α, as constructed, is represented by a countable well-ordered set, namely the set of all ordinals less than α. The set Ω of all countable ordinals itself becomes a well ordered set under the ordering $\alpha \leq \beta$ if there is an order preserving embedding from α to β. Note that, under this ordering, countable ordinals are of two types: limit ordinals like $\omega, 2\omega, \omega^2, \omega^5 + 3\omega^2$ which have no immediate predecessor and non-limit ordinals like 7 and $\omega^5 + 3\omega^2 + 4$ which are of the form $\beta + 1$ for β their immediate predecessor. The ordered set Ω is a representative of the smallest uncountable ordinal. Note that, as for countable ordinals, Ω is represented by the ordered set of all previously constructed smaller ordinals.

Exercise 8.7.1 *Give a precise description of the ordered set corresponding to $\omega^5 + 3\omega^2 + 4$.*

Exercise 8.7.2 *Prove that $n + \omega = \omega$ for all natural numbers n. (First, decide what you should mean by $n + \omega$. Hence, addition of ordinals is non-commutative.*

The next theorem, proved by Ulm in 1933, was the first major result in the theory of torsion groups. Coincidentally, Baer proved the first major result in the theory of torsion-free groups, Theorem 8.4.1, during the same year. First we need some definitions refining our notion of height. Since, in this section, T is a p-group, the only relevant height is the p-height.

Definition 8.7.1 *Set $p^\omega T = \cap_{n<\omega} p^n T$. If α is a countable ordinal and $p^\beta T$ has been defined for all $\beta < \alpha$, define $p^\alpha T$ in one of two ways: If $\alpha = \gamma + 1$ for some γ, set $p^\alpha T = p(p^\gamma T)$. If α is a limit ordinal, set $p^\alpha T = \cap_{\beta<\alpha} p^\beta T$.*

Definition 8.7.2 *For $0 \neq t \in T$, say t is of **height** α ($h(t) = \alpha$) if $t \in p^\alpha T$ but $t \notin p^{\alpha+1} T$. Put $h(0) = \infty$.*

For a slight simplification of notation write T_α for $p^\alpha T$.

Note that if $T_\alpha = T_{\alpha+1} = pT_\alpha$, then $T_\alpha = 0$, since T_α is divisible and T is reduced. Furthermore, we cannot have $T_{\alpha+1} \subsetneqq T_\alpha$ for all countable ordinals α, since T is countable and Ω is an uncountable set. Hence, the set $N = \{\beta \in \Omega : T_\beta = 0\}$ is nonempty. Let λ, called the **length** of T, be the least ordinal in N.

Recall that $T[p] = \{t \in T : pt = 0\}$.

Definition 8.7.3 *Let α be a countable ordinal and T a countable reduced p-group. Then the α-th **Ulm Invariant** of T ($f_\alpha(T)$) is the Z/pZ dimension of the vector space $T_\alpha[p]/T_{\alpha+1}[p]$.*

Recall that we previously used the finite Ulm Invariants, $f_k^p(T) = \dim_{Z/pZ}(p^k T[p]/(p^{k+1} T[p]))$, to show that, for T a direct sum of cyclics, the number of $Z(p^{k+1})$ summands in any decomposition of T was an invariant. Here, of course, T need not be a direct sum of cyclics but we can still determine its structure with this finer set of invariants at our disposal. We are ready to prove Ulm's Theorem. We follow the proof presented in [Ka].

Theorem 8.7.1 *(**structure theorem for countable p-groups**) Let T be a countable reduced p-group of length λ. Then the set $\{f_\alpha(T) : \alpha < \lambda\}$ is a complete set of isomorphism invariants for T.*

Proof. It's reasonably clear that the cardinal numbers $f_\alpha(T)$ are isomorphism invariants. (See Problem 8.7.1 below.) So suppose that T, T' are countable reduced p-groups of length λ and that $f_\alpha(T) = f_\alpha(T')$ for all $\alpha < \lambda$. If $S \leq T$ call an element $t \in T$ **proper** with respect to S if $h(t) \geq h(t + s)$ for all $s \in S$. In other words, t is an element of maximal height in its coset mod S. If S is finite there certainly will be a proper element in each coset.

We set up a somewhat unmotivated isomorphism. For $S \leq T$, let $S_\alpha = (p^\alpha T) \cap S$. Write $S_\alpha^* = S_\alpha \cap p^{-1} T_{\alpha+2}$. By this we mean the collection of all elements $s \in S$ such that $h(s) \geq \alpha$ and $h(ps) \geq \alpha + 2$. Certainly, by definition of height, if $s \in S_\alpha$, $h(ps) \geq \alpha + 1$. But, even in a finite p-group, we can have $h(px) > h(x) + 1$ (and $px \neq 0$). (See Problem 8.7.2.)

If $x \in S_\alpha^*$ we can write $px = py$ with $y \in S_{\alpha+1}$. The element y is not uniquely determined by x. However, if y' is another such element, then $(y - y') \in S_{\alpha+1}[p] \subset T_{\alpha+1}[p]$. Consider the element $x - y$. Since $x \in S_\alpha$, $y \in S_{\alpha+1}$ we have $x - y \in S_\alpha$. Also $p(x - y) = 0$ so that $(x - y) \in S_\alpha[p] \subset T_\alpha[p]$. Thus, the map $x \to x - y$ is a well defined homomorphism from S_α^* into $T_\alpha[p]/T_{\alpha+1}[p]$. The kernel of this homomorphism is precisely $S_{\alpha+1}$. Hence, we have an embedding $\theta : S_\alpha^*/S_{\alpha+1} \to T_\alpha[p]/T_{\alpha+1}[p]$ given by $\theta(S_{\alpha+1} + x) = T_{\alpha+1}[p] + (x - y)$. We now interrupt the proof of the theorem to present a technical lemma. ∎

Lemma 8.7.1 *The following two statements are equivalent: (1) The map θ is not onto. (2) There exists in $T_\alpha[p]$ an element of height exactly α proper with respect to S.*

Proof. $(1) \Longrightarrow (2)$ Let $t \in T_\alpha[p]$ be such that $T_{\alpha+1}[p] + t$ is not in the range of θ. Then, since $t \notin T_{\alpha+1}$, the height of t is precisely α. We show by contradiction that t is proper with respect to S. If not, there exists $s \in S$ with $h(s - t) > \alpha$. Write $s - t = pv$ with $v \in T_\alpha$. Since $pt = 0$ we have $ps = p^2v$ so that $s \in S_\alpha^*$. A simple calculation shows that $\theta(S_{\alpha+1} + s) = T_{\alpha+1}[p] + t$, a contradiction. Hence, t is proper with respect to S.

$(2) \Longrightarrow (1)$ Suppose that $t \in T_\alpha[p]$ is an element of height α proper with respect to S. Then $T_{\alpha+1}[p] + t$ is not in the range of θ. For if it were, there would exist elements $x \in S, y \in T_{\alpha+1}$ with $p(x - y) = 0$ and $T_{\alpha+1}[p] + t = T_{\alpha+1}[p] + (x - y)$. Then $T_{\alpha+1}[p] + (t - x) = T_{\alpha+1}[p] - y$ so that $h(t - x) > \alpha$, contradicting the assumption that t is proper with respect to S.

We return to the proof of the theorem. Before we finally give the proof, some remarks are in order. (a) The isomorphisms $\Phi : T \cong T'$ and $\Phi^{-1} : T' \cong T$ will be built up, step by step, from partial isomorphisms $\Phi_n : S_n \cong S_n'$, $\Phi_n^{-1} : S_n' \cong S_n$, where S_n, S_n' are finite subgroups of T, T'. We will go back and forth from T to T'. Put $S_0 = S_0' = <0>$ and $\Phi_0 = 0$. Starting at step one, we extend each map with even index $\Phi_{2m} : S_{2m} \cong S_{2m}'$ to a map $\Phi_{2m+1} : S_{2m+1} \cong S_{2m+1}'$. Then at each even step $2m \geq 2$, we extend the map with odd index $\Phi_{2m-1}^{-1} : S_{2m-1}' \cong S_{2m-1}$ to a map $\Phi_{2m}^{-1} : S_{2m}' \cong S_{2m}$. (b) Before we begin, we enumerate the elements of T and of T', $T = \{t_i : i \geq 1\}, T' = \{t_i' : i \geq 1\}$. At step $2m+1$ we will make sure that $t_m \in S_{2m+1}$; at step $2m$ we will make sure that $t_m' \in S_{2m}'$. Then, when we take $\Phi = \cup \Phi_n$ we will have an isomorphism from all of T to all of T'. Of course, to carry out this procedure we need the assumption of countability. (c) Each Φ_j, in order to be extended to an isomorphism from T to T', must be **height-preserving**; that is it must have the property that $h[\Phi_j(s)] = h(s)$ for all

$s \in S_j$. Similarly, each Φ_j^{-1} must be height-preserving. We emphasize that all heights are being computed in the overlying groups T', T.

In view of the remarks, we may assume the following situation: the map $\Phi_n : S_n \cong S'_n$, $n = 2m \geq 0$, is a height-preserving isomorphism, where S_n, S'_n are finite subgroups of T, T'. Our goal is to extend Φ_n to a height-preserving isomorphism $\Phi_{2m+1} : S_{2m+1} \cong S'_{2m+1}$ where $S_{2m+1} =<$ $S_{2m}, t_m >$. If we can do this, a symmetric argument will show that $\cup \Phi_n :$ $T \cong T'$. Replacing t_m by $p^{k-1} t_m$, where $p^k t_m \in S_{2m}$, we can assume $p t_m \in$ S_{2m}. (Then we repeat our extension k times to pick up t_m.) By adding, if necessary, an element of S_{2m} to t_m, it also does no harm to assume that t_m is proper with respect to S_{2m}, that is $\alpha = h(t_m) \geq h(t_m + s), s \in S_{2m}$. Our final normalizing assumption is that $h(p t_m)$ is maximal in the finite set $\{h[p(t_m + s)] : s \in S_{2m}, t_m + s \text{ is proper with respect to } S_{2m}\}$.

With these normalizations, let $\Phi_{2m}(p t_m) = z \in S'_{2m}$. Our problem is reduced to this: Find $w \in T'_\alpha$ with $pw = z$ and w proper with respect to S'_{2m}. We then can define $\Phi_{2m+1}(k t_m + s) = kw + \Phi_{2m}(s)$ for $0 \leq k <$ $p, s \in S_{2m}$. It is easy to check that, since w is proper, Φ_{2m+1} will be a height preserving isomorphism. We will have to consider two cases.

Case I. $h(z) = \alpha + 1$ Let w be any element of T'_α such that $pw = z$. Since $h(z) = \alpha + 1$, we have $h(w) = \alpha$. We claim that $w \notin S'_{2m}$. To see this, suppose $w = \Phi_{2m}(y), y \in S_{2m}$. Then $p t_m = py$ since both map onto z. Note that $h(t_m - y) = \alpha$ since t_m is proper with respect to S_{2m}. We have: $(i) : h[p(t_m - y)] = h(0) = \infty$. But, since Φ_{2m} is height preserving:

$$(ii) : h(p t_m) = h(py) = h\Phi_{2m}(py) = h(pw) = h(z) = \alpha + 1.$$

Equations (i) and (ii) contradict the maximality of $h(p t_m)$ mod S_{2m}. Thus, $w \notin S'_{2m}$.

We next show, again by contradiction, that w is proper with respect to S'_{2m}. Suppose that $h[w + \Phi_{2m}(s)] \geq \alpha + 1$. Then:

$$h[pw + p\Phi_{2m}(s)] = h[z + \Phi_{2m}(ps)] = h\Phi_{2m}[(p(t_m + s)] \geq \alpha + 2,$$

so that $h(t_m + s) \geq \alpha + 1$. This contradicts the fact that t_m is of height α and proper with respect to S_{2m}. The proof in Case I is complete. Note that in this case, we didn't need to use our hypothesis, that T and T' have the same Ulm invariants. We'd better use it in Case II, since it's certainly not true that any two countable reduced p-groups are isomorphic.

Case II $h(z) > \alpha + 1$ Now $h(p t_m) > \alpha + 1$ means that $p t_m = pv$ with $v \in T_{\alpha+1}$. Then $t_m - v$ has order p and has height α. (See Problem 8.7.4 below.) Also, $t_m - v$ is proper with respect to S_{2m} since v will not interfere with computations of any heights less than or equal to α. We apply Lemma 8.7.1 to conclude that the embedding θ is not onto. Since the Z/pZ dimension of $(S_{2m})^*_\alpha/(S_{2m})_{\alpha+1}$ is finite,

$$\dim_{Z/pZ}(S_{2m})^*_\alpha/(S_{2m})_{\alpha+1} < \dim_{Z/pZ} T_\alpha[p]/T_{\alpha+1}[p] = f_\alpha(T).$$

Now $\Phi_{2m} : S_{2m} \cong S'_{2m}$ is height preserving, so it maps (employing the obvious notation) $(S_{2m})^*_\alpha$ onto $(S'_{2m})^*_\alpha$ and $(S_{2m})_{\alpha+1}$ onto $(S'_{2m})_{\alpha+1}$. Hence Φ_{2m} induces an isomorphism $(S_{2m})^*_\alpha/(S_{2m})_{\alpha+1} \cong (S'_{2m})^*_\alpha/(S'_{2m})_{\alpha+1}$. Thus,

$$\dim_{Z/pZ}(S'_{2m})^*_\alpha/(S'_{2m})_{\alpha+1} < f_\alpha(T') = f_\alpha(T).$$

We now employ Lemma 8.7.1 in the reverse direction to conclude that T' contains an element w_1 of height α and order p proper with respect to S'_{2m}. Since $h(z) > \alpha + 1$, $z = pw_2$ with $w_2 \in T'_{\alpha+1}$. Let $w = w_1 + w_2$. Then $h(w) = h(w_1 + w_2) = \alpha$ (Problem 8.7.4). Also $pw = pw_2 = z$. Finally, w is proper with respect to S'_{2m}, since w_1 is and $h(w_2) \geq \alpha + 1$. The proof of Ulm's Theorem is complete. ∎

Fifty-fourth Problem Set

Problem 8.7.1 Let T, T' be p-groups and $\Phi : T \cong T'$. Prove: (1) if $x \in p^\alpha T$ then $\Phi(x) \in p^\alpha T'$. A symmetric argument shows that $\Phi : p^\alpha T \cong p^\alpha T'$. It follows directly that the Ulm Invariants are isomorphism invariants. To prove (1) use the method of **transfinite induction**. Check that (1) holds for $\alpha = 0$ (a triviality). Then suppose (1) holds for all ordinals $\gamma < \alpha$ and prove (1) holds for α. You will have to distinguish between the two cases when $\alpha = \beta + 1$ and when α is a limit ordinal.

Problem 8.7.2 Find a finite p-group T and $t \in T$ such that $h(x) + 1 < h(px) < \infty$.

Problem 8.7.3 Let T be the torsion subgroup of the group $\prod_{k=1}^{\infty} Z(p^k)$. (a) Describe the elements of T and explain why T is uncountable. Hint: We can assume the set of real numbers is uncountable. Every real corresponds to a decimal expansion in base p. (b) Show that $f_k(T) = 1$ for $0 \leq k < \omega$ and that $f_\omega(T) = 0$. (c) Find a p-group T' of length ω such that $f_k(T) = f_k(T')$ for all $k < \omega$ but T, T' are not isomorphic. This shows the hypothesis of countability is necessary for Ulm's Theorem.

Problem 8.7.4 Prove that if $h(x) = \alpha$ and $h(v) > \alpha$, then $h(x + v) = \alpha$.

Further reading: As already mentioned, one can't do better than [F]. For a pleasant and compact introduction to the ideas of abelian group theory, either [Ka] (if you can find it) or [G] are very nice.

References

[B], A. Blass, Existence of bases implies the axiom of choice, Axiomatic Set Theory, Contemp. Math. 31, A.M.S. 1984.

[F] L.Fuchs, Infinite Abelian Groups I, Academic Press, 1970.

[Fr] J. Fraleigh, Abstract Algebra, Addison-Wesley, 1999.

[FR] B. Fine and G. Rosenberger, The Fundamental Theorem of Algebra, UTM, Springer-Verlag, 1997.

[GLS] D. Gorenstein, R.Lyons and R.Solomon, Finite Simple Groups, A.M.S. Monograph 40.4, Amer. Math. Soc., 1999.

[H] I.Herstein, Noncommutative Rings, Carus Mathematical Monographs 15, Mathematical Association of America, 1968.

[HK] K. Hoffman and R. Kunze, Linear Algebra, Prentice Hall,1971.

[J] N. Jacobson, Lectures in Abstract Algebra II (Linear Algebra), Van Nostrand, 1953.

[J2] _____, Basic Algebra I, W.H. Freeman, 1974.

[K] J. Kelley, General Topology, D. Van Nostrand, 1955.

[R] J.Rotman, An Introduction to Homological Algebra, Academic Press, 1979.

[R2] _____, A First Course in Abstract Algebra, 2nd e., Prentice Hall, 2000.

[W] E. Walker, Introduction to Abstract Algebra, Random House/Birkhauser Mathematics Series, 1987.

Index

9 780367 394417